T0192203

A COURSE IN MATHEMATICAL ANALYSIS
Volume III: Complex Analysis, Measure and Integration

The three volumes of *A Course in Mathematical Analysis* provide a full and detailed account of all those elements of real and complex analysis that an undergraduate mathematics student can expect to encounter in the first two or three years of study. Containing hundreds of exercises, examples and applications, these books will become an invaluable resource for both students and instructors.

Volume I focuses on the analysis of real-valued functions of a real variable. Volume II goes on to consider metric and topological spaces, and functions of a vector variable, and includes an introduction to the theory of manifolds in Euclidean space. This third volume develops the classical theory of functions of a complex variable. It carefully establishes the properties of the complex plane, including a proof of the Jordan curve theorem. Lebesgue measure is introduced, and is used as a model for other measure spaces, where the theory of integration is developed. The Radon–Nikodym theorem is proved, and the differentiation of measures is discussed.

D. J. H. GARLING is Emeritus Reader in Mathematical Analysis at the University of Cambridge and Fellow of St. John's College, Cambridge. He has fifty years' experience of teaching undergraduate students in most areas of pure mathematics, but particularly in analysis.

A COURSE IN MATHEMATICAL ANALYSIS

Volume III
Complex Analysis,
Measure and Integration

D. J. H. GARLING

Emeritus Reader in Mathematical Analysis,
University of Cambridge, and
Fellow of St John's College, Cambridge

CAMBRIDGE
UNIVERSITY PRESS

Shaftesbury Road, Cambridge CB2 8EA, United Kingdom

One Liberty Plaza, 20th Floor, New York, NY 10006, USA

477 Williamstown Road, Port Melbourne, VIC 3207, Australia

314–321, 3rd Floor, Plot 3, Splendor Forum, Jasola District Centre, New Delhi – 110025, India

103 Penang Road, #05–06/07, Visioncrest Commercial, Singapore 238467

Cambridge University Press is part of Cambridge University Press & Assessment,
a department of the University of Cambridge.

We share the University's mission to contribute to society through the pursuit of
education, learning and research at the highest international levels of excellence.

www.cambridge.org
Information on this title: www.cambridge.org/9781107663305

© D. J. H. Garling 2014

This publication is in copyright. Subject to statutory exception and to the provisions
of relevant collective licensing agreements, no reproduction of any part may take
place without the written permission of Cambridge University Press & Assessment.

First published 2014
Reprinted 2014

A catalogue record for this publication is available from the British Library

Library of Congress Cataloging-in-Publication data
Garling, D. J. H.
Foundations and elementary real analysis / D. J. H. Garling.
pages   cm. – (A course in mathematical analysis; volume 1)
Includes bibliographical references and index.
ISBN 978-1-107-03202-6 (hardback) – ISBN 978-1-107-61418-5 (paperback)
1. Mathematical analysis.   I. Title.
QA300.G276   2013
515–dc23   2012044420

ISBN   978-1-107-03204-0   Hardback
ISBN   978-1-107-66330-5   Paperback

Cambridge University Press & Assessment has no responsibility for the persistence
or accuracy of URLs for external or third-party internet websites referred to in this
publication and does not guarantee that any content on such websites is, or will
remain, accurate or appropriate.

Contents

Introduction

This book is the third and final volume of a full and detailed course in the elements of real and complex analysis that mathematical undergraduates may expect to meet. Indeed, I have based it on those parts of analysis that undergraduates at Cambridge University meet, or used to meet, in their first two years. I have however found it desirable to go rather further in certain places, in order to give a rounded account of the material.

In Part Five, we develop the theory of functions of a complex variable. To begin with, we consider holomorphic functions (functions which are complex-differentiable) and analytic functions (functions which can be defined by power series), and the results seem similar to those of real case. Things change when path-integrals are introduced. To use these, a good understanding of the topology of the plane is needed. We give a careful account of this, including a proof of the Jordan curve theorem (every simple closed curve has an inside and an outside). With this in place, various forms of Cauchy's theorem and Cauchy's integral formula are proved. These lead on to many magical results. Chapter 25 is geometric. A single-valued holomorphic function is conformal (that is, it preserves angles and orientations). We consider the problem of mapping one domain conformally onto another, and end by proving the celebrated Riemann mapping theorem, which says that if U and V are domains in the complex plane which are proper subsets of the plane and are simply-connected (there are no holes) then there exists a conformal mapping of U onto V. In Chapter 26, we apply the theory that we have developed to various problems, some of which were first introduced in Volume I.

In Volume I, we developed properties of the Riemann integral. This is very satisfactory when we wish to integrate continuous or monotonic functions, and is a useful precursor for the complex path integrals that we consider in Part Five, but it has serious shortcomings. In Part Six, we introduce

Lebesgue measure on the real line. Abstract measure theory is a large and important subject, but the topological properties of the real line make the construction of Lebesgue measure on the real line rather straightforward. With this example in place, we introduce the notion of a measure space, and the corresponding space of measurable functions. This then leads on easily to the theory of integration, and the space L^p of p-th power integrable functions. These results are used to construct Lebesgue measure in higher dimensions, using Fubini's theorem. Properties of the Hilbert space L^2 are then used to give von Neumann's proof of the Radon–Nikodym theorem, and this is used to establish differentiability properties of measures and functions on $\mathbf{R}^d$. Almost all measures that arise in practice are defined on topological spaces, and we establish regularity properties, which show that such measures are rather well behaved. A final chapter uses the theory that we have established to obtain further results, largely concerning Fourier series (first considered in Volume I), and the boundary behaviour of harmonic functions on the unit disc.

The text includes plenty of exercises. Some are straightforward, some are searching, and some contain results needed later. All help develop an understanding of the theory: do them!

I am again extremely grateful to Zhuo Min 'Harold' Lim, who read the proofs and found many errors. Any remaining errors are mine alone. Corrections and further comments can be found on a web page on my personal home page at `www.dpmms.cam.ac.uk`.

Part Five

Complex analysis

20

Holomorphic functions and analytic functions

20.1 Holomorphic functions

Suppose that f is a continuous complex-valued function defined on an open subset U of the complex plane $\mathbf{C}$. Recall that the set U is the union of countably many connected components, each of which is an open subset of U (Volume II, Proposition 16.1.15 and Corollary 16.1.18). The behaviour of f on each component does not depend on its behaviour on the other components. For this reason, we restrict our attention to functions defined on a connected open subset of $\mathbf{C}$; such a set is called a *domain*.

We begin by considering differentiability: the definition is essentially the same as in the real case. Suppose that f is a complex-valued function on a domain U, and that $z \in U$. Then f is *differentiable* at z, with *derivative* $f'(z)$, if whenever $\epsilon > 0$ there exists $\delta > 0$ such that the open neighbourhood $N_\delta(z) = \{w : |w - z| < \delta\}$ of z is contained in U and such that if $0 < |w - z| < \delta$ then

$$\left| \frac{f(w) - f(z)}{w - z} - f'(z) \right| < \epsilon.$$

In other words,

$$\frac{f(w) - f(z)}{w - z} \to f'(z) \text{ as } w \to z.$$

Thus if f is differentiable at z, then the derivative $f'(z)$ is uniquely determined. The derivative $f'(z)$ is also denoted by $\frac{df}{dz}(z)$.

Proposition 20.1.1 *Suppose that f is a complex-valued function on a domain U, that $N_\delta(z) \subseteq U$, and that $l \in \mathbf{C}$. The following statements are equivalent.*

(i) f is differentiable at z, with derivative l.

(ii) There is a complex-valued function r on $N_\delta^*(0) = N_\delta(0) \setminus \{0\}$ such that

$$f(z+w) = f(z) + lw + r(w) \text{ for } 0 < |w| < \delta$$

for which $r(w)/w \to 0$ as $w \to 0$.

(iii) There is a complex-valued function s on $N_\delta(0)$ such that

$$f(z+w) = f(z) + (l + s(w))w \text{ for } |w| < \delta$$

for which $s(0) = 0$ and s is continuous at 0.

If so, then f is continuous at z.

Proof   This corresponds to Volume I, Proposition 7.1.1, and the easy proof is essentially the same.   □

If f is differentiable at every point of U, then we say that f is *holomorphic* on U. If $U = \mathbf{C}$, then we say that f is an *entire* function. Although the form of the definition of differentiability that we have just given is exactly the same as the form of the definition in the real case, we shall see that holomorphic functions are very different from differentiable functions on an open interval of $\mathbf{R}$.

Example 20.1.2   Let $f(z) = 1/z$ for $z \in \mathbf{C} \setminus \{0\}$. Then f is holomorphic on $\mathbf{C} \setminus \{0\}$, with derivative $-1/z^2$.

For if $0 < |w| < |z|$, then $z + w \neq 0$ and

$$\frac{f(z+w) - f(z)}{w} - \frac{-1}{z^2} = \frac{z^2 - (z+w)z + w(z+w)}{wz^2(z+w)} = \frac{w}{z^2(z+w)} \to 0$$

as $w \to 0$.

Proposition 20.1.3   *Suppose that f and g are complex-valued functions defined on a domain U, and that f and g are differentiable at z. Suppose also that $\lambda, \mu \in \mathbf{C}$.*

(i) $\lambda f + \mu g$ is differentiable at z, with derivative $\lambda f'(z) + \mu g'(z)$.

(ii) The product fg is differentiable at z, with derivative $f'(z)g(z) + f(z)g'(z)$.

Proof   An easy exercise for the reader.   □

Theorem 20.1.4 (The chain rule)   *Suppose that f is a complex-valued function defined on a domain U, that h is a complex-valued function defined* on a domain V and that $f(U) \subseteq V$. Suppose that f is differentiable at z and that h is differentiable at $f(z)$. Then the composite function $h \circ f$ is differentiable at z, with derivative $h'(f(z)).f'(z)$.

Proof   There are two possibilities. First, there exists $\delta > 0$ such that $N_\delta(z) \subseteq U$ and $f(z + w) \neq f(z)$ for $0 < |w| < \delta$. If $0 < |w| < \delta$ then

$$\frac{h(f(z+w)) - h(f(z))}{w} = \left(\frac{h(f(z+w)) - h(f(z))}{f(z+w) - f(w)} \right) . \left(\frac{f(z+w) - f(z)}{w} \right).$$

Since f is continuous at z, $f(z+w) - f(z) \to 0$ as $w \to 0$, and so

$$\frac{h(f(z+w)) - h(f(z))}{f(z+w) - f(z)} \to h'(f(z)) \text{ as } w \to 0.$$

Since $(f(z+w) - f(z))/w \to f'(z)$ as $w \to 0$, the result follows.

Secondly, z is the limit point of a sequence $(z_n)_{n=1}^\infty$ in $U \setminus \{z\}$ for which $f(z_n) = f(z)$. In this case it follows that $f'(z) = 0$, and we must show that $(h \circ f)'(z) = 0$. We use Proposition 20.1.1. Let $b = f(z)$. There exist $\eta > 0$ such that $N_\eta(f(z)) \subseteq V$ and a function t on $N_\eta(0)$, with $t(0) = 0$, such that $h(b + k) = h(b) + (h'(b) + t(k))k$ for $k \in N_\eta(0)$ and such that t is continuous at 0. Similarly, there exist $\delta > 0$ such that $N_\delta(z) \subseteq U$ and a function s on $N_\delta(0)$, with $s(0) = 0$, such that $f(z + w) = b + s(w)w$ for $h \in N_\delta(0)$ and such that s is continuous at 0. Since f is continuous at z, we can suppose that $f(N_\delta(z)) \subseteq N_\eta(b)$. If $0 < |w| < \delta$ then

$$h(f(z+w)) = h(b + s(w)w) = h(b) + (h'(b) + t(s(w)w))s(w)w$$

so that

$$\frac{h(f(z+w)) - h(f(z))}{w} = (h'(b) + t(s(w)w))s(w) \to 0 \text{ as } w \to 0,$$

since $s(w) \to 0$ and $t(s(w)w) \to 0$ as $w \to 0$. □

This is essentially the same proof as in the real case. But, as we shall see (Theorem 23.1.1), the second case can only arise if f is constant on U: complex differentiation is in fact very different from real differentiation.

Corollary 20.1.5   *Suppose that g is a complex-valued function on U, which is differentiable at z. If $g(z) \neq 0$ then there is a neighbourhood $N_\delta(z) \subseteq U$ such that $g(w) \neq 0$ for $w \in N_\delta(z)$. The function $1/g$ on*

$N_\delta(z)$ is differentiable at z, with derivative $-g'(z)/g(z)^2$. Furthermore f/g is differentiable at z, with derivative

$$\left(\frac{f}{g}\right)'(z) = \frac{f'(z)g(z) - f(z)g'(z)}{(g(z))^2}.$$

Proof Since g is continuous at z, there is a neighbourhood $N_\delta(z) \subseteq U$ such that $g(w) \neq 0$ for $w \in N_\delta(z)$. Then $g(N_\delta(z)) \subseteq \mathbf{C} \setminus \{0\}$. Let $h(z) = 1/z$ for $z \in \mathbf{C} \setminus \{0\}$. Then the first result follows from the chain rule, and the second from Proposition 20.1.3. □

For example, if $p(z) = a_0 + \cdots + a^n z^n$ is a polynomial function, then p is an entire function, and $p'(z) = a_1 + 2a_2 z + \cdots + na_n z^{n-1}$. Similarly, if p and q are polynomials, and U is an open set in which q has no zeros then the rational function $r(z) = p(z)/q(z)$ is holomorphic on U, and

$$r'(z) = \frac{q(z)p'(z) - q'(z)p(z)}{q(z)^2}.$$

Exercises

20.1.1 Suppose that f is a holomorphic function on $N_1(i)$ and that $(f(z))^5 = z$ for $z \in N_1(i)$. What is $f'(i)$?

20.1.2 Suppose that f is a holomorphic function on $\mathbf{D} = \{z \in \mathbf{C} : |z| < 1\}$. Show that the set $\{n \in \mathbf{N} : f(1/(n+1)) = 1/n\}$ is finite.

20.2 The Cauchy–Riemann equations

Suppose that f is a complex-valued function on a domain U, and that $z = x + iy \in U$. We can write $f(z)$ as $u(x,y) + iv(x,y)$, where $u(x,y)$ and $v(x,y)$ are the real and imaginary parts of $f(z)$. The functions u and v are real-valued functions of two real variables. How are differentiability properties of f related to differentiability properties of u and v?

Let us make this more explicit. Let $k : \mathbf{R}^2 \to \mathbf{C}$ be defined by setting $k((x,y)) = x + iy$; k is a linear isometry of $\mathbf{R}^2$ onto $\mathbf{C}$, considered as a real vector space. Let $j : \mathbf{C} \to \mathbf{R}^2$ be the inverse mapping. If f is a complex-valued function on U, let $\tilde{f} = j \circ f \circ k$; $\tilde{f}$ is a mapping from the open set $j(U)$ into $\mathbf{R}^2$. If $\tilde{f}(x,y) = (u(x,y), v(x,y))$, then $f(x + iy) = u(x,y) + iv(x,y)$:

$$
\begin{array}{ccc}
x + iy & \xrightarrow{\ f\ } & f(x+iy) = u(x,y) + iv(x,y) \\[4pt]
{\scriptstyle k}\Big\uparrow & & \Big\downarrow{\scriptstyle j} \\[4pt]
(x,y) & \xrightarrow{\ \tilde{f}\ } & (u(x,y), v(x,y))
\end{array}
$$

Theorem 20.2.1   *Suppose that f is a complex-valued function on a domain U, and that $z_0 = x_0 + iy_0 \in U$. With the notation described above, the following are equivalent:*

(i) *f is differentiable at z_0;*
(ii) *the function $\tilde{f} : (x,y) \to (u(x,y), v(x,y))$ from $j(U)$ to $\mathbf{R}^2$ is differentiable at (x_0, y_0), and the partial derivatives satisfy the Cauchy–Riemann equations:*

$$\frac{\partial u}{\partial x}(x_0, y_0) = \frac{\partial v}{\partial y}(x_0, y_0) \text{ and } \frac{\partial u}{\partial y}(x_0, y_0) = -\frac{\partial v}{\partial x}(x_0, y_0).$$

If so, then

$$\frac{df}{dz}(z_0) = \frac{\partial u}{\partial x}(x_0, y_0) + i\frac{\partial v}{\partial x}(x_0, y_0) = \frac{\partial v}{\partial y}(x_0, y_0) - i\frac{\partial u}{\partial y}(x_0, y_0).$$

Proof   Suppose first that f is differentiable at z_0. Then

$$\frac{df}{dz}(z_0) = \lim_{x \to 0} \frac{f(z_0 + x) - f(z_0)}{x}$$

$$= \lim_{x \to 0} \frac{u(x_0 + x, y_0) - u(x_0, y_0)}{x} + i \lim_{x \to 0} \frac{v(x_0 + x, y_0) - v(x_0, y_0)}{x},$$

so that the partial derivatives $(\partial u / \partial x)(x_0, y_0)$ and $(\partial v / \partial x)(x_0, y_0)$ exist, and

$$\frac{df}{dz}(z_0) = \frac{\partial u}{\partial x}(x_0, y_0) + i\frac{\partial v}{\partial x}(x_0, y_0).$$

But also

$$\frac{df}{dz}(z_0) = \lim_{y \to 0} \frac{f(z_0 + iy) - f(z_0)}{iy}$$

$$= -i \lim_{y \to 0} \frac{u(x_0, y_0 + y) - u(x_0, y_0)}{y} + \lim_{y \to 0} \frac{v(x_0, y_0 + y) - v(x_0, y_0)}{y}$$

$$= \frac{\partial v}{\partial y}(x_0, y_0) - i\frac{\partial u}{\partial y}(x_0, y_0),$$

so that the partial derivatives $(\partial u / \partial y)(x_0, y_0)$ and $(\partial v / \partial y)(x_0, y_0)$ exist, and

$$\frac{df}{dz}(z_0) = \frac{\partial v}{\partial y}(x_0, y_0) - i\frac{\partial u}{\partial y}(x_0, y_0).$$

Thus the partial derivatives satisfy the Cauchy–Riemann equations.

Suppose that $z \in U$. Using these equations, we see that the real part of $(z - z_0)f'(z_0)$ is

$$(x - x_0)\frac{\partial u}{\partial x}(x_0, y_0) + i(y - y_0)(-i\frac{\partial u}{\partial y}(x_0, y_0))$$

$$= (x - x_0)\frac{\partial u}{\partial x}(x_0, y_0) + (y - y_0)\frac{\partial u}{\partial y}(x_0, y_0),$$

so that if we set

$$r(x, y) = u(x, y) - u(x_0, y_0) - (x - x_0)\frac{\partial u}{\partial x}(x_0, y_0) - (y - y_0)\frac{\partial u}{\partial y}(x_0, y_0)$$

then $r(x, y)$ is the real part of $f(z) - f(z_0) - (z - z_0)f'(z_0)$. Consequently, u is differentiable at (x_0, y_0). An exactly similar argument shows that the same is true for v.

Conversely, suppose that (ii) holds. Let

$$g = \frac{\partial u}{\partial x}(x_0, y_0) + i\frac{\partial v}{\partial x}(x_0, y_0) = \frac{\partial v}{\partial y}(x_0, y_0) - i\frac{\partial u}{\partial y}(x_0, y_0).$$

Suppose that $z \in U$. Let $f(z) - f(z_0) - (z - z_0)g = h(z) + ik(z)$. Then easy calculations show that

$$h(x + iy) = u(x, y) - u(x_0, y_0) - (x - x_0)\frac{\partial u}{\partial x}(x_0, y_0) - (y - y_0)\frac{\partial u}{\partial y}(x_0, y_0),$$

$$k(x + iy) = v(x, y) - v(x_0, y_0) - (x - x_0)\frac{\partial v}{\partial x}(x_0, y_0) - (y - y_0)\frac{\partial v}{\partial y}(x_0, y_0),$$

so that

$$\frac{f(z) - f(z_0)}{z - z_0} - g = \frac{h(z) + ik(z)}{z - z_0} \to 0$$

as $z \to z_0$; hence f is differentiable at z_0, with derivative g. $\qquad\square$

Corollary 20.2.2 *If f is holomorphic and twice continuously differentiable on U then u and v are harmonic functions; that is*

$$\frac{\partial^2 u}{\partial x^2} + \frac{\partial^2 u}{\partial y^2} = \frac{\partial^2 v}{\partial x^2} + \frac{\partial^2 v}{\partial y^2} = 0.$$

Proof For

$$\frac{\partial^2 u}{\partial x^2} = \frac{\partial^2 v}{\partial x \partial y} = \frac{\partial^2 v}{\partial y \partial x} = -\frac{\partial^2 u}{\partial x^2},$$

$$\frac{\partial^2 v}{\partial x^2} = -\frac{\partial^2 u}{\partial x \partial y} = -\frac{\partial^2 u}{\partial x \partial y} = -\frac{\partial^2 v}{\partial x^2}. \qquad \square$$

We shall see later that every holomorphic function is infinitely differentiable. Harmonic functions in Euclidean space were considered in Volume II, Section 19.8.

This result suggests a rather different approach. Suppose that $\tilde{f}$ is differentiable at (x_0, y_0). Let $\check{f} = f \circ k$, so that $\check{f}(x, y) = f(x + iy)$. We set

$$\partial f = \frac{1}{2}\left(\frac{\partial \check{f}}{\partial x} - i\frac{\partial \check{f}}{\partial y}\right), \quad \bar{\partial} f = \frac{1}{2}\left(\frac{\partial \check{f}}{\partial x} + i\frac{\partial \check{f}}{\partial y}\right).$$

Then

$$\bar{\partial} f = \frac{1}{2}\left(\left(\frac{\partial u}{\partial x} - \frac{\partial v}{\partial y}\right) + i\left(\frac{\partial v}{\partial x} + \frac{\partial u}{\partial y}\right)\right),$$

so that f is differentiable at z_0 if and only if $\bar{\partial} f(z_0) = 0$. If this is so, then

$$\partial f(z_0) = \frac{1}{2}\left(\left(\frac{\partial u}{\partial x} + \frac{\partial v}{\partial y}\right)(x_0, y_0) + i\left(\frac{\partial v}{\partial x} - \frac{\partial u}{\partial y}\right)(x_0, y_0)\right) = f'(z_0).$$

We can use the Cauchy–Riemann equations and the differentiable inverse mapping theorem to prove an inverse mapping theorem for holomorphic functions. An injective holomorphic function on a domain U is said to be *univalent*: that is, it takes each value at most once on U.

Theorem 20.2.3 *Suppose that f is a univalent function on a domain U, with continuous derivative f', and suppose that $f'(z) \neq 0$ for all $z \in U$. Then $f(U)$ is an open subset of $\mathbf{C}$, the mapping $f : U \to f(U)$ is a homeomorphism, the inverse mapping $f^{-1} : f(U) \to U$ is holomorphic, and if $f(z) = w$ then $(f^{-1})'(w) = 1/f'(z)$.*

Proof Suppose that $z = x + iy \in U$. Let $r = |f'(z)|$. Since

$$f'(z) = \frac{\partial u}{\partial x}(x, y) + i\frac{\partial v}{\partial x}(x, y),$$

it follows that

$$r^2 = \left(\frac{\partial u}{\partial x}(x,y)\right)^2 + \left(\frac{\partial v}{\partial x}(x,y)\right)^2,$$

so that there exists $0 \le \theta < 2\pi$ such that

$$\frac{\partial u}{\partial x}(x,y) = \frac{\partial v}{\partial y}(x,y) = r\cos\theta, \quad \frac{\partial u}{\partial y}(x,y) = -\frac{\partial v}{\partial x}(x,y) = -r\sin\theta.$$

Hence $f'(z) = r(\cos\theta + i\sin\theta)$. Thus the Jacobian $J(\widetilde{f})$ of the mapping $\widetilde{f}$ from $j(U)$ to $j(f(U))$ is

$$\det \begin{bmatrix} r\cos\theta & -r\sin\theta \\ r\sin\theta & r\cos\theta \end{bmatrix} = r^2 > 0.$$

By the differentiable inverse mapping theorem (Volume II, Theorem 17.4.1), $j(f(U))$ is an open subset of $\mathbf{R}^2$, and the inverse mapping $\widetilde{f}^{-1} : j(f(U)) \to j(U)$ is differentiable, with derivative

$$(D\widetilde{f}^{-1})_{(u(x,y),v(x,y))} = ((D\widetilde{f})_{(x,y)})^{-1} = \begin{bmatrix} r^{-1}\cos\theta & r^{-1}\sin\theta \\ -r^{-1}\sin\theta & r^{-1}\cos\theta \end{bmatrix}.$$

Consequently, the Cauchy–Riemann equations are satisfied by f^{-1}, and f^{-1} is holomorphic; if $w = s + it = f(z) \in f(U)$ then

$$(f^{-1})'(w) = \frac{\partial \widetilde{f}^{-1}}{\partial s}(s,t) + i\frac{\partial \widetilde{f}^{-1}}{\partial t}(s,t) = \frac{\cos\theta - i\sin\theta}{r} = \frac{1}{f'(z)}. \qquad \square$$

At first sight, this looks like a strong and useful result. In fact, as we shall see, two of the hypotheses are redundant. First, the derivative of a holomorphic function on a domain is always continuous (Corollary 22.6.6), and secondly, if f is a univalent function on a domain U, then its derivative cannot take the value 0 on U (Theorem 23.6.8).

Exercises

20.2.1 Why was the chain rule not used to prove the Cauchy–Riemann equations?

20.2.2 Suppose that f is holomorphic on a domain U and that $|f|$ is constant on U. By considering $|f|^2$ and using the Cauchy–Riemann equations, show that f is constant on U.

20.2.3 Suppose that f is a non-constant holomorphic function on a domain U. Show that if $c \in \mathbf{R}$ then $\{z \in U : |f(z)| = c\}$ has an empty interior.

20.2.4 Suppose that f is an entire function and that u and v are its real and imaginary parts. Show that if $u(z) + v(z) \geq 0$ for $z \in \mathbf{C}$, then f is constant. Show that if $u(z)v(z) \geq 0$ for $z \in \mathbf{C}$, then f is constant.

20.2.5 Suppose that u is a harmonic twice-differentiable function on an open neighbourhood $N_r(x_0, y_0)$ of (x_0, y_0) in $\mathbf{R}^2$. If $(x, y) \in N_r(x_0, y_0)$ let

$$v(x, y) = - \int_{x_0}^{x} \frac{\partial u}{\partial y}(s, y_0)\, ds + \int_{y_0}^{y} \frac{\partial u}{\partial x}(x, t)\, dt.$$

Show that u and v satisfy the Cauchy–Riemann equations. If $z_0 = x_0 + iy_0$ and $z = x + iy \in N_r(z_0)$, let $f(z) = u(x, y) + iv(x, y)$. Then f is a holomorphic function on $N_r(z_0)$. The function v is the *harmonic conjugate* of u.

20.3 Analytic functions

So far, we only have a meagre supply of examples of holomorphic functions. When we considered functions of a real variable, we used a power series to define the exponential function and the circular functions. We shall see that power series not only enable us to do the same in the complex case, but also play a fundamental role in the theory of functions of a complex variable. First we consider power series quite generally.

Recall that a *complex power series* is an expression of the form $\sum_{n=0}^{\infty} a_n(z - z_0)^n$, where $(a_n)_{n=0}^{\infty}$ is a sequence of complex numbers, z_0 is a complex number, and z is a complex number, which we also allow to vary. Here are some of the fundamental results that were established in Volume I, Sections 4.7 and 6.6.

- Suppose that $\sum_{n=0}^{\infty} a_n(z - z_0)^n$ is a complex power series. There exists $0 \leq R \leq \infty$ (the *radius of convergence*) such that $\sum_{n=0}^{\infty} a_n(z - z_0)^n$ converges locally absolutely uniformly on $\{z : |z - z_0| < R\}$ to a continuous function f on $\{z : |z - z_0| < R\}$; that is, if $0 < S < R$ then $\sum_{n=0}^{\infty} a_n(z - z_0)^n$ converges absolutely uniformly to f on $\{z : |z - z_0| \leq S\}$. If $|z - z_0| > R$, then the sequence $(a_n(z - z_0)^n)_{n=0}^{\infty}$ is unbounded, so that the series certainly does not converge. All sorts of things can happen on the *circle of convergence* $\{z : |z - z_0| = R\}$.
- Suppose that $\sum_{n=0}^{\infty} a_n(z - z_0)^n$ is a power series with radius of convergence R. Let $\Lambda = \limsup |a_n|^{1/n}$. If $\Lambda = 0$ then $R = \infty$. If $\Lambda = \infty$ then $R = 0$. Otherwise, $R = 1/\Lambda$.
- If $\sum_{n=0}^{\infty} a_n(z - z_0)^n$ and $\sum_{n=0}^{\infty} b_n(z - z_0)^n$ are power series, we can form the formal product $\sum_{n=0}^{\infty} c_n(z - z_0)^n$, where $c_n = \sum_{j=0}^{n} a_j b_{n-j}$.

If $\sum_{n=0}^{\infty} a_n(z - z_0)^n$ has radius of convergence R and $\sum_{n=0}^{\infty} b_n(z - z_0)^n$ has radius of convergence R', then the power series $\sum_{n=0}^{\infty} c_n(z - z_0)^n$ has radius of convergence greater than or equal to $\min(R, R')$. If $|z - z_0| < \min(R, R')$ then

$$\left(\sum_{n=0}^{\infty} a_n(z - z_0)^n \right) \left(\sum_{n=0}^{\infty} b_n(z - z_0)^n \right) = \sum_{n=0}^{\infty} c_n(z - z_0)^n.$$

- Provided that their radii of convergence are positive, different power series define different functions. Suppose that each of the power series $\sum_{n=0}^{\infty} a_n(z - z_0)^n$ and $\sum_{n=0}^{\infty} b_n(z - z_0)^n$ has radius of convergence greater than or equal to $R > 0$. Let $f(z) = \sum_{n=0}^{\infty} a_n(z - z_0)^n$ and $g(z) = \sum_{n=0}^{\infty} b_n(z - z_0)^n$, for $|z - z_0| < R$. Suppose that $(w_k)_{k=1}^{\infty}$ is a null sequence of non-zero complex numbers in $\{z : |z| < R\}$ such that $f(z_0 + w_k) = g(z_0 + w_k)$ for all $k \in \mathbf{N}$. Then $a_n = b_n$ for all $n \in \mathbf{Z}^+$. This means that if we obtain two power series for the same function, we can 'equate coefficients'.
- Suppose that the power series $\sum_{n=0}^{\infty} a_n(z - z_0)^n$ has positive radius of convergence R; if $|z - z_0| < R$, let $f(z) = \sum_{n=0}^{\infty} a_n(z - z_0)^n$. Suppose that $0 < S \leq R$, and that f has no zeros in $\{z : |z - z_0| < S\}$. Then there exists a power series $\sum_{n=0}^{\infty} c_n(z - z_0)^n$ with positive radius of convergence T such that, if $g(z) = \sum_{n=0}^{\infty} c_n(z - z_0)^n$ for $|z - z_0| < T$, then $f(z)g(z) = 1$ for $|z - z_0| < \min(S, T)$.

What about the differentiability of power series?

Theorem 20.3.1 *Suppose that the power series $\sum_{n=1}^{\infty} a_n(z - z_0)^n$ has radius of convergence R. Then the power series $\sum_{n=1}^{\infty} na_n(z - z_0)^{n-1}$ has radius of convergence R. If $f(z) = \sum_{n=0}^{\infty} a_n(z - z_0)^n$ for $|z - z_0| < R$, then f is differentiable on the set $\{z \in \mathbf{C} : |z - z_0| < R\}$, and*

$$f'(z) = \sum_{n=1}^{\infty} na_n(z - z_0)^{n-1} = \sum_{n=0}^{\infty} (n + 1)a_{n+1}(z - z_0)^n.$$

In other words, we can differentiate a power series term by term within its circle of convergence.

Proof   We can clearly suppose that $z_0 = 0$. Since $(\log(n + 1))/n \to 0$ as $n \to \infty$, $(n + 1)^{1/n} \to 1$ as $n \to \infty$, and so

$$\limsup((n + 1)|a_{n+1}|)^{1/n} = \limsup |a_n|^{1/n}.$$

Thus the power series $\sum_{n=0}^{\infty} (n + 1)a_{n+1}z^n$ has radius of convergence R.

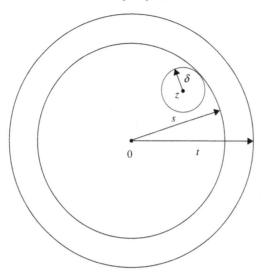

Figure 20.3.

Suppose that $|z| = r < R$. Choose s and t with $r < s < t < R$, and let $M = \sup_n |a_n| t^n$; then $M < \infty$. Let $\delta = s - r$ and let $B_\delta = \{w \in \mathbf{C} : |w - z| \leq \delta\}$; $B_\delta \subseteq \{w : |w| \leq s\}$.

We use the identity

$$a^n - b^n = (a - b)(a^{n-1} + ba^{n-2} + \cdots + b^{n-2}a + b^{n-1}).$$

If $w \in B_\delta$, let

$$q_1(w) = 1,$$
$$q_n(w) = (z + w)^{n-1} + z(z + w)^{n-2} + \cdots + z^{n-2}(z + w) + z^{n-1}$$

for $n > 1$. Then $q_n(0) = nz^{n-1}$, and $q_n(w) = ((z + w)^n - z^n)/w$ for $w \neq 0$, so that

$$\frac{f(z + w) - f(z)}{w} = \sum_{n=1}^{\infty} a_n q_n(w).$$

Now if $|w| \leq \delta$ then $|q_n(w)| \leq n(|z| + |w|)^{n-1} \leq ns^{n-1}$, so that

$$\sup\{|a_n q_n(w)| : w \in B_\delta\} \leq ns^{n-1}\left(\frac{M}{t^n}\right) = \left(\frac{nM}{s}\right)\left(\frac{s}{t}\right)^n.$$

Since $\sum_{n=1}^{\infty} n(s/t)^n < \infty$, the series $\sum_{n=1}^{\infty} a_n q_n(w)$ converges absolutely and uniformly on B_δ. Consequently, the function $\sum_{n=1}^{\infty} a_n q_n$ is continuous on B_δ, and so

$$\frac{f(z+w) - f(z)}{w} = \sum_{n=1}^{\infty} a_n q_n(w) \to \sum_{n=1}^{\infty} a_n q_n(0) = \sum_{n=1}^{\infty} n a_n z^{n-1}$$

as $w \to 0$.   □

Corollary 20.3.2 $f(z)$ *is infinitely differentiable on the set* $\{z \in \mathbf{C} : |z - z_0| < R\}$, *and*

$$f^{(k)}(z) = \sum_{j=0}^{\infty} (k+1)(k+2) \cdots (k+j) a_{k+j} z^j = \sum_{j=0}^{\infty} \frac{(k+j)!}{j!} a_{k+j} z^j.$$

Proof   For we can apply the result inductively to f', and to the higher derivatives of f.   □

Corollary 20.3.3 $a_n = f^{(n)}(z_0)/n!$, *so that*

$$f(z) = \sum_{n=0}^{\infty} \frac{f^{(n)}(z_0)}{n!} (z - z_0)^n$$

is the Taylor series expansion of f.

Corollary 20.3.4   *If* $|z - z_0| < R$ *then*

$$f^{(k)}(z) = \sum_{j=0}^{\infty} \frac{f^{(k+j)}(z_0)}{j!} z^j.$$

   Thus if a power series has positive radius of convergence, it defines a holomorphic function within the radius of convergence, and this function is infinitely differentiable.

   This leads to the following definition. Suppose that f is a complex-valued function on a domain U. f is *analytic* on U if for each $w \in U$ there exists a power series $\sum_{n=0}^{\infty} a_n(w)(z - w)^n$ with positive radius of convergence $R(w)$ such that $f(z) = \sum_{n=0}^{\infty} a_n(w)(z - w)^n$ for all $z \in N_{R(w)}(w) \cap U$.

Corollary 20.3.5   *If* f *is analytic on a domain* U, *then* f *is holomorphic, and indeed is infinitely differentiable on* U.

   We shall see later (Theorem 22.6.5) that the converse holds: a holomorphic function on a domain is analytic.

Theorem 20.3.6   *Let* $A(U)$ *denote the set of all analytic functions on a domain* U. *If* $f, g \in A(U)$, *then* $f + g \in A(U)$ *and* $fg \in A(U)$. *If* $f \in A(U)$ *and* $f(z) \neq 0$ *for* $z \in U$ *then* $1/f \in A(U)$.

Proof These results follow directly from the properties of power series listed at the beginning of this section. □

Corollary 20.3.7 *The function $J(z) = 1/z$ is analytic on $\mathbf{C} \setminus \{0\}$.*

Proof For the function $f(z) = z$ is analytic on $\mathbf{C} \setminus \{0\}$, and $f(z) \neq 0$ on $\mathbf{C} \setminus \{0\}$. □

It is instructive to obtain the power series expansion of J.

Proposition 20.3.8 *If $z_0 \neq 0$ then*

$$\frac{1}{z_0 + w} = \frac{1}{z_0} - \frac{w}{z_0^2} + \cdots + \frac{(-w)^n}{z_0^{n+1}} + \cdots,$$

for $|w| < |z_0|$, and the power series has radius of convergence $|z_0|$.

Proof Suppose that $|w| < |z_0|$. Using the formula

$$(1 - y)(1 + y + \cdots + y^n) = 1 - y^{n+1},$$

with $y = -w/z_0$, we find after a little manipulation that

$$\frac{1}{z_0 + w} = \frac{1}{z_0} + \frac{-w}{z_0^2} + \cdots + \frac{(-w)^n}{z_0^{n+1}} + \frac{(-w)^{n+1}}{z_0^{n+1}(z_0 + w)}.$$

Now

$$\left| \frac{(-w)^{n+1}}{z_0^{n+1}(z_0 + w)} \right| \leq \frac{1}{|z_0| - |w|} \left(\frac{|w|}{|z_0|} \right)^{n+1} \to 0 \text{ as } n \to \infty,$$

and so the series converges. It follows directly from the definition that the radius of convergence of the power series is $|z_0|$. □

Proposition 20.3.9 *Suppose that f is an analytic function on a domain U, and that there exists $z_0 \in U$ such that $f^{(k)}(z_0) = 0$ for all $k \in \mathbf{N}$. Then f is constant on U.*

Proof We use the connectedness of U. Let

$$A = \{z \in U : f^{(k)}(z) = 0 \text{ for all } k \in \mathbf{N}\}.$$

If $k \in \mathbf{N}$ then $f^{(k)}$ is continuous on U, so that $\{z \in U : f^{(k)}(z) = 0\}$ is closed in U. Since $A = \cap_{k \in \mathbf{N}} \{z \in U : f^{(k)}(z) = 0\}$, A is closed in U. If $w \in A$, there exists $R > 0$ such that $N_R(w) \subseteq U$ and

$$f(z) = \sum_{n=0}^{\infty} \frac{f^{(n)}(w)}{n!}(z - w)^n = f(w) \text{ for } z \in N_R(w).$$

Thus $f^{(k)}(z) = 0$ for $z \in N_R(w)$ and $k \in \mathbf{N}$. Hence $N_R(w) \subset A$, and A is open. Since U is connected and A is not empty, it follows that $A = U$, and that f is constant on U. $\square$

This means that an analytic function on a domain U is determined by its values near an arbitrary point of U.

Corollary 20.3.10 *Suppose that f and g are analytic functions on a domain U, and that there exists $z_0 \in U$ such that $f^{(k)}(z_0) = g^{(k)}(z_0)$ for all $k \in \mathbf{Z}^+$. Then $f = g$.*

Proof   Apply the proposition to $f - g$. $\square$

We now show that a power series with positive radius of convergence R defines an analytic function within its circle of convergence. The Taylor series expression suggests that it is convenient to consider power series of the form $\sum_{n=0}^{\infty} c_n (z - z_0)^n / n!$.

Theorem 20.3.11 *Suppose that the power series $\sum_{n=1}^{\infty} c_n (z - z_0)^n / n!$ has positive radius of convergence R; for $|z - z_0| < R$ let $f(z) = \sum_{n=1}^{\infty} c_n (z - z_0)^n / n!$. Suppose that $|w - z_0| = r < R$. Then the power series $\sum_{k=0}^{\infty} f^{(k)}(w)(z - w)^k / k!$ has radius of convergence at least $R - r$, and if $|z| < R - r$ then*

$$f(w + z) = \sum_{k=0}^{\infty} \frac{f^{(k)}(w)}{k!} z^k.$$

Proof   We can clearly suppose that $z_0 = 0$. First,

$$f^{(k)}(w) = \sum_{n=k}^{\infty} \frac{n(n-1)\ldots(n-k+1)}{n!} c_n w^{n-k} = \sum_{n=k}^{\infty} \frac{c_n}{(n-k)!} w^{n-k}.$$

We consider absolute values, and change the order of summation. If $|z| < R - |w|$ then

$$\sum_{k=0}^{\infty} \frac{|f^{(k)}(w)||z|^k}{k!} \leq \sum_{k=0}^{\infty} \left(\sum_{n=k}^{\infty} \frac{|c_n|}{(n-k)!} |w|^{n-k} \right) \frac{|z|^k}{k!}$$

$$= \sum_{n=0}^{\infty} \frac{|c_n|}{n!} \left(\sum_{k=0}^{n} \frac{n!}{(n-k)!k!} |w|^{n-k} |z|^k \right)$$

$$= \sum_{n=0}^{\infty} \frac{|c_n|}{n!} (|w| + |z|)^n < \infty.$$

Thus the radius of convergence of the power series $\sum_{k=0}^{\infty} f^{(k)}(w)z^k/k!$ is at least $R - |w|$, and the double sum

$$\sum_{k=0}^{\infty} \left(\sum_{n=k}^{\infty} \frac{c_n}{(n-k)!} w^{n-k} \right) \frac{z^k}{k!}$$

is absolutely convergent for $|z| < R - |w|$. We can therefore change the order of summation:

$$\sum_{k=0}^{\infty} \frac{f^{(k)}(w)z^k}{k!} = \sum_{k=0}^{\infty} \left(\sum_{n=k}^{\infty} \frac{c_n}{(n-k)!} w^{n-k} \right) \frac{z^k}{k!}$$

$$= \sum_{n=0}^{\infty} \frac{c_n}{n!} \left(\sum_{k=0}^{n} \frac{n!}{(n-k)!k!} w^{n-k} z^k \right)$$

$$= \sum_{n=0}^{\infty} \frac{c_n}{n!} (w+z)^n = f(w+z).$$

$\square$

Exercises

20.3.1 Suppose that the power series $\sum_{n=0}^{\infty} a_n z^n$ has positive radius of convergence R, and that $f(z) = \sum_{n=0}^{\infty} a_n z^n$ for $|z| < R$. Show that there exists an analytic function F on $|z| < R$ such that $F'(z) = f(z)$ for $|z| < R$.

20.4 The exponential, logarithmic and circular functions

In Volume I, we used power series to define the real-valued exponential and circular functions on the real line. We now consider their complex-valued counterparts, defined on $\mathbf{C}$. First, the power series

$$e^z = \exp(z) = 1 + \frac{z}{1!} + \frac{z^2}{2!} + \cdots + \frac{z^n}{n!} + \cdots$$

has infinite radius of convergence and so defines an entire function. Of course, the restriction of exp to $\mathbf{R}$ is real-valued, and is the function that we considered in Volume I, Section 7.4. Differentiating term by term, we see that $de^z/dz = e^z$, and multiplying the series for e^z and e^w, we see that

$e^{z+w} = e^z e^w$. Consequently, if $z = x + iy$ then $e^z = e^x e^{iy}$. Since $-iy = \overline{iy}$ it follows that $e^{-iy} = \overline{e^{iy}}$. Thus

$$|e^{iy}|^2 = e^{iy}\overline{e^{iy}} = e^{iy}e^{-iy} = e^{iy-iy} = 1,$$

so that $|e^{iy}| = 1$.

The power series

$$\cos z = \sum_{n=0}^{\infty}(-1)^n \frac{z^{2n}}{(2n)!} \text{ and } \sin z = \sum_{n=0}^{\infty}(-1)^n \frac{z^{2n+1}}{(2n+1)!}$$

also have infinite radii of convergence, and inspection shows that

$$\cos z = \frac{e^{iz} + e^{-iz}}{2} \text{ and } \sin z = \frac{e^{iz} - e^{-iz}}{2i},$$

so that $e^{iz} = \cos z + i\sin z$. In particular, if $x \in \mathbf{R}$ then $\cos x$ and $\sin x$ are the real and imaginary parts of e^{ix}. Many of the results about the real-valued circular functions can be deduced from this.

Proposition 20.4.1   *The mapping $t \to e^{it} = \cos t + i\sin t$ from $\mathbf{R}$ to $\mathbf{T} = \{z : |z| = 1\}$ is a continuous homomorphism of the additive group $(\mathbf{R}, +)$ onto the multiplicative group $(\mathbf{T}, .)$, with kernel $2\pi\mathbf{Z}$.*

Proof   The mapping is certainly continuous, and is a homomorphism into $(\mathbf{T}, .)$. If $z = x+iy \in \mathbf{T}$ then $-1 \le x \le 1$; by the intermediate value theorem, there exists $s \in [0, \pi]$ such that $x = \cos s$. Then $y^2 = 1 - \cos^2 s = \sin^2 s$. If $y = \sin s$ take $t = s$, and if $y = -\sin s$ take $t = -s$. Then $e^{it} = z$, and so the mapping is surjective. Finally, $e^{it} = \cos t + i\sin t = 1$ if and only if $\cos t = 1$ and $\sin t = 0$, and this happens if and only if $t = 2\pi k$, for some $k \in \mathbf{Z}$. $\square$

Recall that $\mathbf{C}^*$, the punctured plane, is the set $\mathbf{C} \setminus \{0\}$.

Corollary 20.4.2   *The mapping $\exp : z \to e^z$ is a continuous homomorphism of the additive group $(\mathbf{C}, +)$ onto the multiplicative group $(\mathbf{C}^*, .)$, with kernel $\{2\pi ki : k \in \mathbf{Z}\}$.*

Proof   Again, the mapping is certainly continuous, and is a homomorphism into $(\mathbf{C}^*, .)$. If $w \in \mathbf{C}^*$ and $r = |w|$ then $w/r \in \mathbf{T}$, so that there exists $y \in \mathbf{R}$ such that $w/r = e^{iy}$. Let $x = \log r$. Then $r = e^x$, and so $w = e^x e^{iy} = e^z$, where $z = x + iy$; the mapping is surjective. Since $e^z = 1$ if and only if $z = 2\pi ki$ for some $k \in \mathbf{Z}$, its kernel is $\{2\pi ki : k \in \mathbf{Z}\}$. $\square$

Thus if $w \in \mathbf{C}^*$, we can write $w = re^{i\theta}$ with $r = |w|$ and $\theta \in \mathbf{R}$; this is the *polar form* of w. The number θ is not unique; we set $\text{Arg } w = \{\theta \in \mathbf{R} : w =$

$|w|e^{i\theta}\}$. The set $\mathrm{Arg}\, w$ is called the *argument* of w, and elements of $\mathrm{Arg}\, w$ are called *values* of the argument. There is a unique $\theta \in \mathrm{Arg}\, w \cap (-\pi, \pi]$; this is the *principal value* of the argument, and is denoted by $\arg w$. Then $\mathrm{Arg}\, w = \{\arg w + 2k\pi : k \in \mathbf{Z}\}$.

In the same way, if $w \in \mathbf{C}^*$ we set $\mathrm{Log}\, w = \{z \in \mathbf{C} : e^z = w\}$, so that $\mathrm{Log}\, w = \log|w| + i\mathrm{Arg}\, w$. Thus Log is a set-valued function on $\mathbf{C}^*$: an element of $\mathrm{Log}\, w$ is called a *value* of $\mathrm{Log}\, w$. We define the *principal logarithm* of w to be $\log|w| + i\arg w$, where $\arg$ is the principal value of the argument.

The *strip* $\{z = x + iy : -\pi < y < \pi\}$ is a connected open subset of $\mathbf{C}$, and the restriction of $\exp$ to the strip is a univalent map of the strip onto the *cut complex plane*

$$\mathbf{C}_0 = \mathbf{C} \setminus (-\infty, 0] = \{w = re^{i\theta} : r > 0, -\pi < \theta < \pi\}.$$

Then the restriction of $\log$ to $\mathbf{C}_0$ is the inverse mapping from the cut complex plane $\mathbf{C}_0$ onto the strip $\{z = x + iy : -\pi < y < \pi\}$. It is also a univalent mapping, and $\log' w = 1/w$, as in the real case.

Proposition 20.4.3 *If $|z| < 1$ then $\log(1 + z) = \sum_{n=1}^{\infty}(-1)^{n+1}z^n/n$.*

Proof For the power series on the right-hand side has radius of convergence 1; if $|z| < 1$, let $l(z) = \sum_{n=1}^{\infty}(-1)^{n+1}z^n/n$. Then

$$\frac{d}{dz}(\log(1 + z) - l(z)) = \frac{1}{1 + z} - \sum_{n=0}^{\infty}(-z)^n = 0,$$

so that $\log(1 + z) - l(z)$ is constant on $\{z : |z| < 1\}$. But $\log 1 = 0 = l(0)$, and so $\log(1 + z) = l(z)$, for $|z| < 1$. $\qquad\square$

The complex function $\log$ and the real function $\arg$ cannot be extended to continuous functions on $\mathbf{C}^*$, or on $\mathbf{T}$, since

$$\lim_{y\searrow 0} \log(-r + iy) = \log r + i\pi \text{ and } \lim_{y\nearrow 0} \log(-r + iy) = \log r - i\pi.$$

We can cut the complex plane in other ways. If $-\pi < \beta \le \pi$, let

$$\mathbf{C}_\beta = \mathbf{C} \setminus \{-re^{i\beta} : 0 \le r < \infty\} = \{w = re^{i\theta} : r > 0, \beta - \pi < \theta < \beta + \pi\}.$$

$\mathbf{C}_\beta$ is a cut plane, cut along a ray in the direction *opposite* to $e^{i\beta}$. If $w \in \mathbf{C}_\beta$ there exists a unique $\theta \in \mathrm{Arg}\, w \cap (\beta - \pi, \beta + \pi)$, which we denote by $\arg_\beta w$. Similarly, if $w \in \mathbf{C}_\beta$, we set $\log_{(\beta)} w = \log|w| + i\arg_\beta w$. Then $\log_{(\beta)}$ is a holomorphic function on $\mathbf{C}_\beta$.

Care is needed when working with principal values. If $w_1 = e^{\log w_1}$ and $w_2 = e^{\log w_2}$ are in $\mathbf{C}^*$, then

$$w_1 w_2 = e^{\log w_1} e^{\log w_2} = e^{\log w_1 + \log w_2},$$

so that $\log w_1 + \log w_2 \in \mathrm{Log}\,(w_1 w_2)$, but $\log w_1 + \log w_2$ need not equal $\log(w_1 w_2)$. For example, if $w = i - 1$ then

$$w = \sqrt{2} e^{3\pi i/4}, \text{ so that } 2\log w = \log 2 + 3\pi i/2,$$

whereas

$$w^2 = -2i, \text{ so that } \log(w^2) = \log 2 - \pi i/2 \neq 2\log w.$$

We can use the exponential and logarithmic functions to define complex powers. If $w \in \mathbf{C}^*$ and $\alpha \in \mathbf{C}$, we define $\{w^\alpha\}$ to be the set $\{\exp(\alpha z) : z \in \mathrm{Log}\,w\}$; any element of $\{w^\alpha\}$ is then a *value* of w^α. The *principal value* of w^α is obtained by taking the principal value of $\log w$. If $-\pi < \beta \leq \pi$, and $w \in \mathbf{C}_\beta$, we set $w^\alpha_{(\beta)} = \exp(\alpha \log_{(\beta)} w)$. Then the function $w \to w^\alpha_{(\beta)}$ is a holomorphic function on $\mathbf{C}_\beta$.

Exercises

20.4.1 The functions $\cosh z$ and $\sinh z$ are defined as

$$\cosh z = \frac{e^z + e^{-z}}{2} \text{ and } \sinh z = \frac{e^z - e^{-z}}{2}.$$

Write down their power series. Show that $\cos z = \cosh iz$ and $i\sin z = \sinh iz$, and prove the inequalities

$$|\sinh y| \leq |\sin z| \leq \cosh y, \qquad |\sinh y| \leq |\cos z| \leq \cosh y$$

for $z = x + iy \in \mathbf{C}$.

20.4.2 Find the zeros of $\cosh z$ and $\sinh z$, and of $\cos z + \sin z$.

20.4.3 Suppose that f is a complex-valued function on a domain U which does not contain 0, and that $f(re^{i\theta}) = u(r, \theta) + iv(r, \theta)$. Show that the Cauchy-Riemann equations become

$$\frac{\partial u}{\partial r} = \frac{1}{r}\frac{\partial v}{\partial \theta} \text{ and } \frac{\partial v}{\partial r} = -\frac{1}{r}\frac{\partial u}{\partial \theta}.$$

20.4.4 Suppose that $w \in \mathbf{C}^*$, and that $\arg w = -\beta$. Find the power series for $\log_{(\beta)}(w + z)$ in a neighbourhood of w. What is its radius of convergence?

20.4.5 Evaluate i^i.

20.4.6 Define the function z^z on the cut complex plane $\mathbf{C}_0$, and show that it is holomorphic. What is its derivative? What happens as $z \to 0$?

20.5 Infinite products

We now use properties of the complex functions exp and log to consider infinite products of the form $\prod_{j=1}^{\infty}(1+a_j)$, where $a_j \in \mathbf{C}$ and $|a_j| < 1$ for $j \in \mathbf{N}$. We need the following inequality.

Proposition 20.5.1 *Suppose that $z_1, \ldots, z_k$ are complex numbers for which $\sum_{j=1}^{k}|z_j| \le \sigma < \frac{1}{2}$. Then*

$$\left| \prod_{j=1}^{k}(1+z_j) - 1 \right| < \frac{\sigma}{1-2\sigma}.$$

Proof   If $|z| < 1$ then

$$|\log(1+z)| = \left| -\sum_{j=1}^{\infty}\frac{(-z)^j}{j} \right| \le \sum_{j=1}^{\infty}\frac{|z|^j}{j} = -\log(1-|z|) \le \sum_{j=1}^{\infty}|z|^j = \frac{|z|}{1-|z|},$$

and

$$|e^z - 1| = \left| \sum_{j=1}^{\infty}\frac{z^j}{j!} \right| \le \sum_{j=1}^{\infty}\frac{|z|^j}{j!} = e^{|z|} - 1 \le \sum_{j=1}^{\infty}|z|^j = \frac{|z|}{1-|z|}.$$

Thus

$$\left| \prod_{j=1}^{k}(1+z_j) - 1 \right| = \left| \exp\left(\sum_{j=1}^{k}\log(1+z_j) \right) - 1 \right|$$

$$\le \exp\left(\left| \sum_{j=1}^{k}\log(1+z_j) \right| \right) - 1$$

$$\le \exp\left(\sum_{j=1}^{k}|\log(1+z_j)| \right) - 1$$

$$\le \exp\left(\sum_{j=1}^{k}\frac{|z_j|}{1-|z_j|} \right) - 1 \le \exp(\sigma/(1-\sigma)) - 1$$

$$\le \sigma/(1-2\sigma).$$

$\square$

The following result corresponds to Corollary 14.2.10 of Volume II.

Proposition 20.5.2   (Weierstrass' uniform M-test for complex products)
*Suppose that (X, τ) is a topological space and that $(f_j)_{j=1}^{\infty}$ is a sequence
of bounded continuous complex-valued functions on X. If $\|f_j\|_{\infty} \leq M_j$ for
each $j \in \mathbf{N}$, and $\sum_{j=1}^{\infty} M_j < \infty$, then the infinite product $\prod_{j=1}^{\infty}(1 + f_j(x))$
converges uniformly to a bounded continuous function on X.*
  Further, if $\sum_{j=1}^{\infty} M_j = \sigma < \frac{1}{4}$ then $|\prod_{j=1}^{\infty}(1 + f_j(x)) - 1| \leq 2\sigma$.

Proof   First we show that the finite products are uniformly bounded. If
$k \in \mathbf{N}$ and $x \in X$, then

$$\left| \prod_{j=1}^{k}(1 + f_j(x)) \right| \leq \prod_{j=1}^{k}(1 + |f_j(x)|)$$

$$\leq \prod_{j=1}^{k}(1 + M_j) \leq \prod_{j=1}^{\infty}(1 + M_j) < \infty.$$

Thus there exists $K \geq 1$ such that $|\prod_{j=1}^{k}(1 + f_j(x))| \leq K$ for all $k \in \mathbf{N}$ and
all $x \in X$.
  Suppose that $0 < \epsilon < 1/2$. There exists j_0 such that $\sum_{j=j_0}^{\infty} M_j < \epsilon/2K$.
By Proposition 20.5.1, if $j_0 \leq j < l$ then

$$\left| \prod_{j=k+1}^{l}(1 + f_j(x)) - 1 \right| \leq \frac{\epsilon}{K},$$

and so

$$\left| \prod_{j=1}^{l}(1 + f_j(x)) - \prod_{j=1}^{k}(1 + f_j(x)) \right|$$

$$= \left| \prod_{j=1}^{k}(1 + f_j(x)) \right| \cdot \left| 1 - \prod_{j=k+1}^{l}(1 + f_j(x)) \right| < \epsilon.$$

Thus the products converge uniformly on X.
  The final statement follows by applying Proposition 20.5.1 to the product
of k terms, and then letting k tend to infinity.   □

20.6 The maximum modulus principle

Theorem 20.6.1 (The maximum modulus principle)   *Suppose that U is
a bounded domain and that f is a non-constant continuous function on $\overline{U}$*

whose restriction to U is analytic. If $z_0 \in U$ then

$$|f(z_0)| < \sup\{|f(z)| : z \in \partial U\}.$$

Proof   Since $\overline{U}$ is compact and $|f|$ is continuous on $\overline{U}$, $|f|$ is bounded on $\overline{U}$. Let $M = \sup\{|f(z)| : z \in \overline{U}\}$. Then $L = \{z \in \overline{U} : |f(z)| = M\}$ is a closed non-empty subset of $\overline{U}$.

We must show that $L \cap U$ is empty. Suppose not, and suppose that $z_0 \in L \cap U$. There exists $\delta > 0$ such that $N_\delta(z_0) \subseteq U$. Since f is analytic on U, there exists a power series $\sum_{n=0}^\infty a_n(z - z_0)^n$ which converges to $f(z)$ for $z \in N_\delta(z_0)$. Note that $a_0 = f(z_0)$, so that $|a_0| = M$. By Proposition 20.3.9, there is a least index k in $\mathbf{N}$ for which $a_k \neq 0$. Thus

$$f(z) = a_0 + a_k(z - z_0)^k + s(z)(z - z_0)^k, \text{ where } s(z) = \sum_{n=1}^\infty a_{k+n}(z - z_0)^n,$$

for $z \in N_\delta(z_0)$. Then $s(z) \to 0$ as $z \to z_0$, and so there exists $0 < \eta < \delta$ such that $|s(z)| < |a_k|/2$ for $|z - z_0| \leq \eta$. Now consider $f(z_0 + \eta e^{it})$, for $t \in [0, 2\pi)$:

$$f(z_0 + \eta e^{it}) = a_0 + \eta^k e^{ikt} a_k + \eta^k s(z_0 + \eta e^{it}),$$

so that

$$|f(z_0 + \eta e^{it})| \geq |a_0 + \eta^k e^{ikt} a_k| - \eta^k |a_k|/2.$$

Now $\mathrm{Arg}(\eta^k e^{ikt} a_k) = (\mathrm{Arg}\, a_k) + kt$, and so we can choose $t \in [0, 2\pi)$ such that $\mathrm{Arg}(\eta^k e^{ikt} a_k) = \mathrm{Arg}\, a_0$; a_0 and $\eta^k e^{ikt} a_k$ point in the same direction. Thus

$$|a_0 + \eta^k e^{ikt} a_k| = |a_0| + |\eta^k e^{ikt} a_k| = |a_0| + \eta^k |a_k|,$$

and so $|f(z_0 + \eta e^{it})| \geq |a_0| + \eta^k |a_k|/2 > |a_0| = M$. This gives a contradiction, and so $L \cap U = \emptyset$.   □

In particular, if f is not constant, $|f|$ has no local maxima in U.

Corollary 20.6.2   *If $|f(z)|$ is constant on ∂U, then f has a zero in U: there exists $z_0 \in U$ for which $f(z_0) = 0$.*

Proof   Let c be the value of $|f(z)|$ on ∂U. By the maximum modulus principle, $c > 0$. Suppose that $f(z) \neq 0$ for $z \in U$. Then the function $g = 1/f$ is continuous on $\overline{U}$, and its restriction to U is analytic. But if $z_0 \in U$ then $|g(z_0)| > 1/c = \sup\{|g(z)| : z \in \partial U\}$, contradicting the maximum modulus principle.   □

We shall improve on this in Corollary 20.6.6. As an application, let us give the first of several proofs of the fundamental theorem of algebra.

Corollary 20.6.3 (The fundamental theorem of algebra) *Suppose that* $p(z) = a_0 + \cdots + a_n z^n$ *is a non-constant complex polynomial function on* $\mathbf{C}$. *Then there exists* $z_1 \in \mathbf{C}$ *such that* $p(z_1) = 0$.

Proof   Suppose not. Then $a_0 = p(0) \neq 0$. Since p is not constant, $n > 0$, and we can suppose that $a_n \neq 0$. If $z \neq 0$ let

$$ p(z) = z^n \left(a_n + \frac{a_{n-1}}{z} + \cdots + \frac{a_0}{z^n} \right) = z^n h(z). $$

Then $h(z) \to a_n$ as $z \to \infty$, and so $|p(z)| \to \infty$ as $z \to \infty$. Thus there exists R such that $|p(z)| \geq 2|a_0|$ for $|z| \geq R$. Let

$$ V = \{ z \in \mathbf{C} : |p(z)| < 2|a_0| \}. $$

V is a non-empty bounded open subset; let U be a connected component. Then $|p(z)| = 2|a_0|$ for $z \in \partial U$ (justify this!). Then p has a zero in U, by the previous corollary. □

Corollary 20.6.4   *If* $a_n \neq 0$ *there exist* $z_1, \ldots, z_n$ *such that*

$$ p(z) = a_n(z - z_1) \ldots (z - z_n). $$

Proof   A straightforward induction argument. □

Theorem 20.6.5 (The open mapping theorem)   *If* f *is a non-constant analytic function on a domain* U, *and if* V *is an open subset of* U, *then* $f(V)$ *is an open subset of* $\mathbf{C}$.

Proof   Suppose that $z_0 \in V$. Let $g(z) = f(z) - f(z_0)$. Arguing as in Theorem 20.6.1, there exists $\delta > 0$ and $m > 0$ such that $M_\delta(z_0) = \{ z : |z - z_0| \leq \delta \} \subseteq U$ and such that $|g(z)| \geq m$ for $|z - z_0| = \delta$. Suppose now that $|\zeta - f(z_0)| < m/2$. Let $h(z) = f(z) - \zeta$. Then $|h(z)| > m/2$ for $z \in \mathbf{T}_\delta(z_0) = \{ z : |z - z_0| = \delta \}$, and $|h(z_0)| < m/2$. Let $W = \{ z \in M_\delta(z_0) : |h(z)| < m/2 \}$. Then W is a non-empty open set, and $\overline{W} \cap \mathbf{T}_\delta(z_0) = \emptyset$, so that $W \subseteq N_\delta(z_0)$. Let X be a connected component of W. As before, $|h(z)| = m/2$ for $z \in \partial X$, and so, by Corollary 20.6.2, there exists $w \in X$ such that $h(w) = 0$. Thus $f(w) = \zeta$, and so $f(N_\delta(z_0)) \supset N_{m/2}(f(z_0))$. This implies that $f(V)$ is open. □

Corollary 20.6.6   *Suppose that* U *is a bounded domain and that* f *is a non-constant continuous complex-valued function on* $\overline{U}$ *which is analytic on* U. *If* $|f(z)|$ *takes the constant value* c *on* ∂U, *then*

$$ f(U) = \{ z : |z| < c \} \text{ and } f(\partial U) = \{ z : |z| = c \}. $$

Proof   The set $f(\overline{U})$ is compact, and is therefore closed in $\mathbf{C}$. Since $f(U) = f(\overline{U}) \cap \{w : |w| < c\}$, $f(U)$ is closed in $\{w : |w| < c\}$ in the subspace topology. But $f(U)$ is open in $\{w : |w| < c\}$, by the open mapping theorem. The set $\{z : |z| < c\}$ is connected, and so $f(U) = \{z : |z| < c\}$. Since $f(\overline{U})$ is closed, it contains $\overline{f(U)} = \{z : |z| \leq c\}$ and so

$$f(\partial U) = \{z : |z| = c\}.$$

□

Compare the open mapping theorem with Theorem 20.2.3. We give another proof of the open mapping theorem in Section 23.5.

Exercises

20.6.1 Suppose that f is analytic on a bounded domain U and that

$$\limsup\nolimits_{(z \to w : z \in U)} |f(z)| \leq K$$

for each $w \in \partial U$. Show that $|f(z)| \leq K$ for each $z \in U$. [Hint: Show that if $L > K$ and $V = \{z \in U : |f(z)| > L\}$ then $\overline{V}$ is a compact subset of U.]

20.6.2 Let $U = \{z = x + iy : -\pi/2 < y < \pi/2\}$ and let $f(z) = \exp(e^z)$, for $z \in U$. (You may assume that f is analytic.) Show that

$$\limsup\nolimits_{(z \to w : z \in U)} |f(z)| = 1$$

for each $w \in \partial U$, but that f is unbounded on U.

20.6.3 Suppose that f is analytic on an unbounded domain U, that

$$\limsup\nolimits_{(z \to w : z \in U)} |f(z)| \leq K$$

for each $w \in \partial U$ and that $\limsup_{z \to \infty} |f(z)| \leq K$. Show that $|f(z)| \leq K$ for each $z \in U$.

20.6.4 Let $A_{r,R}$ be the annulus $\{z \in \mathbf{C}; r < |z| < R\}$. Show that if p is a polynomial then $\sup\{|p(z) - 1/z| : z \in A_{r,R}\} \geq \frac{1}{2}(1/r - 1/R)$. The function $J(z) = 1/z$ cannot be approximated uniformly on $A_{r,R}$ by polynomials.

21

The topology of the complex plane

21.1 Winding numbers

The complex analysis that we have so far developed is essentially a straight-forward development of ideas from real analysis. In the next chapter, we consider path integrals, and things will change dramatically. For this, we need to establish some of the topological properties of the complex plane $\mathbf{C}$. Since the mapping $(x, y) \to x + iy$ is an isometry of $\mathbf{R}^2$ onto $\mathbf{C}$, these properties correspond to topological properties of $\mathbf{R}^2$.

Suppose that (X, τ) is a topological space and that f is a continuous mapping from X into $\mathbf{C}^*$. A *continuous branch* of Arg f on X is a continuous mapping θ of (X, τ) into $\mathbf{R}$ such that $\theta(x) \in$ Arg $f(x)$ for each $x \in X$. We shall be concerned with the question of when continuous branches exist. Note that continuous branches are functions on X, and not on $f(X)$. As a particular case, if $X \subseteq \mathbf{C}^*$ and $f(z) = z$, then a *continuous branch*of Arg z on X is a continuous branch of the inclusion mapping of X into $\mathbf{C}^*$.

For example, the principal value mapping $z \to$ arg z is a continuous branch of Arg z on the cut plane $\mathbf{C}_0$. Similarly, the mapping $z \to \arg_\alpha z$ is a continuous branch of Arg z on the cut plane $\mathbf{C}_\alpha$.

Proposition 21.1.1 *Suppose that $f : (X, \tau) \to \mathbf{C}^*$ is continuous and that θ is a continuous branch of Arg f on (X, τ), that $x_0 \in X$ and that $t_0 \in$ Arg $f(x_0)$.*

(i) *If g is a continuous mapping of a topological space (Y, σ) into (X, τ), then $\theta \circ g$ is a continuous branch of Arg $(f \circ g)$ on Y. In particular, if Y is a subset of X, then the restriction of θ to Y is a continuous branch of Arg f on Y.*

(ii) *There exists a continuous branch θ_0 of Arg f on (X, d) with $\theta_0(x_0) = t_0$.*

(iii) If (X, d) is connected, the continuous branch θ_0 of Arg f on (X, d) with $\theta_0(x_0) = t_0$ is unique.

Proof (i) follows directly from the definition. For (ii), let $\theta_0 = \theta + (t_0 - \theta(x_0))$; θ_0 satisfies (ii). If θ_1 also satisfies (ii), then $(\theta_0 - \theta_1)/2\pi$ is a continuous integer-valued function, which vanishes at x_0; thus if X is connected, then $(\theta_0 - \theta_1)/2\pi = 0$, so that $\theta_0 = \theta_1$. □

Corollary 21.1.2 *There is no continuous branch of Arg z on $\mathbf{T} = \{z : |z| = 1\}$.*

Proof Suppose that a continuous branch existed on $\mathbf{T}$. Since $\mathbf{T}$ is connected, there would be a unique branch a on $\mathbf{T}$ with $a(1) = 0$. But $\mathbf{T} \setminus \{-1\}$ is connected, and so the restriction of a would be the principal value of the argument. But, as we saw in Section 20.4, arg has no continuous extension to $\mathbf{T}$. □

Recall that a *path* γ in $\mathbf{C}$ is a continuous mapping from a closed interval $[a, b]$ into C. $\gamma(a)$ is the *initial point* of the path, and $\gamma(b)$ is its *final point*, and γ is a path from a to b. The image $\gamma([a, b])$ is called the *track* from $\gamma(a)$ to $\gamma(b)$, and is denoted by $[\gamma]$. A path γ is *closed* if $\gamma(a) = \gamma(b)$; we return to our starting point. A path $\gamma : [a, b] \to \mathbf{C}$ is *simple* if γ is an injective mapping from $[a, b]$ into $\mathbf{C}$. A *simple closed path* $\gamma : [a, b] \to \mathbf{C}$ is a closed path whose restriction to $[a, b)$ is injective. If $\gamma : [a, b] \to X$ and $\delta : [c, d] \to X$ are paths, and $\gamma(b) = \delta(c)$, the *juxtaposition* $\gamma \vee \delta$ is the path from $[a, b + (d - c)]$ into X defined by $\gamma \vee \delta(x) = \gamma(x)$ for $x \in [a, b]$ and $\gamma \vee \delta(x) = \delta(x + (c - b))$ for $x \in [b, b + (d - c)]$. If $\gamma : [a, b] \to X$ is a path, the *reverse* $\gamma^{\leftarrow}(t)$ is defined as $\gamma^{\leftarrow}(t) = \gamma(a + b - t)$ for $t \in [a, b]$. If $\gamma : [a, b] \to X$ and $\delta : [c, d] \to X$ are paths, γ and δ are *similar paths*, or *equivalent paths*, if there exists a homeomorphism $\phi : [c, d] \to [a, b]$ such that $\phi(c) = a$, $\phi(d) = b$ and $\delta = \gamma \circ \phi$. Properties of paths are established in Volume II, Section 16.2.

Theorem 21.1.3 *If $\gamma : [a, b] \to \mathbf{C}^*$ is a path in $\mathbf{C}^*$ then there exists a continuous branch of Arg γ on $[a, b]$.*

If α and α' are two such continuous branches, then $\alpha(b) - \alpha(a) = \alpha'(b) - \alpha'(a)$.

Proof (i) We use the fact that $[a, b]$ is connected. If $s, t \in [a, b]$, set $s \sim t$ if there is a continuous branch of Arg γ on $[s, t]$. We show that this is an equivalence relation on $[a, b]$. Clearly $t \sim t$, and $t \sim s$ if $s \sim t$. Suppose that $s \sim t$ and $t \sim u$, and that θ is a continuous branch of Arg γ on $[s, t]$, θ' a continuous branch of Arg γ on $[t, u]$. Let $k = \theta(t) - \theta'(t)$, and let

$\theta(v) = \theta'(v) + k$, for $v \in [t, u]$. Then θ is a continuous branch of Arg γ on $[s, u]$, so that $s \sim u$.

Suppose that $s \in [a, b]$, and that $\arg \gamma(s) = \alpha$. Then $\gamma(s) \in C_\alpha$. Since γ is continuous, there exists $\delta > 0$ such that $\gamma(N_\delta(s)) \cap [a, b] \subseteq C_\alpha$. Then $\arg_\alpha \circ \gamma$ is a continuous branch of Arg γ on $N_\delta(s) \cap [a, b]$, and so the equivalence classes of $\sim$ are open. Since $[a, b]$ is connected, there is just one equivalence class, namely $[a, b]$, and so there exists a continuous branch of Arg γ on $[a, b]$.

(ii) The function $(\alpha - \alpha')/2\pi$ is a continuous integer-valued function on the connected set $[a, b]$, and is therefore constant.   □

Suppose that $\gamma : [a, b] \to C$ is a path and that w does not belong to the track $[\gamma]$ of γ. Then $\gamma - w$ is a path in C^*, and so there exists a continuous branch θ of Arg $(\gamma - w)$ on $[a, b]$. The *winding number* $n(\gamma, w)$ *of* γ *about* w is defined to be

$$n(\gamma, w) = \frac{\theta((\gamma - w)(b)) - \theta((\gamma - w)(a))}{2\pi} = \frac{\theta(\gamma(b) - w) - \theta(\gamma(a) - w)}{2\pi}.$$

It follows from Theorem 21.1.3 that this is well defined. In fact, we shall be principally concerned with the case where γ is a closed path, so that $\gamma(a) - w = \gamma(b) - w$, and $n(\gamma, w)$ is an integer.

As an easy example, let $\gamma(t) = w + re^{ikt}$ for $t \in [0, 2\pi]$, where $r > 0$ and $k \in Z$. Then kt is a continuous branch of Arg$(\gamma - w)$ on $[0, 2\pi]$, and so $n(\gamma, w) = k$. This accords with common sense: the path winds k times round w. But note that k can be positive, negative or zero; if $k > 0$ then γ winds k times round w in an anti-clockwise sense, and if $k < 0$ then γ winds $|k|$ times round w in a clockwise sense.

Here are some basic properties of winding numbers.

Proposition 21.1.4 *Suppose that* $\gamma : [a, b] \to C$ *is a path, and that* $w \notin [\gamma]$.

(i) *If* γ *is a constant path then* $n(\gamma, w) = 0$.

(ii) *If* $\gamma = \alpha \vee \beta$ *is the juxtaposition of two paths then* $n(\gamma, w) = n(\alpha, w) + n(\beta, w)$.

(iii) *If* $s : [c, d] \to [a, b]$ *is continuous, and* $s(c) = a$, $s(d) = b$ *then* $n(\gamma \circ s, w) = n(\gamma, w)$.

(iv) *If* $s : [c, d] \to [a, b]$ *is continuous, and* $s(c) = b$, $s(d) = a$ *then* $n(\gamma \circ s, w) = -n(\gamma, w)$.

Proof   The easy proofs are left as worthwhile exercises for the reader.   □

Corollary 21.1.5 *If* γ *and* δ *are similar paths, or similar closed paths, then* $n(\gamma, w) = n(\delta, w)$.

Corollary 21.1.6 $n(\gamma^{\leftarrow}, w) = -n(\gamma, w)$.

We can rotate, dilate and translate $\mathbf{C}$ without changing winding numbers.

Proposition 21.1.7 *Suppose that* $\gamma : [a, b] \to \mathbf{C}$ *is a path, and that* $w \notin [\gamma]$.

(i) *If* $\theta \in \mathbf{R}$ *then* $n(e^{i\theta}\gamma, e^{i\theta}w) = n(\gamma, w)$.
(ii) *If* $\lambda > 0$ *then* $n(\lambda\gamma, \lambda w) = n(\gamma, w)$.
(iii) *If* $b \in \mathbf{C}$ *then* $n(\gamma + b, w + b) = n(\gamma, w)$.

Proof   More easy exercises for the reader.   □

Proposition 21.1.8 *Suppose that* $\gamma : [a, b] \to \mathbf{C}$ *is a closed path.*
(i) *If* $w \notin [\gamma]$, *and if there exists* $\alpha \in (-\pi, \pi]$ *such that*

$$-\alpha \notin Arg\ (\gamma(t) - w)\ for\ a \le t \le b,$$

then $n(\gamma, w) = 0$.
(ii) *Suppose that* $\delta : [a, b] \to \mathbf{C}$ *is a closed path for which*

$$|\delta(t) - \gamma(t)| < |\gamma(t) - w| + |\delta(t) - w|\ for\ all\ t \in [a, b].$$

Then $w \notin [\gamma] \cup [\delta]$ *and* $n(\delta, w) = n(\gamma, w)$.

Proof   (i) The track $[\gamma - w]$ is contained in $\mathbf{C}_\alpha$ and $\arg_\alpha$ is a continuous branch of $\mathrm{Arg}\ z$ on $\mathbf{C}_\alpha$. Then

$$n(\gamma, w) = \arg_\alpha(\gamma(b) - w) - \arg_\alpha(\gamma(a) - w) = 0.$$

(ii) Translating and rotating if necessary, we can suppose that $w = 0$ and that $\gamma(a)$ is real and positive. Then the inequality implies that $0 \notin [\gamma] \cup [\delta]$ and that $\delta(t) \ne -\lambda\gamma(t)$, for some $\lambda > 0$. Let θ_γ be a continuous branch of $\mathrm{Arg}\ \gamma$ on $[a, b]$ with $\theta_\gamma(a) = \arg\gamma(a) = 0$, and let θ_δ be a continuous branch of $\mathrm{Arg}\ \delta$ on $[a, b]$ with $\theta_\delta(a) = \arg\delta(a)$. Since $\delta(a)$ is not real and negative, $-\pi < \theta_\delta(a) < \pi$, so that $|\theta_\delta(a) - \theta_\gamma(a)| < \pi$. We claim that $|\theta_\delta(t) - \theta_\gamma(t)| < \pi$ for all $t \in [a, b]$. If not, then by the intermediate value theorem there exists $t_0 \in [a, b]$ with $|\theta_\delta(t_0) - \theta_\gamma(t_0)| = \pi$. But then $\delta(t) = -\lambda\gamma(t)$ for some $\lambda > 0$, giving a contradiction. In particular, $|\theta_\delta(b) - \theta_\gamma(b)| < \pi$; thus

$$|n(\delta, 0) - n(\gamma, 0)| \le (|\theta_\delta(b) - \theta_\gamma(b)| + |\theta_\delta(a) - \theta_\gamma(a)|)/2\pi < 1.$$

Since $n(\delta, 0)$ and $n(\gamma, 0)$ are integers, it follows that $n(\delta, 0) = n(\gamma, 0)$.   □

The track $[\gamma]$ of a closed path γ is a compact subset of $\mathbf{C}$. Its complement $\mathbf{C} \setminus [\gamma]$ is an unbounded open subset of $\mathbf{C}$. It therefore has one unbounded connected component, and finitely many or countably many bounded components.

Corollary 21.1.9 *The function* $n_\gamma : \mathbf{C} \setminus [\gamma] \to \mathbf{Z}$ *defined by* $n_\gamma(w) = n(\gamma, w)$ *is continuous, and so is constant on each of the connected components of* $\mathbf{C} \setminus [\gamma]$. *If* w *is in the unbounded component of* $\mathbf{C} \setminus [\gamma]$ *then* $n(\gamma, w) = 0$.

Proof   Suppose that $w \in \mathbf{C} \setminus [\gamma]$. Let

$$\delta = d(w, [\gamma]) = \inf\{|\gamma(t) - w| : t \in [a, b]\}.$$

Since $[\gamma]$ is closed, $\delta > 0$. If $z \in N_\delta(w)$ and $t \in [a, b]$, then

$$|(\gamma(t) - w) - (\gamma(t) - z)| = |w - z| < |\gamma(t) - w|.$$

Thus

$$n(\gamma, z) = n(\gamma - z, 0) = n(\gamma - w, 0) = n(\gamma, w),$$

and so n_γ is continuous on $\mathbf{C} \setminus [\gamma]$. Since n_γ is integer-valued, it is constant on each of the connected components of $\mathbf{C} \setminus [\gamma]$.

Let $M = \sup\{|\gamma(t)| : t \in [a, b]\}$. If $r > M$ then $-r$ is in the unbounded connected component of $\mathbf{C} \setminus [\gamma]$, and $[\gamma + r] \cap C_0 = \emptyset$, so that $n(\gamma, -r) = n(\gamma + r, 0) = 0$. The result follows, since n_γ is constant on the unbounded connected component.   □

Exercises

21.1.1 Give the details of the proof of Proposition 21.1.7.

21.1.2 Suppose that f is holomorphic on a domain U and that $\arg f$ is constant on U. Show that f is constant on U.

21.1.3 Suppose that $\gamma_1 : [0, 1] \to \mathbf{C}^*$ and $\gamma_2 : [0, 1] \to \mathbf{C}^*$ are closed paths. Let $\gamma(t) = \gamma_1(t)\gamma_2(t)$. Show that $n(\gamma, 0) = n(\gamma_1, 0) + n(\gamma_2, 0)$.

21.1.4 Suppose that $\gamma_1 : [0, 1] \to \mathbf{C}$ and $\gamma_2 : [0, 1] \to \mathbf{C}$ are closed paths, and that $w \notin [\gamma_1]$. By considering the path

$$\eta(t) = 1 - \frac{\gamma_1(t) - \gamma_2(t)}{\gamma_1(t) - w} = \frac{\gamma_2(t) - w}{\gamma_1(t) - w},$$

show that if $|\gamma_1(t) - \gamma_2(t)| < |\gamma_1(t) - w|$ for $t \in [0, 1]$ then $w \notin [\gamma_2]$ and $n(\gamma_1, w) = n(\gamma_2, w)$.

21.1.5 Give an example of a closed rectifiable path γ in $\mathbf{C}$ for which

$$\{n(\gamma, w) : w \notin [\gamma]\} = \mathbf{Z}.$$

21.2 Homotopic closed paths

Proposition 21.1.8 shows that a small perturbation of a path does not change its winding number about a point. We can obtain further results like this. In order to do so, we need to introduce the notion of homotopy of closed paths.

Suppose that $\gamma_0 : [a, b] \to U$ and $\gamma_1 : [a, b] \to U$ are two closed paths in a domain U. Then γ_0 and γ_1 are *homotopic in U* if there is a continuous mapping $\Gamma : [0, 1] \times [a, b] \to U$ such that

(i) $\Gamma(0, t) = \gamma_0(t)$ and $\Gamma(1, t) = \gamma_1(t)$ for $t \in [a, b]$, and
(ii) $\Gamma(s, a) = \Gamma(s, b)$ for $s \in [0, 1]$.

Let us set $\gamma_s(t) = \Gamma(s, t)$. Then γ_s is a closed path in U. As s increases from 0 to 1, γ_s moves continuously from γ_0 to γ_1. The function Γ is called a *homotopy* connecting γ_0 and γ_1.

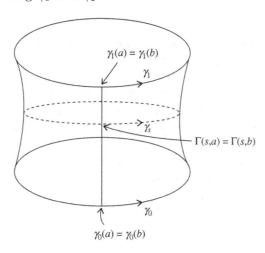

$\gamma_1(a) = \gamma_1(b)$

γ_1

γ_s

$\Gamma(s,a) = \Gamma(s,b)$

γ_0

$\gamma_0(a) = \gamma_0(b)$

Figure 21.2a.

Theorem 21.2.1 *Suppose that $\gamma_0 : [a, b] \to U$ and $\gamma_1 : [a, b] \to U$ are homotopic closed paths in a domain U, and that $w \in \mathbf{C} \setminus U$. Then $n(\gamma_0, w) = n(\gamma_1, w)$.*

Proof   Let Γ be a homotopy connecting γ_0 and γ_1. Since $\Gamma([0, 1] \times [a, b])$ is a compact subset of U, $\delta = d(w, \Gamma([0, 1] \times [a, b])) > 0$. Thus $|\Gamma(s, t) - w| \geq \delta$ for all $(s, t) \in [0, 1] \times [a, b]$. The function Γ is uniformly continuous on $[0, 1] \times [a, b]$, and so there exists $\eta > 0$ such that if $|u - s| < \eta$ then $|\Gamma(u, t) - \Gamma(s, t)| < \delta$ for all $t \in [a, b]$. If so, then

$$|\gamma_u(t) - \gamma_s(t)| < |\gamma_s(t) - w| \text{ for all } t \in [a, b].$$

Thus $n(\gamma_u, w) = n(\gamma_s, w)$, by Proposition 21.1.8, and so $n(\gamma_s, w)$ is a continuous function of s on $[0, 1]$. Since $n(\gamma_s, w)$ is integer-valued and $[0, 1]$ is connected, $n(\gamma_s, w)$ is constant, and so $n(\gamma_0, w) = n(\gamma_1, w)$. □

We say that a closed path γ in an open subset U of $\mathbf{C}$ is *null-homotopic in U* if it is homotopic in U to a constant path.

Corollary 21.2.2 *Suppose that γ is a null-homotopic closed path in U and that $w \notin U$. Then $n(\gamma, w) = 0$.*

Proposition 21.2.3 *If γ is a closed path in U and $\gamma = \gamma_1 \vee \gamma_2$, where γ_1 and γ_2 are null-homotopic closed paths in U, then γ is null-homotopic.*

Proof We can suppose that γ_1 maps $[a, b]$ into U and that γ_2 maps $[b, c]$ into U. Let δ be the constant path taking the value $\gamma_1(b)$. Then γ is homotopic to $\gamma_1 \vee \delta$, which is homotopic to $\delta \vee \delta$, which is homotopic to δ. □

As an example, let us show that a closed path in a domain U is homotopic to a dyadic rectilinear closed path.

Proposition 21.2.4 *Suppose that $\gamma : [0, 1] \to U$ is a closed path in a domain U and that $\delta > 0$. Then there is a dyadic rectilinear path $\beta : [0, 1] \to U$, with $|\beta(t) - \gamma(t)| < \delta$ for $t \in [0, 1]$, which is homotopic to γ.*

Proof We can suppose that $N_\delta([\gamma]) \subseteq U$. By Corollary 16.2.3 of Volume II, there exists a dyadic rectilinear path $\beta : [0, 1] \to U$ with $|\beta(t) - \gamma(t)| < \delta$ for $t \in [0, 1]$. Since γ is closed, we can also suppose that $\beta(0) = \beta(1)$. A homotopy is then given by setting $\Gamma(s, t) = (1 - s)\gamma(t) + s\beta(t)$. □

Let us give some applications of Theorem 21.2.1 and its corollary. Suppose that $w \in \mathbf{C}$ and that $r > 0$. Recall that the circular path $\kappa_r(w) : [0, 2\pi] \to \mathbf{C}$ is defined as $\kappa_r(w)(t) = w + re^{it}$, and that its track $\mathbf{T}_r(w) = \{z \in \mathbf{C} : |z - w| = r\}$ is the *circle* with centre a and radius r. Also $N_r(w) = \{z \in \mathbf{C} : |z - w| < r\}$ is the open r-neighbourhood of w, and $M_r(w) = \{z \in \mathbf{C} : |z - w| \leq r\}$ is the closed r-neighbourhood of w.

Proposition 21.2.5 *Suppose that f is a continuous complex-valued function on $M_r(w)$, and suppose that $z_0 \notin f(\mathbf{T}_r(w))$. If $n(f \circ \kappa_r(w), z_0) \neq 0$ then the equation $f(z) = z_0$ has a solution in $N_r(w)$.*

Proof Suppose not. Let $U = \mathbf{C} \setminus \{z_0\}$. Then f maps $M_r(w)$ into U. Let

$$\Gamma(s, t) = f(w + sre^{it}), \text{ for } (s, t) \in [0, 1] \times [0, 2\pi].$$

Then Γ is a homotopy in U connecting the constant path $f(w)$ to $\gamma_1 = f \circ \kappa_r$. Thus $f \circ \kappa_r(w)$ is null-homotopic in U, and so $n(f \circ \kappa_r(w), z_0) = 0$, giving a contradiction. $\qquad\square$

Notice that this result uses the convexity of $M_r(w)$.

We use this to give a second proof of the fundamental theorem of algebra. As in Corollary 20.6.3, it is enough to show that if

$$p(z) = a_0 + a_1 z + \cdots + a_n z^n, \text{ with } n > 0 \text{ and } a_n \neq 0,$$

then there exists z with $p(z) = 0$. Let $f(z) = a_n z^n$ and let $g(z) = a_0 + a_1 z + \cdots + a_{n-1} z^{n-1}$. Then there exists $R > 0$ such that $|g(z)| < |f(z)|$ for $|z| \geq R$. Let $\gamma(t) = R e^{it}$, for $t \in [0, 2\pi]$, so that $f(\gamma(t)) = a_n R^n e^{int}$. Then $n(f \circ \gamma, 0) = n$. Since

$$|p(\gamma(t)) - f(\gamma(t))| = |g(\gamma(t))| < |f(\gamma(t))| \text{ for } t \in [0, 2\pi],$$

$n(p \circ \gamma, 0) = n \neq 0$, by Proposition 21.1.8. Thus there exists $z \in M_R(0)$ with $p(z) = 0$, by the proposition.

Proposition 21.2.6   *Suppose that $f : M_r(w) \to M_r(w)$ is continuous. Then f has a fixed point: there exists $z \in M_r(w)$ with $f(z) = z$.*

Proof   Without loss of generality, we can suppose that $w = 0$. Let $g(z) = z - f(z)$ for $z \in M_r(0)$. We must show that the equation $g(z) = 0$ has a solution in $M_r(0)$. Suppose not. Let $\gamma(t) = \kappa_r(0)(t) = r e^{it}$. Let $h(t) = e^{-it} g(\gamma(t))$, for $0 \leq t \leq 2\pi$. Then

$$|h(t) - r| = |e^{-it} g(\gamma(t)) - e^{-it} \gamma(t)|$$
$$= |e^{-it} (g(\gamma(t)) - \gamma(t))| = |f(\gamma(t))| \leq r.$$

Also $g(z) \neq 0$ for $z \in M_r(0)$, and so $h(t) \neq 0$ for $t \in [0, 2\pi]$. Thus $[h] \subseteq \{z : \Re z > 0\} \subseteq \mathbf{C}_0$, and so $n(h, 0) = 0$. Let θ_h be a continuous branch of $\operatorname{Arg} h$ on $[0, 2\pi]$; then the function $t \to \theta_h(t) + t$ is a continuous branch of $\operatorname{Arg} g$ on $[0, 2\pi]$, so that

$$n(g, 0) = \frac{(\theta_h(2\pi) + 2\pi) - (\theta_h(0) + 0)}{2\pi} = n(h, 0) + 1 = 1.$$

Thus the equation $g(z) = 0$ must have a solution in $M_r(0)$, by Proposition 21.2.5. $\qquad\square$

Corollary 21.2.7   *Suppose that C is a compact convex body in $\mathbf{R}^2$. If $f : C \to C$ is continuous then it has a fixed point.*

Proof   For C is homeomorphic to $M_1(0)$ (Exercise 18.5.4).   □

A continuous mapping f of a topological space (X, τ) onto a subset Y of X is a *retract* of X onto Y if $f(y) = y$ for $y \in Y$. Suppose that $w \in \mathbf{C}$ and $r > 0$. If $z \in \mathbf{C}$ and $z \neq w$, let

$$\rho(z) = w + \frac{r(z - w)}{|z - w|};$$

$\rho(z)$ is the unique point in $\mathbf{T}_r(w) \cap R_{w,z}$, where

$$R_{w,z} = \{w + \lambda(z - w) : \lambda \geq 0\}$$

is the ray from w that contains z. The mapping $\rho : \mathbf{C} \setminus \{w\} \to \mathbf{T}_r(w)$ is a retract of $\mathbf{C} \setminus \{w\}$ onto $\mathbf{T}_r(w)$; it is the *natural retract* of $\mathbf{C} \setminus \{w\}$ onto $\mathbf{T}_r(w)$. The restriction of ρ to the punctured neighbourhood $M_r^\circ(w) = M_r(w) \setminus \{w\}$ is also a retract, of $M_r^\circ(w)$ onto $\mathbf{T}_r(w)$. We cannot do better.

Proposition 21.2.8   *There does not exist a retract f of $M_r(w)$ onto* $\mathbf{T}_r(w)$.

Proof   If $w + z \in \mathbf{T}_r(w)$, let $t(w + z) = w - z$. t is a homeomorphism of $\mathbf{T}_r(w)$ onto itself, called the *antipodal map*. Suppose that f is a retract of $M_r(w)$ onto $\mathbf{T}_r(w)$. Then $t \circ f$ is a continuous mapping from $M_r(w)$ to itself with no fixed point, giving a contradiction.   □

The next result is intuitively 'obvious', but requires proof.

Proposition 21.2.9   *Let $R = [a, b] \times [c, d]$ be a closed rectangle. Suppose that $h = (h_1, h_2) : [-1, 1] \to R$ and $v = (v_1, v_2) : [-1, 1] \to R$ are paths for which $h_1(-1) = a$ and $h_1(1) = b$, and $v_2(-1) = c$, $v_2(1) = d$. Then $[h] \cap [v]$ is not empty.*

Proof h is a path which joins the left and right sides of the rectangle R and v is a path which joins the bottom and top sides. The proposition says that the paths must meet.

Suppose that they do not, so that $h(s) \neq v(t)$, for $(s, t) \in [-1, 1] \times [-1, 1]$. We may clearly suppose that $R = [-1, 1] \times [-1, 1]$. Then R is the closed unit ball of $\mathbf{R}^2$, with norm $\|(s, t)\|_\infty = \max(|s|, |t|)$. Let

$$g(s, t) = (g_1(s, t), g_2(s, t)) = \frac{h(s) - v(t)}{\|h(s) - v(t)\|_\infty}, \quad \text{for } (s, t) \in [-1, 1] \times [-1, 1].$$

Thus $g(s, t)$ is the unit vector in the direction $h(s) - v(t)$, and so belongs to ∂R. Next, we reflect in the y-axis: let $f = (f_1, f_2) = (-g_1, g_2)$. f is a continuous mapping of R into ∂R. We shall show that f has no fixed point. This contradicts Corollary 21.2.7.

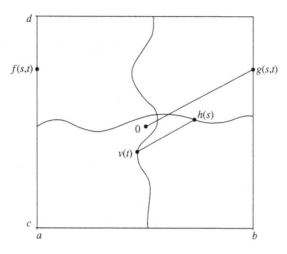

Figure 21.2b.

Suppose that (s_0, t_0) is a fixed point of f. Then $(s_0, t_0) \in \partial R$, so that

$$(s_0, t_0) = (f_1(s_0, t_0), f_2(s_0, t_0)) = (v_1(t_0) - h_1(s_0), h_2(s_0) - v_2(t_0)).$$

Thus
$$s_0 = v_1(t_0) - h_1(s_0) \text{ and } t_0 = h_2(s_0) - v_2(t_0).$$

The point (s_0, t_0) lies on one of the sides of the square ∂R. We consider each case in turn.

If $s_0 = -1$ then $h_1(s_0) = -1$ and $-1 = s_0 = v_1(t_0) + 1 \geq 0$;

if $s_0 = 1$ then $h_1(s_0) = 1$ and $1 = s_0 = v_1(t_0) - 1 \leq 0$;

if $t_0 = -1$ then $v_2(t_0) = -1$ and $-1 = t_0 = h_2(s_0) + 1 \geq 0$;

if $t_0 = 1$ then $v_2(t_0) = 1$ and $1 = t_0 = h_2(s_0) - 1 \leq 0$.

In each case, we obtain a contradiction.   □

Exercises

Homotopy can be defined in a more general setting. In these exercises, some of the basic theory is developed. Suppose that (X, d) is a metric space and that $x_0 \in X$. x_0 is called a *base point*. Let

$$L(X, x_0) = \{\gamma : \gamma \text{ is a closed path in } X \text{ with } \gamma(0) = \gamma(1) = x_0\}.$$

We define *juxtaposition* in a slightly different way. If $\gamma, \delta \in L(X, x_0)$, set

$$(\gamma \vee \delta)(t) = \gamma(2t) \text{ for } 0 \leq t \leq 1/2,$$
$$= \delta(2t - 1) \text{ for } 1/2 \leq t \leq 1.$$

If $\gamma, \delta \in L(X, x_0)$, a *homotopy* connecting γ and δ is a continuous mapping $h : [0,1] \times [0,1] \to X$ such that $h(0, t) = \gamma(t)$ and $h(1, t) = \delta(t)$ for $0 \leq t \leq 1$ and $h(s, 0) = h(s, 1) = x_0$ for $0 \leq s \leq 1$. We set $h_s(t) = h(s, t)$. γ is *null-homotopic* if it is homotopic to the constant map ϵ taking the value x_0.

21.2.1 Set $\gamma \sim \delta$ if there is a homotopy connecting γ and δ. Show that this is an equivalence relation. Let $\Pi_1(X, x_0)$ denote the quotient space of equivalence classes which this defines, and let $\{\gamma\}$ be the equivalence class to which γ belongs.

21.2.2 Suppose that $\gamma_1 \sim \gamma_2$ and $\delta_1 \sim \delta_2$. Show that $\gamma_1 \vee \delta_1 \sim \gamma_2 \vee \delta_2$. Use this to define a law of composition $*$ on $\Pi_1(X, x_0)$.

21.2.3 Prove associativity: show that

$$(\{\gamma_1\} * \{\gamma_2\}) * \{\gamma_3\} = \{\gamma_1\} * (\{\gamma_2\} * \{\gamma_3\}).$$

21.2.4 Let $e = \{\epsilon\}$ be the equivalence class of null-homotopic maps. Show that e is an identity element: $\{\gamma\} * e = e * \{\gamma\} = \{\gamma\}$.

21.2.5 Show that $\{\gamma\} * \{\gamma^{\leftarrow}\} = \{\gamma^{\leftarrow}\} * \{\gamma\} = e$. ($\{\gamma\}$ has an inverse). Thus $(\Pi_1(X, x_0), *)$ is a group, the *homotopy group* of (X, d) relative to x_0.

21.2.6 Suppose that (X, d) is path-connected and that $x_1 \in X$. Show that $(\Pi_1(X, x_0), *)$ and $(\Pi_1(X, x_1), *)$ are isomorphic.

21.2.7 The next few exercises show that the homotopy group need not be commutative. Let $U = \mathbf{R}^2 \setminus \{(1,0), (-1,0)\}$. Let

$$\gamma(t) = (1 - \cos 2\pi t, \sin 2\pi t) \text{ and } \delta(t) = (-1 + \cos 2\pi t, \sin 2\pi t)$$

for $0 \leq t \leq 1$, and let $\beta = \gamma \vee \delta \vee \gamma^{\leftarrow} \vee \delta^{\leftarrow}$. Show that $n(\beta, w) = 0$ for all $w \notin [\beta]$.

21.2.8 Show that there is a retract of U onto $[\beta]$. Deduce that if β is null-homotopic then there is a homotopy connecting β to a constant map taking values in $[\beta]$.

21.2.9 Suppose that h is such a homotopy. Let $C = \{s \in [0, 1] : [h_s] = [\beta]\}$. Use a compactness argument to show that C is closed.

21.2.10 Use the intermediate value theorem and a compactness argument to show that C is open.

21.2.11 Deduce that β is not null-homotopic, and that $\Pi_1(U, (0,0))$ is not commutative.

21.3 The Jordan curve theorem

The results of the previous section now enable us to prove one of the famous results of mathematics. To conform with tradition, we shall call a simple closed path in $\mathbf{C}$ a *Jordan curve*, although, in the terminology used in Volume II, it need not be a curve.

Theorem 21.3.1 (The Jordan curve theorem)   *Suppose that γ is a Jordan curve. Then $\mathbf{C} \setminus [\gamma]$ has exactly two connected components. One is unbounded (the 'outside') and one is bounded (the 'inside').*

We denote the outside of γ by $out[\gamma]$, and the inside by $in[\gamma]$. We denote the closure $[\gamma] \cup in[\gamma]$ of $in[\gamma]$ by $\overline{in}[\gamma]$. A point in the bounded connected component of $\mathbf{C} \setminus [\gamma]$ is said to be *inside* $[\gamma]$, and a point in the unbounded connected component to be *outside* $[\gamma]$. The theorem is intuitively true, but it needs to be proved. Bolzano was the first to observe this. Jordan gave a proof in 1887, but this was considered to be incomplete. A complete proof was given by Veblen in 1905, and many proofs have been given since then. We shall present the proof given by Ryuji Maehara[1] in 1984, and shall to a large extent use his notation.

Before proving the Jordan curve theorem, we prove two results, of interest in their own right.

Theorem 21.3.2   *Suppose that $\gamma : [a, b] \to \mathbf{C}$ is a simple path in $\mathbf{C}$ and that U is a non-empty bounded open subset of $\mathbf{C} \setminus [\gamma]$. Then ∂U is not contained in $[\gamma]$.*

Proof   Suppose that $\partial U \subseteq [\gamma]$. Let $w \in U$, and let $M_R(w)$ be a closed disc which contains $U \cup [\gamma] = \overline{U} \cup [\gamma]$. The mapping $\gamma^{-1} : [\gamma] \to [a, b]$ is a homeomorphism. By Tietze's extension theorem (Volume II, Theorem 14.4.3), there exists a continuous mapping $f : M_R(w) \to [a, b]$ which extends γ^{-1}. Thus if $r = \gamma \circ f$, r is a retract of $M_R(w)$ onto $[\gamma]$. Let $q(z) = r(z)$ for $z \in \overline{U}$ and let $q(z) = z$ for $z \in M_R(w) \setminus \overline{U}$. (Note that $q(z) = r(z) = z$, for $z \in \partial U$.) Then q is continuous on each of the closed sets $\overline{U}$ and $M_R(w) \setminus U$, and their union is $M_R(w)$, and so q is a continuous mapping of $M_R(w)$ onto $M_R(w) \setminus U$. Thus if ρ is the natural retract of $\mathbf{C} \setminus \{w\}$ onto $\mathbf{T}_R(w)$, $\rho \circ q$ is a continuous mapping of $M_R(w)$ onto $\mathbf{T}_R(w)$ which fixes the points of $\mathbf{T}_R(w)$, contradicting Proposition 21.2.8.   □

Corollary 21.3.3   *If γ is a simple path in $\mathbf{C}$ then $\mathbf{C} \setminus [\gamma]$ is connected.*

---

[1] *American Mathematical Monthly* 91 (1984) 641–643.

Proof   Suppose, if possible, that U is a bounded connected component of $\mathbf{C} \setminus [\gamma]$. Then $\partial U \subseteq [\gamma]$, giving a contradiction. Thus every connected component of $\mathbf{C} \setminus [\gamma]$ is unbounded. Since $[\gamma]$ is bounded, there can be only one unbounded connected component, and so $\mathbf{C} \setminus [\gamma]$ is connected.   □

Theorem 21.3.4   *Suppose that γ is a Jordan curve for which $\mathbf{C} \setminus [\gamma]$ has at least two connected components. If O is any one of these, then $\partial O = [\gamma]$.*

Proof   First suppose that O is a bounded connected component of $\mathbf{C} \setminus [\gamma]$. Then $\partial O \subseteq [\gamma]$. Suppose that $\partial O \neq [\gamma]$, and that $z_0 \in [\gamma] \setminus \partial O$. We can parametrize $[\gamma]$ as a closed path starting and finishing at z_0; there is a simple closed path $\beta : [0, 1] \to \mathbf{C}$ such that $[\beta] = [\gamma]$ and $\beta(0) = \beta(1) = z_0$. Since ∂O is closed, there exists $\delta > 0$ such that $\partial O \subseteq \beta([\delta, 1 - \delta])$. This contradicts Theorem 21.3.2.

Next, suppose that O is the unbounded connected component of $\mathbf{C} \setminus [\gamma]$, and let O' be a bounded connected component. Without loss of generality, we can suppose that $0 \in O'$. Let $j : \mathbf{C}^* \to \mathbf{C}^*$ be the inversion mapping $z \to 1/z$; j is a homeomorphism of $\mathbf{C}^*$ onto itself. Then $j \circ \gamma$ is a Jordan curve in $\mathbf{C}^*$ with path $j([\gamma])$, $j(O) \cup \{0\}$ is a bounded connected component of $\mathbf{C} \setminus j([\gamma])$ and $j(O' \setminus \{0\})$ is the unbounded connected component of $\mathbf{C} \setminus j([\gamma])$. By the result that we have just proved, $\partial(j(O) \cup \{0\}) = j([\gamma])$. Thus $\partial O = [\gamma]$.   □

Before proving the Jordan curve theorem, let us recall some notation, introduce some more, and set the scene. If $x, y \in \mathbf{C}$, we denote by $\sigma(x, y)$ the linear path from x to y: $\sigma(x, y)(t) = (1 - t)x + ty$ for $t \in [0, 1]$, and we denote its track by $[x, y]$. If γ is a simple path in $\mathbf{C}$ and $u = \gamma(r)$ and $v = \gamma(s)$ are points on its track, we denote by $\gamma(u, v)$ the restriction of γ to the part connecting u and v: if $r < s$ then $\gamma(u, v)$ is the restriction of γ to $[r, s]$; if $r > s$ then $\gamma(u, v)(t) = \gamma^{\leftarrow}(t) = \gamma(-t)$ for $t \in [-r, -s]$; if $r = s$, so that $u = v$, then $\gamma(u, v)(t) = u$ for $t \in [0, 1]$.

We shall work within the rectangle with vertices $\pm 1 \pm 2i$. We label certain points of ∂R as follows:

$$N = 2i, \ S = -2i, \ E = 1, \ W = -1,$$

$$NE = 1 + 2i, \ SE = 1 - 2i, \ NW = -1 + 2i, \ SW = -1 - 2i.$$

We call $[NW, NE]$ the *top* of ∂R, and $[SW, SE]$ the *bottom* of ∂R. The set ∂R is the track of a Jordan curve, and of course the Jordan curve theorem holds for it; let U be the inside of ∂R and V the outside. If $\gamma : [a, b] \to \mathbf{C}$ is a path with $\gamma(a) \in U$ and $\gamma(b) \in V$ then $\gamma^{-1}(U)$ and $\gamma^{-1}(V)$ are disjoint non-empty open subsets of $[a, b]$. Since $[a, b]$ is connected, it follows that

$\gamma^{-1}(\partial R)$ is a non-empty closed subset of $[a, b]$. If t is the least element of $\gamma^{-1}(\partial R)$ then $\gamma(t)$ is called the *exit point* of γ.

We now prove the Jordan curve theorem.

Proof   Suppose that γ is a Jordan curve. $[\gamma]$ is a compact subset of $\mathbf{C}$, and so there exist points a and b in $[\gamma]$ such that $|a - b| = \sup\{|c - d| : c, d \in [\gamma]\}$. By scaling, rotation and translation, we may suppose that $a = W$ and $b = E$. Then $[\gamma] \subseteq R$ and $[\gamma] \cap \partial R = \{W, E\}$. Further, we can split $[\gamma]$ into two: there exist simple paths $\gamma_1, \gamma_2 : [0, 1] \to [\gamma]$ such that

$$\gamma = \gamma_1 \vee \gamma_2^{\leftarrow}, \ \gamma_1(0) = \gamma_2(0) = W \text{ and } \gamma_1(1) = \gamma_2(1) = E.$$

We now consider $[\gamma] \cap [N, S]$. By Proposition 21.2.9, $[\gamma_1] \cap [N, S]$ and $[\gamma_2] \cap [N, S]$ are not empty. Let $l = \sup\{t \in [-2, 2] : it \in [\gamma]\}$, and let $L = il$. By relabelling if necessary, we can suppose that $L \in [\gamma_1]$. Next, let $m = \inf\{t \in [-2, 2] : it \in [\gamma_1]\}$, and let $M = im$. (It may well be that $L = M$.) Thus $[\gamma_1] \cap [N, S] \subseteq [L, M]$.

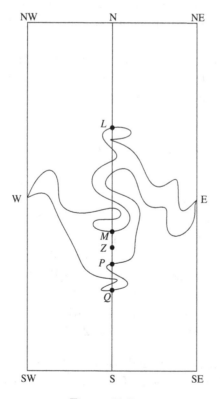

Figure 21.3a.

Now consider the path $\delta = \sigma(N, L) \vee \gamma_1(L, M) \vee \sigma(M, S)$. This connects the top and bottom of ∂R, and so $[\gamma_2] \cap [\delta]$ is not empty. But $[\gamma_2] \cap [N, L]$ is empty, and so is $[\gamma_2] \cap [\gamma_1(L, M)]$, and so $[\gamma_2] \cap [M, S]$ is not empty. Let $p = \sup\{t \in [-2, m] : it \in [\gamma_2]\}$ and let $q = \inf\{t \in [-2, m] : it \in [\gamma_2]\}$, Then $m > p \geq q > -2$. Let $P = ip$, $Q = iq$. Then $P \neq M$, but it may well be that $P = Q$. Finally, let $Z = \frac{1}{2}(M + P)$. See Figure 21.3a.

The point Z does not belong to $[\gamma]$. Let O be the connected component of $\mathbf{C} \setminus [\gamma]$ to which it belongs. First we show that O is bounded. If not, there exists a path ϵ in O from Z to a point outside R. Let $X = x + iy$ be the exit point of ϵ. Then $y \neq 0$, since $[\epsilon] \cap [\gamma] = \emptyset$. Suppose that $y < 0$. Then there exists a simple path ζ in ∂R from X to S. Now let

$$\eta = \sigma(N, L) \vee \gamma_1(L, M) \vee \sigma(M, Z) \vee \epsilon(Z, X) \vee \zeta.$$

Then η is a path joining the top and bottom of ∂R whose track is disjoint from $[\gamma_2]$; this contradicts Proposition 21.2.9. Suppose that $y > 0$. Then there exists a simple path θ in ∂R from N to X, so that $\theta \vee \epsilon(X, Z) \vee \sigma(Z, S)$ is a path joining the top and bottom of ∂R whose track is disjoint from $[\gamma_1]$, again giving a contradiction. Thus O is bounded. Since $\mathbf{C} \setminus [\gamma]$ has an unbounded connected component O_∞, there are at least two connected components of $\mathbf{C} \setminus [\gamma]$.

Since O_∞ is the only unbounded connected component of $\mathbf{C} \setminus [\gamma]$, it is enough to show that there are no more bounded connected components. Suppose, if possible, that O' is another bounded connected component. Since $O_\infty \supset V$, $O' \subseteq U$. Let

$$\iota = \sigma(N, L) \vee \gamma_1(L, M) \vee \sigma(M, P) \vee \gamma_2(P, Q) \vee \sigma(Q, S).$$

ι is a path joining the top and bottom of ∂R. Since neither W nor E is in $[\iota]$, there are neighbourhoods $N_\delta(W)$ and $N_\delta(E)$ disjoint from $[\iota]$. Now $[\gamma_1(L, M)] \cup [\gamma_2(P, Q)] \subseteq [\gamma] = \partial O$ by Theorem 21.3.4, $[N, L]$ and $[Q, S]$ are contained in $\overline{O}_\infty$, and $[M, P] \subseteq \overline{O}$ (since $Z \in O$). Consequently, $[\iota]$ is disjoint from O'. Since there are at least two connected components, $\partial O' = [\gamma]$, so that $W, E \in \overline{O}'$, and there are points $W' \in N_\delta(W) \cap O'$ and $E' \in N_\delta(E) \cap O'$. Since O' is path-connected, there is a path λ in O' joining W' and E'. Then the path $\lambda = \sigma(W, W') \vee \lambda \vee \sigma(E', E)$ is a path from W to E disjoint from $[\iota]$. Once again, this contradicts Proposition 21.2.9; the proof is complete. $\square$

We can say more about the inside of a Jordan curve γ. If w is outside $[\gamma]$ then $n(\gamma, w) = 0$. What happens if w is inside $[\gamma]$? Certainly the winding number is constant on the inside of γ.

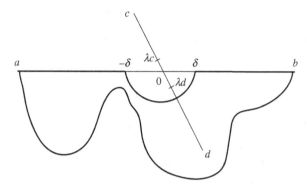

Figure 21.3b.

Theorem 21.3.5   *If z_0 is inside a Jordan curve γ, then $n(\gamma, z_0) = \pm 1$.*

Proof   First we consider the case where $[\gamma]$ contains a straight line segment $[a, b]$. Since we shall need this result later, we state it separately.

Proposition 21.3.6   *Suppose that γ is a Jordan curve whose track contains a straight line segment $[a, b]$. Suppose that $c, d \notin [\gamma]$ and that $[\gamma] \cap [c, d] = \{e\}$, where e is an interior point of the segment $[a, b]$. Then $|n(\gamma, c) - n(\gamma, d)| = 1$, so that one of $\{c, d\}$ is inside $[\gamma]$ and the other is outside.*

Proof   First we make some simplifications. By scaling, rotation and translation, we can suppose that $[a, b] \subseteq \mathbf{R}$, that $e = 0$ and that $a < 0 < b$. There exists a simple path β such that $\gamma = \sigma(a, b) \vee \beta$. Let $2\delta = \inf\{|z| : z \in [\beta]\}$. Since $[\beta]$ is a compact subset of $\mathbf{C}$, $\delta > 0$. There exists $\lambda > 0$ such that $|\lambda c| < \delta$ and $|\lambda d| < \delta$. Since $[\lambda c, c] \cap [\gamma] = \emptyset$, λc and c are in the same connected component of $\mathbf{C} \setminus [\gamma]$ and so $n(\gamma, c) = n(\gamma, \lambda c)$; similarly, $n(\gamma, d) = n(\gamma, \lambda d)$. This means that we can suppose that $|c| < \delta$ and $|d| < \delta$. Since $0 \in [c, d]$, we can suppose that the imaginary part of c is positive and that the imaginary part of d is negative. We reparametrize γ to start at $-\delta$; we can suppose that

$$\gamma = \sigma(-\delta, \delta) \vee \sigma(\delta, b) \vee \beta \vee \sigma(a, -\delta) = \sigma(-\delta, \delta) \vee \epsilon, \text{ say.}$$

Now let $\theta(t) = \delta e^{it}$ for $t \in [-\pi, 0]$; θ is a simple semicircular path from $-\delta$ to δ. Let $\eta = \theta \vee \epsilon$, and let $\kappa = \theta \vee \sigma(\delta, -\delta)$. Since $|c| < \delta$ and $|d| < \delta$, $[c, d] \cap [\eta] = \emptyset$, so that $n(\eta, c) = n(\eta, d)$. But

$$n(\eta, c) = n(\kappa \vee \gamma, c) = n(\kappa, c) + n(\gamma, c) = 0 + n(\gamma, c)$$
$$\text{and } n(\eta, d) = n(\kappa \vee \gamma, d) = n(\kappa, d) + n(\gamma, d) = 1 + n(\gamma, d),$$

so that $n(\gamma, c) - n(\gamma, d) = 1$.   □

Inspection of the proof shows that we have in fact shown the following.

Corollary 21.3.7 *Suppose that $a, b \in \mathbf{R}$ and that $a < b$. Suppose that $\gamma = \sigma(a, b) \vee \beta$ is a simple closed path, where β is a simple path from b to a whose track is in the upper half-space: if $x + iy \in [\beta]$ then $y \geq 0$. Then $n(\gamma, z) = 1$ for z inside $[\gamma]$.*

Proof Let $e = (a + b)/2$. Then $d = e + iy$ is outside $[\gamma]$ for negative y, and $c = e + iy$ is inside $[\gamma]$ for small positive y. The proof then shows that $n(\gamma, c) - n(\gamma, d) = 1$.   □

Now let us return to the proof of the theorem, and consider the general case. We can suppose that we are in the situation described in the proof of the Jordan curve theorem, and that $\gamma = \gamma_1 \vee \gamma_2^{\leftarrow}$.

We consider two Jordan curves. Let

$$\epsilon = \gamma_1(W, M) \vee \sigma(M, P) \vee \gamma_2^{\leftarrow}(P, W)$$
$$\text{and } \zeta = \gamma_2^{\leftarrow}(E, P) \vee \sigma(P, M) \vee \gamma_1(M, E).$$

Note that if $w \notin [\epsilon] \cup [\zeta]$ then $n(\gamma, w) = n(\epsilon, w) + n(\zeta, w)$. There exists $\eta > 0$ such that $N_\eta(W) \cap [\zeta] = \emptyset$. Since $W \in \overline{in}[\epsilon]$, there exists $W' \in N_\eta(W) \cap in[\epsilon]$. Thus $n(\epsilon, W') = \pm 1$ by Proposition 21.3.6. On the other hand, W is outside ζ, and $N_\eta(W)$ is connected, so that W' is outside $[\zeta]$ and $n(\zeta, W') = 0$. Thus $n(\gamma, W') = \pm 1$. This implies that W' is inside $[\gamma]$. Since $n(\gamma, z)$ is constant on the inside of $[\gamma]$, the result follows.   □

Thus if γ is a simple closed path and z is inside $[\gamma]$ then γ winds round z once, either in a clockwise sense or in an anti-clockwise sense. If $n(\gamma, z) = 1$ for z inside γ, we say that γ is *positively oriented*; if $n(\gamma, z) = -1$ for z inside γ, we say that γ is *negatively oriented*. If γ is positively oriented, then $\gamma^{\leftarrow}$ is negatively oriented. A positively oriented rectifiable simple closed path is called a *contour*.

Exercises

21.3.1 Suppose that γ_1, γ_2 and γ_3 are simple paths in $\mathbf{C}$ from a to b, with no points other than a and b in common. Thus the paths $\delta_1 = \gamma_2 \vee \gamma_3^{\leftarrow}$, $\delta_2 = \gamma_3 \vee \gamma_1^{\leftarrow}$ and $\delta_3 = \gamma_1 \vee \gamma_2^{\leftarrow}$ are Jordan curves. Let z_j be a point in $[\gamma_j] \setminus \{a, b\}$, for $j = 1, 2, 3$. Show that there is exactly one j such that z_j is inside $[\delta_j]$. [Hint: consider the proof of Theorem 21.3.1.]

21.3.2 Three utilities, gas, water and electricity, have plants at distinct points G, W and E, and wish to provide supplies to each of three distinct houses X, Y, Z. The plants and houses lie in a plane, and supplies are delivered along a simple path. Show that at least two paths must cross. What is the minimal number of crossings?

21.4 Surrounding a compact connected set

We now show that a compact connected subset of $\mathbf{C}$ can be squeezed between some simple closed dyadic rectilinear paths. We need a certain amount of notation. We begin with the set $\mathbf{Z} + i\mathbf{Z} = \{m + in : m, n \in \mathbf{Z}\}$. If $w = m + in \in \mathbf{Z} + i\mathbf{Z}$, there are linear paths to the four nearest elements of $\mathbf{Z} + i\mathbf{Z}$:

$$E_{m,n} = \sigma(w, w + 1), \quad N_{m,n} = \sigma(w, w + i),$$
$$W_{m,n} = \sigma(w, w - 1), \quad S_{m,n} = \sigma(w, w - i).$$

The path $E_{m,n} \vee N_{m+1,n} \vee W_{m+1,n+1} \vee S_{m,n+1}$ is then a simple closed path, the *square path* $sq_{m,n}$. Its inside is the *open square* $Q_{m,n}$, and its closure $Q_{m,n} \cup [sq_{m,n}]$ is the *closed square* $\overline{Q}_{m,n}$. We say that two squares are *adjacent* if they have an edge in common. For example, the closed squares $\overline{Q}_{m,n}$ and $\overline{Q}_{m+1,n}$ have an edge $[(m + 1) + in, (m + 1) + i(n + 1)]$ in common. Notice though that

$$[(m + 1) + in, (m + 1) + i(n + 1)] = [N_{m+1,n}] \subseteq [sq_{m,n}]$$
$$\text{and } [(m + 1) + i(n + 1), (m + 1) + in] = [S_{m+1,n+1}] \subseteq [sq_{m+1,n}];$$

the paths $sq_{m,n}$ and $sq_{m+1,n}$ traverse the edge in opposite directions. This elementary fact is of critical importance, since it leads to essential cancellation results.

We now scale all of the above by a factor 2^{-k}, where $k \in \mathbf{Z}$. We consider the set $(\mathbf{Z} + i\mathbf{Z})/2^k = \{(m + in)/2^k : m, n \in \mathbf{Z}\}$. Elements of $(\mathbf{Z} + i\mathbf{Z})/2^k$ are called *k-points*. If $m + in \in \mathbf{Z} + i\mathbf{Z}$, we set

$$E_{m,n}^{(k)} = E_{m,n}/2^k, \quad N_{m,n}^{(k)} = N_{m,n}/2^k, \quad W_{m,n}^{(k)} = W_{m,n}/2^k, \quad S_{m,n}^{(k)} = S_{m,n}/2^k.$$

Similarly, we set $sq_{m,n}^{(k)} = sq_{m,n}/2^k$ and $Q_{m,n}^{(k)} = Q_{m,n}/2^k$. Paths $E_{m,n}^{(k)}, N_{m,n}^{(k)}, W_{m,n}^{(k)}$ and $S_{m,n}^{(k)}$ are called *elementary k-paths*, and paths obtained by juxtaposing elementary k-paths are called *k-rectilinear paths*. The track $[\gamma]$ of

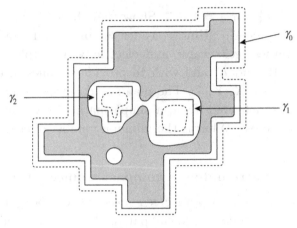

Figure 21.4.

a k-rectilinear path γ is the union of a finite number of rectilinear line seg-
ments of length $1/2^k$, which join a finite number of *vertices* $v_0, \ldots, v_n$. Two
vertices v_i and v_j are *adjacent* if $|i - j| = 1$. In particular the k-*square
path* $sq_{m,n}^{(k)}$ is a k-rectilinear path. Notice that a k-rectilinear path is also a
$(k+1)$-rectilinear path. The set $Q_{m,n}^{(k)}$ is an *open* k-*square* and $\overline{Q}_{m,n}^{(k)}$ is a
closed k-*square*.

Theorem 21.4.1   *Suppose that K is a non-empty compact connected sub-
set of* $\mathbf{C}$, *and that $\delta > 0$. Let $N_\delta(K) = \cup\{N_\delta(k) : k \in K\}$ be the open
δ-neighbourhood of K. Then there is a finite sequence $(\gamma_0, \ldots, \gamma_j)$ (here j
may be 0) of simple closed dyadic rectilinear paths in $N_\delta(K)$ such that*

(i) *K is inside γ_0;*
(ii) *K is outside γ_i for $1 \le i \le j$;*
(iii) *$\overline{in}[\gamma_r]$ is inside γ_0, for $1 \le r \le j$;*
(iv) *$\overline{in}[\gamma_r] \cap \overline{in}[\gamma_s] = \emptyset$ for $1 \le r < s \le j$.*

Proof   The idea of the proof is simple: we cover K with a finite collection
of small dyadic squares in such a way that the boundary of their union is
the track of finitely many disjoint closed dyadic rectilinear paths.
   There exists $l \in \mathbf{Z}$ such that $2^{-l} < \delta/2$. Then the set F of closed l-squares
which have a non-empty intersection with K is finite; list F as $(\overline{Q}_1, \ldots, \overline{Q}_w)$,
and let $G = \cup_{u=1}^{w} \overline{Q}_u$. Then G is a closed set, and $K \subseteq G \subseteq N_\delta(K)$. Suppose
that $k \in K$ belongs to the boundary of a l-square. Then k is an interior
point of the union of the l-squares to which it belongs, and so $k \notin \partial G$. Thus

K is contained in the interior of G. There therefore exists $m \geq l + 2$ such that $2^{-m} < d(\partial G, K)$.

The boundary ∂G is a finite union of some of the edges of the l-squares in F. Suppose that e is an edge contained in ∂G, and that e is an edge of $\overline{Q}_u = Q_{a,b}^{(l)}$. Then e is a subset of the l-square path $sq_{a,b}^{(l)}$. We use the orientation of $sq_{a,b}^{(l)}$ to orient e; it has a beginning point and an end point.

Let us now consider an element v of ∂G which is the corner of an l-square. It may belong to one, two or three l-squares in F. If it belongs to one or three l-squares in F, or to two adjacent l-squares in F, then it belongs to exactly two edges in ∂G, and is the beginning of one and the end of the other. If it belongs to two non-adjacent squares $\overline{Q}_r$ and $\overline{Q}_s$ in F, then it belongs to four edges in ∂G. We remove an m-square containing v from each of $\overline{Q}_r$ and $\overline{Q}_s$, to obtain disjoint closed sets H_r and H_s. We orient the edges of ∂H_r in such a way that each corner of ∂H_r is the beginning of an edge of ∂H_r and the end of an edge of ∂H_r, and so that the orientation of the edges of ∂Q_r which have not been changed are preserved; similarly for H_s.

We carry out this procedure for each vertex of this kind. As a consequence, we obtain closed sets $H_1, \ldots, H_w$, each a finite union of closed m-squares, such that $H_u \cap K \neq \emptyset$, for $1 \leq u \leq w$. Further, if $H = \cup_{u=1}^{w} H_u$, then

$$K \subseteq H^\circ \subset H \subseteq G \subseteq N_\delta(K).$$

The boundary ∂H is the union of edges of m-squares, and each element v of ∂H which is the corner of an m-square is the beginning of just one edge in ∂H and the end of just one other.

We now show that ∂H is the track of finitely many disjoint closed m-rectilinear paths. Suppose that $v_0 = (m_0 + in_0)/2^m$ is a vertex in ∂H for which $m_0 + n_0$ is as small as possible, and suppose that $v_0 \in H_{u_0}$. Then the edge $e_0 = [(m_0 + in_0)/2^m, ((m_0 + 1) + in_0)/2^m]$ is contained in ∂H, and v_0 is the beginning of e_0; let $v_1 = (m_0 + 1) + in_0)/2^m$ be its end. Then v_1 is the beginning of just one edge in ∂H; let v_2 be its end. We iterate this procedure until we reach a vertex which has already been listed. This must be v_0, since each vertex is the end of exactly one edge in ∂H. Thus we obtain a simple closed m-rectilinear path γ_0 in ∂H, with vertices $v_0, v_1, \ldots, v_p$. If $[\gamma_0] = \partial H$, the construction is finished. If not, choose $v_0^{(1)}$ a vertex in $\partial H \setminus [\gamma_0]$. It is the beginning of an edge in ∂H; let $v_1^{(1)}$ be its end, and repeat the procedure to obtain a simple closed m-rectilinear path γ_1 in ∂H, with $[\gamma_1]$ disjoint from $[\gamma_0]$. Repeat this procedure until all the vertices have been used. Thus we have simple closed m-rectilinear paths $\gamma_0, \gamma_1, \ldots, \gamma_j$ with disjoint tracks such that $\cup_{k=0}^{j}[\gamma_k] = \partial H$. We show that these paths satisfy the conclusions of the theorem.

First note that $P = (m_0 + \frac{1}{2} + i(n_0 - \frac{1}{2}))/2^m$ is outside each of the tracks $[\gamma_r]$, for $0 \le r \le j$. On the other hand, $Q = (m_0 + \frac{1}{2} + i(n_0 + \frac{1}{2}))/2^m$ is in the interior of H_{u_0}, and the line segment $[P, Q]$ meets e_0. Thus Q is inside $[\gamma_0]$. But there exists $k_0 \in K \cap H_{u_0}^{\circ}$, and $H_{u_0}^{\circ}$ is connected, and so k_0 is inside $[\gamma_0]$. Since K is connected, $K \subseteq in[\gamma_0]$. Thus (i) is satisfied. Since each H_u is connected, it follows that $H^{\circ} = in[\gamma_0]$.

If $1 \le r \le j$ then $[P, Q] \cap [\gamma_r] = \emptyset$, so that Q is outside $[\gamma_r]$. Then, arguing as for $[\gamma_0]$, we see that $K \subseteq out[\gamma_r]$. Thus (ii) is satisfied. Again, it follows that $H^{\circ} \subseteq out[\gamma_r]$.

Suppose now that $[v^{(r)}, v^{(r)} + 1/2^m]$ is a horizontal edge in $[\gamma_r]$. There exists $\lambda_r = \pm 1$ such that $Q_r = v^{(r)} + (1 + i\lambda_r)/2^{m+1} \in H^{\circ}$ and $P_r = v^{(r)} + (1 - i\lambda_r)/2^{m+1} \notin H^{\circ}$. Then Q_r is inside $[\gamma_0]$ and outside $[\gamma_s]$ for $1 \le s \le j$. But $[P_r, Q_r]$ meets $[\gamma_r]$ and none of the other paths, so that P_r is inside $[\gamma_r]$ and $[\gamma_0]$, and is outside $[\gamma_s]$ for $s \ne 0, r$. Conditions (iii) and (iv) follow from this. $\qquad \square$

Exercises

21.4.1 Prove the following generalization of Theorem 21.4.1.

Suppose that K is a non-empty compact subset of $\mathbf{C}$, and that $\delta > 0$. Let $N_\delta(K) = \cup\{N_\delta(k) : k \in K\}$. Then there is a finite sequence $(\gamma_1, \ldots, \gamma_j)$ of disjoint simple closed dyadic rectilinear paths in $N_\delta(K)$ and, for each $1 \le i \le j$ a finite set Δ_i of disjoint simple closed dyadic rectilinear paths in $N_\delta(K)$ such that

(i) $K \subseteq \cup_{i=1}^{j} in[\gamma_i]$;
(ii) $\overline{in}[\gamma_h] \cap \overline{in}[\gamma_i] = \emptyset$ for $1 \le h < i \le j$;
and, for each $1 \le i \le j$,
(iii) $[\delta]$ is inside $[\gamma_i]$ and K is outside $[\delta]$, for each $\delta \in \Delta_i$;
(iv) $\overline{in}[\delta] \cap \overline{in}[\delta'] = \emptyset$ for distinct $\delta, \delta' \in \Delta_i$.

21.5 Simply connected sets

A domain U is said to be *simply connected* if every closed path in U is null-homotopic.

Theorem 21.5.1 *Suppose that U is a domain. The following are equivalent:*

(i) *U is simply connected;*
(ii) *Every simple closed dyadic rectilinear path γ in U is null-homotopic;*

(iii) $n(\gamma, w) = 0$ *for all closed paths* γ *in* U *and all* $w \notin U$;

(iv) $n(\gamma, w) = 0$ *for all simple closed dyadic rectilinear paths* γ *in* U *and all* $w \notin U$;

(v) *If* γ *is a simple closed path in* U *then* $in[\gamma] \subseteq U$;

(vi) *If* γ *is a simple closed dyadic rectilinear paths in* U *then* $in[\gamma] \subseteq U$.

Proof   Clearly (i) implies (ii), (iii) implies (iv), and (v) implies (vi). It follows from Corollary 21.2.2 that (i) implies (iii), and (ii) implies (iv). If γ is a simple closed path in U and $w \in in[\gamma]$ then $n(\gamma, w) = \pm 1$, by Theorem 21.3.5. Thus (iii) implies (v) and (iv) implies (vi). It is therefore sufficient to show that (ii) implies (i) and that (vi) implies (ii).

Suppose that (ii) holds, and that γ is a simple closed path in U. There exists $\delta > 0$ such that $N_\delta([\gamma]) \subseteq U$, and by Proposition 21.2.4 there exists a closed dyadic rectilinear paths β in U which is homotopic to γ. The path β may not be simple, but we can suppose that $\beta = \beta_1 \vee \cdots \vee \beta_k$, where each β_j is simple. Then each β_j is null-homotopic, and so β is null-homotopic, by Proposition 21.2.3. Thus γ is null-homotopic, and (i) holds.

Suppose that (vi) holds. Suppose that γ is a simple closed k-dyadic rectilinear path in U. We prove that γ is null-homotopic by induction on the number $n_k(\gamma)$ of k-squares in $in[\gamma]$. If $n_k(\gamma) = 1$ then γ is a square path in U with $\overline{in}[\gamma] \subseteq U$, and so γ is clearly null-homotopic. Suppose that the result holds for all simple closed k-dyadic rectilinear paths in U with $n_k(\gamma) < n$, and that γ is a simple closed k-dyadic rectilinear path in U with $n_k(\gamma) = n$. There exists a vertex $v_0 = (m_0 + in_0)/2^k$ in $[\gamma]$ for which $m_0 + n_0$ is minimal, so that $((m_0 + 1) + in_0)/2^k$ and $(m_0 + i(n_0 + 1))/2^k$ are the two adjacent vertices. Let γ' be the path obtained by replacing v_0 by $v_0' = ((m_0 + 1) + i(n_0 + 1))/2^k$. Then γ and γ' are homotopic in U. There are now two possibilities. First, γ' is simple. Then $n_k(\gamma') = n - 1$, and so γ' is null-homotopic. Secondly, γ' is not simple. then $\gamma' = \delta \vee \epsilon$, where δ an ϵ are simple closed k-dyadic rectilinear paths in U with $n_k(\delta) < n$ and $n_k(\epsilon) < n$. Thus δ and ϵ are null-homotopic, and so therefore is γ', by Proposition 21.2.3. Thus γ is null-homotopic, and (ii) holds.   □

There is another important characterization of simply connected sets, this time for bounded sets. First we need an easy result.

Proposition 21.5.2   *Suppose that* K *is a compact subset of a domain* U. *Then there exists a compact connected subset* L *of* U *which contains* K.

Proof   There exists $\delta > 0$ such that $N_\delta(K) \subseteq U$. Since K is compact, there exists a finite subset $\{k_1, \ldots, k_n\}$ of K such that $K \subseteq \cup_{m=1}^{n} N_{\delta/2}(k_m)$. Since U is path-connected there exists, for each $2 \le m \le n$, a path γ_m in U from

k_1 to k_m. Let

$$L = \left(\cup_{m=1}^{n} M_{\delta/2}(k_m)\right) \cup \left(\cup_{m=2}^{n}[\gamma_m]\right).$$

Then L is a compact connected subset of U which contains K.   □

Corollary 21.5.3   *There exists a simple closed dyadic rectilinear path γ in U such that $K \subseteq in[\gamma]$.*

Proof   For there exists such a path for which $L \subseteq U$, by Theorem 21.4.1.
                        □

Theorem 21.5.4   *A bounded domain U is simply connected if and only if $\mathbf{C} \setminus U$ is connected.*

Proof   Suppose first that $\mathbf{C} \setminus U$ is connected and that γ is a path in U. Then $n(\gamma, w)$ is constant on $\mathbf{C}\setminus U$. Since $\mathbf{C}\setminus U$ is unbounded, it follows that $n(\gamma, w) = 0$ for all $w \notin U$, and U is simply connected.

Suppose next that $\mathbf{C} \setminus U$ is not connected. Let $E \cup F$ be a splitting of $\mathbf{C}\setminus U$, where E and F are disjoint non-empty closed subsets of $\mathbf{C}$. Since U is bounded, $\mathbf{C}\setminus U$ has just one unbounded connected component. Suppose that this is contained in E. Let $V = \mathbf{C}\setminus E = F \cup U$. Then V is a bounded open set, and F is a compact subset of V. There exists a connected component W of V such that $F \cap W \neq \emptyset$. Since W is closed in V, $F \cap W$ is a compact subset of W, and, by the preceding proposition there exists a compact connected subset L of W such that $F \cap W \subseteq L$. By Theorem 21.4.1, there exists a closed path γ_0 in $W \setminus L$ such that $L \subseteq in[\gamma_0]$. Since $[\gamma_0] \subseteq V = F \cup U$ and $[\gamma_0] \cap F = \emptyset$, $[\gamma_0] \subseteq U$. If $k \in F \cap W$ then $k \notin U$ and $n(\gamma_0, k) \neq 0$. Thus U is not simply connected.   □

Corollary 21.5.5   *If γ is a simple closed path then $in[\gamma]$ is simply connected.*

Proof   For $\mathbf{C} \setminus in[\gamma] = [\gamma] \cup out[\gamma] = \overline{out}[\gamma]$. The set $out[\gamma]$ is connected, and so therefore is $\overline{out}[\gamma]$, by Volume II, Corollary 16.1.7. Thus $in[\gamma]$ is simply connected.   □

What more can we say about $in[\gamma]$?

Theorem 21.5.6   *If γ is a simple closed path, there is a homeomorphism of the open unit disc $N_1(0)$ onto $in[\gamma]$.*

This is a consequence of the Riemann mapping theorem (Theorem 25.8.1), which we shall prove much later.

Exercises

21.5.1 Give an example of a domain U which is not simply connected, but for which $\mathbf{C} \setminus U$ is connected.

21.5.2 Give an example of a domain U which is simply connected, but for which $\mathbf{C} \setminus U$ is not connected.

21.5.3 Show that the set $\{z \in \mathbf{C} : r < |z| < R\}$ is not simply connected.

21.5.4 Let U be the domain $\mathbf{C} \setminus \{-1, 1\}$. Give an example of a closed path γ in U for which $n(\gamma, w) = 0$ for $w \notin U$, but which is not null-homotopic in U. (You need not prove that γ is not null-homotopic.)

21.5.5 Suppose that K is a non-empty compact connected subset of $\mathbf{C}$. Show that the unbounded connected component of $\mathbf{C} \setminus K$ is not simply connected, but that every bounded connected component of $\mathbf{C} \setminus K$ is simply connected.

22

Complex integration

22.1 Integration along a path

Suppose that $\gamma : [a, b] \to \mathbf{C}$ is a path. Recall that its length $l(\gamma)$ is defined as

$$l(\gamma) = \sup\{\sum_{j=1}^{n} |\gamma(t_j) - \gamma(t_{j-1})| : n \in \mathbf{N}, a = t_0 < \cdots < t_n = b\},$$

and that γ is rectifiable if $l(\gamma) < \infty$. Properties of rectifiable paths are considered in Volume II, Section 16.6. We now consider the integral of a continuous complex-valued function f along a rectifiable path γ in $\mathbf{C}$. Suppose that $D = (a = t_0 < \cdots < t_n = b)$ is a dissection of $[a, b]$. We set

$$S_D(f; \gamma) = \sum_{j=1}^{n} f(\gamma(t_j))(\gamma(t_j) - \gamma(t_{j-1})).$$

Note the similarity to the approximating sum of a Riemann integral; the increment $t_j - t_{j-1}$ is replaced by the change $\gamma(t_j) - \gamma(t_{j-1})$ in the path between t_{j-1} and t_j.

Recall that the mesh size of D is $\max\{|t_j - t_{j-1}| : 1 \leq j \leq n\}$. We want to show that as the mesh size of D tends to 0 these finite sums converge to an element of $\mathbf{C}$, the *path integral* $\int_\gamma f(z)\, dz$ *of f along γ*. We begin with a preliminary result.

Theorem 22.1.1 *Suppose that $\gamma : [a, b] \to \mathbf{C}$ is a rectifiable path and that f is a continuous complex-valued function on $[\gamma]$. Then given $\epsilon > 0$ there exists $\delta > 0$ such that if $D = (a = t_0 < \cdots < t_n = b)$ is a dissection of $[a, b]$ with mesh size less than δ and if $D' = (a = s_0 < \cdots < s_m = b)$ is a refinement of D then $|S_{D'}(f; \gamma) - S_D(f; \gamma)| < \epsilon$.*

Proof Since $f \circ \gamma$ is uniformly continuous on $[a, b]$, there exists $\delta > 0$ such that if $|s - t| < \delta$ then $|f(\gamma(s)) - f(\gamma(t))| < \epsilon/l(\gamma)$. Suppose that $D = (a = t_0 < \cdots < t_n = b)$ is a dissection of $[a, b]$ with mesh size less than δ and than $D' = (a = s_0 < \cdots < s_m = b)$ is a refinement of D. Then there exist $0 = i_0 < \cdots < i_n = m$ such that $t_j = s_{i_j}$ for $0 \le j \le n$. Now

$$S_{D'}(f; \gamma) = \sum_{j=1}^{n} \left(\sum_{i=i_{j-1}+1}^{i_j} f(\gamma(s_i))(\gamma(s_i) - \gamma(s_{i-1})) \right),$$

and

$$\left| \left(\sum_{i=i_{j-1}+1}^{i_j} f(\gamma(s_i))(\gamma(s_i) - \gamma(s_{i-1})) \right) - f(\gamma(t_j))(\gamma(t_j) - \gamma(t_{j-1})) \right|$$

$$= \left| \sum_{i=i_{j-1}+1}^{i_j} (f(\gamma(s_i)) - f(\gamma(t_j)))(\gamma(s_i) - \gamma(s_{i-1})) \right|$$

$$\le \sum_{i=i_{j-1}+1}^{i_j} |f(\gamma(s_i)) - f(\gamma(t_j))| . |\gamma(s_i) - \gamma(s_{i-1})|$$

$$\le \frac{\epsilon}{l(\gamma)} \sum_{i=i_{j-1}+1}^{i_j} |\gamma(s_i) - \gamma(s_{i-1})|,$$

so that

$$|S_{D'}(f; \gamma) - S_D(f; \gamma)| < \frac{\epsilon}{l(\gamma)} \sum_{j=1}^{n} \left(\sum_{i=i_{j-1}+1}^{i_j} |\gamma(s_i) - \gamma(s_{i-1})| \right) \le \epsilon.$$

□

Corollary 22.1.2 *Suppose that $\gamma : [a, b] \to \mathbf{C}$ is a rectifiable path and that f is a continuous complex-valued function on $[\gamma]$. Then there exists a unique complex number $I_\gamma(f)$ with the property that if $\epsilon > 0$ then there exists $\delta > 0$ such that if $D = (a = t_0 < \cdots < t_n = b)$ is a dissection of $[a, b]$ with mesh size less than δ then*

$$|I_\gamma(f) - S_D(f; \gamma)| \le 2\epsilon.$$

Proof Let D_n be the dissection of $[a, b]$ into 2^n intervals of equal length. Then D_m is a refinement of D_n, for $m > n$, and the mesh size of D_n tends to 0 as $n \to \infty$. It follows from the theorem that $(S_{D_n}(f; \gamma))_{n=1}^{\infty}$ is a Cauchy sequence in $\mathbf{C}$. Let $I_\gamma(f)$ be its limit. If D is a dissection of $[a, b]$ with mesh size less than δ, then

$$|S_{D_n}(f;\gamma) - S_D(f;\gamma)|$$
$$\le |S_{D_n}(f;\gamma) - S_{D_n \vee D}(f;\gamma)| + |S_{D_n \vee D}(f;\gamma) - S_{D_n}(f;\gamma)| < 2\epsilon,$$

so that $|I_\gamma(f) - S_D(f;\gamma)| \le 2\epsilon$.   □

We denote $I_\gamma(f)$ by $\int_\gamma f(z)\,dz$. Then $\int_\gamma f(z)\,dz$ is uniquely determined. It is a *path integral*, the *integral of f along the path γ*. The quantities $\{S_D(f,\gamma) : D$ a dissection of $[a, b]\}$ are approximating sums to the integral.

Proposition 22.1.3   *Suppose that $\gamma : [a, b] \to \mathbf{C}$ is a rectifiable path and that f is a continuous complex-valued function on $[\gamma]$. Then*

$$\left| \int_\gamma f(z)\,dz \right| \le \|f\|_\infty . l(\gamma).$$

Proof   For $|S_D(f;\gamma)| \le \|f\|_\infty . l(\gamma)$ for any dissection D of $[a, b]$.   □

The path integral does not depend upon the parametrization of the path.

Corollary 22.1.4   *Suppose that the path γ is similar to the path γ' : $[c, d] \to E$. Then $\int_{\gamma'} f(z)\,dz = \int_\gamma f(z)\,dz$.*

Proof   Suppose that $\epsilon > 0$. There exists $\delta > 0$ such that if D' is a dissection of $[c, d]$ with mesh size less than δ, and if D is a dissection of $[a, b]$ with mesh size less than δ then

$$|S_{D'}(f;\gamma') - \int_{\gamma'} f(z)\,dz| < \epsilon/2 \text{ and } |S_D(f;\gamma) - \int_\gamma f(z)\,dz| < \epsilon/2.$$

There is a strictly increasing continuous map ϕ of $[c, d]$ onto $[a, b]$ such that $\gamma' = \gamma \circ \phi$. Since ϕ is uniformly continuous on $[c, d]$, there exists $0 < \eta \le \delta$ such that if $s, s' \in [c, d]$ and $|s - s'| < \eta$ then $|\phi(s) - \phi(s')| < \delta$. If D' is a dissection of $[c, d]$ with mesh size less than η then the image dissection $\phi(D')$ has mesh size less than δ. Since $S_{D'}(f;\gamma') = S_{\phi(D')}(f;\gamma)$ it follows that

$$\left| \int_{\gamma'} f(z)\,dz - \int_\gamma f(z)\,dz \right|$$
$$\le \left| \int_\gamma f(z)\,dz - S_{\phi(D')}(f;\gamma) \right| + \left| \int_{\gamma'} f(z)\,dz - S_{D'}(f;\gamma') \right| < \epsilon.$$

Since ϵ is arbitrary, the result follows.   □

Here are some straightforward results.

Proposition 22.1.5 *Suppose that γ and γ' are rectifiable paths in $\mathbf{C}$, and that the final point of γ is the initial point of γ'. Suppose that f and g are continuous functions on $[\gamma]$ and that $\alpha \in \mathbf{C}$.*

(i) $\int_\gamma (f(z) + g(z))\, dz = \int_\gamma f(z)\, dz + \int_\gamma g(z)\, dz.$
(ii) $\int_\gamma \alpha f(z)\, dz = \alpha \int_\gamma f(z)\, dz.$
(iii) $\int_{\gamma \vee \gamma'} f(z)\, dz = \int_\gamma f(z)\, dz + \int_{\gamma'} f(z)\, dz.$

Proof   The proofs are left as an exercise for the reader.   □

A path $\gamma : [a, b] \to \mathbf{C}$ is *piecewise smooth* if there is a dissection $D = (a = t_0 < \cdots < t_n = b)$ of $[a, b]$ such that γ is continuously differentiable on $[t_{j-1}, t_j]$ (with one-sided derivatives at t_{j-1} and t_j), for $1 \le j \le n$.

Theorem 22.1.6   *A piecewise smooth path $\gamma : [a, b] \to \mathbf{C}$ is rectifiable, and $l(\gamma) = \int_a^b |\gamma'(t)|\, dt.$*

Proof   We can clearly suppose that γ is continuously differentiable on $[a, b]$. Suppose that $D = (a = t_0 < \cdots < t_n = b)$ is a dissection of $[a, b]$. Then

$$\sum_{j=1}^n |\gamma(t_j) - \gamma(t_{j-1})| = \sum_{j=1}^n \left| \int_{t_{j-1}}^{t_j} \gamma'(t)\, dt \right|$$

$$\le \sum_{j=1}^n \int_{t_{j-1}}^{t_j} |\gamma'(t)|\, dt = \int_a^b |\gamma'(t)|\, dt.$$

On the other hand, suppose that $\epsilon > 0$. Since γ' is uniformly continuous, there exists $\delta > 0$ such that if $|s - t| < \delta$ then $|\gamma'(s) - \gamma'(t)| < \epsilon / 2((b-a)+1)$. There exists a dissection $D = (a = t_0 < \cdots < t_n = b)$ of $[a, b]$ with mesh size less than δ for which

$$\left| \int_a^b |\gamma'(t)|\, dt - \sum_{j=1}^n |\gamma'(t_j)|(t_j - t_{j-1}) \right| < \frac{\epsilon}{2(b - a)}.$$

But

$$\left| \sum_{j=1}^n |\gamma(t_j) - \gamma(t_{j-1})| - \sum_{j=1}^n |\gamma'(t_j)|(t_j - t_{j-1}) \right|$$

$$\le \sum_{j=1}^n \Big| |\gamma(t_j) - \gamma(t_{j-1})| - |\gamma'(t_j)|(t_j - t_{j-1}) \Big|$$

$$\leq \sum_{j=1}^{n} |\gamma(t_j) - \gamma(t_{j-1}) - \gamma'(t_j)(t_j - t_{j-1})|$$

$$= \sum_{j=1}^{n} \left| \left(\int_{t_{j-1}}^{t_j} (\gamma'(t) - \gamma'(t_j)) \, dt \right) \right|$$

$$\leq \sum_{j=1}^{n} \int_{t_{j-1}}^{t_j} |\gamma'(t) - \gamma'(t_j)| \, dt$$

$$\leq \sum_{j=1}^{n} \frac{\epsilon(t_j - t_{j-1})}{2(b-a)} = \epsilon/2.$$

Hence

$$\left| \sum_{j=1}^{n} |\gamma(t_j) - \gamma(t_{j-1})| - \int_{a}^{b} |\gamma'(t)| \, dt \right| < \epsilon,$$

so that $l(\gamma) > \int_{a}^{b} |\gamma'(t)| \, dt - \epsilon$. Since ϵ is arbitrary, the result follows. $\qquad \square$

In fact, almost all path integrals that arise are path integrals along a piecewise smooth path. Such integrals can be expressed as the integral of a complex function of a real variable.

Theorem 22.1.7 *Suppose that $\gamma : [a, b] \to E$ is a piecewise smooth path in $\mathbf{C}$, and that f is a continuous complex-valued function on $[\gamma]$. Then*

$$\int_{\gamma} f(z) \, dz = \int_{a}^{b} f(\gamma(t)) \gamma'(t) \, dt.$$

Proof For

$$\left| \int_{a}^{b} f(\gamma(t)) \gamma'(t) \, dt - S_D(f; \gamma) \right|$$

$$= \left| \sum_{j=1}^{n} \left(\int_{t_{j-1}}^{t_j} f(\gamma(t)) \gamma'(t) \, dt - f(\gamma(t_j))(\gamma(t_j) - \gamma(t_{j-1})) \right) \right|$$

$$= \left| \sum_{j=1}^{n} \left(\int_{t_{j-1}}^{t_j} (f(\gamma(t)) - f(\gamma(t_j))) \gamma'(t) \, dt \right) \right|$$

$$\le \sum_{j=1}^{n} \int_{t_{j-1}}^{t_j} |f(\gamma(t)) - f(\gamma(t_j))| \cdot |\gamma'(t)| \, dt$$

$$\le \frac{\epsilon}{2l(\gamma)} \sum_{j=1}^{n} \int_{t_{j-1}}^{t_j} |\gamma'(t)| \, dt = \epsilon/2.$$

Thus

$$\left| \int_{\gamma} f(z) \, dz - \int_{a}^{b} f(\gamma(t)) \gamma'(t) \, dt \right| < \epsilon.$$

Since ϵ is arbitrary, the result follows. □

Example 22.1.8 Suppose that $\gamma : [0, 1] \to [z_0, z_1]$ is the linear path from z_0 to z_1, defined as $\gamma(t) = (1 - t)z_0 + tz_1$. Then

$$\int_{\gamma} f(z) \, dz = (z_1 - z_0) \int_{0}^{1} f((1 - t)z_0 + tz_1) \, dt.$$

For $\gamma'(t) = z_1 - z_0$.

In particular, if $z_0 = x_0 + iy$ and $z_1 = x_1 + iy$, then, changing variables,

$$\int_{\gamma} f(z) \, dz = \int_{x_0}^{x_1} f(s + iy) \, ds \text{ if } x_0 < x_1$$

$$= - \int_{x_1}^{x_0} f(s + iy) \, ds \text{ if } x_0 > x_1.$$

Similarly, if $z_0 = x + iy_0$ and $z_1 = x + iy_1$ then

$$\int_{\gamma} f(z) \, dz = i \int_{y_0}^{y_1} f(x + it) \, dt \text{ if } y_0 < y_1$$

$$= -i \int_{y_1}^{y_0} f(x + it) \, dt \text{ if } y_0 > y_1.$$

Example 22.1.9 Suppose that f is a continuous function on a circle $\mathbf{T}_r(w)$. Then

$$\int_{\kappa_r(w)} f(z) \, dz = ir \int_{0}^{2\pi} f(w + re^{it}) e^{it} \, dt$$

and $\int_{\kappa_r(w)} \frac{f(z)}{z - w} \, dz = i \int_{0}^{2\pi} f(w + re^{it}) \, dt,$

where $\kappa_r(w)$ is the circular path defined by $\kappa_r(w)(t) = w + re^{it}$, for $t \in [0, 2\pi]$.

For $\kappa_r'(w)(t) = ire^{it}$ and $z - w = re^{it}$.

Exercises

22.1.1 Evaluate the integrals

$$\int_{\kappa_1(0)} z \, dz, \quad \int_{\kappa_1(1)} z \, dz, \quad \int_{\kappa_1(0)} \bar{z} \, dz \quad \text{and} \quad \int_{\kappa_1(1)} \bar{z} \, dz.$$

22.1.2 Suppose that $\gamma : [a,b] \to \mathbf{C}$ is a rectifiable path in $\mathbf{C}$ and that f is a continuous function on $[\gamma]$. Suppose that $\alpha(t)$ and $\beta(t)$ are the real and imaginary parts of $\gamma(t)$. Show that α and β are rectifiable paths in $\mathbf{R}$. Suppose that α and β are strictly monotonic. Show that there are continuous real-valued functions u_r and v_r on $[\alpha]$ and u_i and v_i on $[\beta]$ such that

$$f(\gamma(t)) = u_r(\alpha(t)) + iv_r(\alpha(t)) = u_i(\beta(t)) + iv_i(\beta(t)),$$

for $t \in [a,b]$. Show that

$$\int_\gamma f(z) \, dz = \left(\int_\alpha u_r(z) \, dz - \int_\beta v_i(z) \, dz \right)$$

$$+ i \left(\int_\alpha v_r(z) \, dz + \int_\beta u_i(z) \, dz \right).$$

Is the result true if the word 'strictly' is omitted?

22.1.3 Let γ be the square path with corners 1, i, -1 and $-i$. Calculate $\int_\gamma dz/z$.

22.1.4 Suppose that f is a continuous function on $\mathbf{T}$. Show that

$$\int_{\kappa_1(0)} \frac{f(z)}{z^{n+1}} \, dz = i \int_0^{2\pi} f(e^{i\theta}) e^{-in\theta} \, d\theta.$$

22.1.5 Let $\gamma(t) = t + it \sin(1/t)$ for $0 < t \le 1$ and let $\gamma(0) = 0$. Show that γ is a continuous path in $\mathbf{C}$ which is not rectifiable, but that the restriction γ_ϵ of γ to $[\epsilon, 1]$ is rectifiable, for $0 < \epsilon < 1$. Suppose that f is a continuous function on $[\gamma]$. Show that $\int_{\gamma_\epsilon} f(z) \, dz$ tends to a limit as $\epsilon \searrow 0$.

22.2 Approximating path integrals

We have defined path integrals of continuous functions along rectifiable paths. It is useful to approximate these by integrals along polygonal and rectilinear paths. The proofs use rather standard approximation arguments.

Theorem 22.2.1 *Suppose that* $\gamma : [a, b] \to U$ *is a rectifiable path in a domain* U, *that* f *is a continuous complex-valued function on* U, *that* H *is a dense subset of* U *and that* $\epsilon > 0$. *Then there exists a polygonal path* $\beta : [a, b] \to U$ *with vertices in* H *such that* $|\beta(t) - \gamma(t)| < \epsilon$ *for* $t \in [a, b]$ *and* $|\int_\beta f(z) \, dz - \int_\gamma f(z) \, dz| < \epsilon$. *If* γ *is a closed path, then* β *can be chosen to be a closed path, homotopic to* γ *in* U.

Proof Since $[\gamma]$ is compact, f is uniformly continuous on $[a, b]$ and, if $U \neq \mathbf{C}$, $d([\gamma], \mathbf{C} \setminus U) > 0$. There therefore exist $0 < \delta < \epsilon$ and a partition $D = (t_0 = a < t_1 < \cdots < t_k = b)$ such that

(i) $N_\delta([\gamma]) \subseteq U$;
(ii) $|\int_\gamma f(z) \, dz - S_D(f; \gamma)| < \epsilon/4$;
(iii) if $t \in [a, b]$ and $|w - \gamma(t)| < \delta$ then $|f(w) - f(\gamma(t))| < \epsilon/4l(\gamma)$;
(iv) $|\gamma(t) - \gamma(t_j)| < \delta/4$ for all $t \in [t_{j-1}, t_j]$ and $1 \leq j \leq k$.

Let $\eta = \min(\delta/4, l(\gamma)/2k)$. Since H is dense in U, there exist $h_j \in U$ with $|h_j - \gamma(t_j)| < \eta$, for $0 \leq j \leq k$; if γ is closed, we can take $h_k = h_0$. For $1 \leq j \leq k$, let $\sigma_j : [t_{j-1}, t_j] \to [h_{j-1}, h_j]$ be the linear path from h_{j-1} to h_j, parametrized by the interval $[t_{j-1}, t_j]$, and let b be the polygonal path $\sigma_1 \vee \cdots \vee \sigma_k$. We shall show that β satisfies the conditions of the theorem.
If $t \in [t_{j-1}, t_j]$, then

$$|\beta(t) - \gamma(t)| \leq |\beta(t) - h_j| + |h_j - \gamma(t_j)| + |\gamma(t_j) - \gamma(t)|$$
$$\leq \delta/4 + \delta/4 + \delta/4 < \epsilon.$$

Further,

$$l(\beta) = \sum_{j=1}^{k} |h_j - h_{j-1}|$$

$$\leq \sum_{j=1}^{k} (|h_j - \gamma(t_j)| + |\gamma(t_j) - \gamma(t_{j-1})| + |\gamma(t_{j-1}) - h_{j-1}|)$$

$$\leq l(\gamma) + 2k\eta \leq 2l(\gamma).$$

Now

$$\left| \int_{\sigma_j} f(z) \, dz - (h_j - h_{j-1}) f(\gamma(t_j)) \right|$$

$$= \left| (h_j - h_{j-1}) \int_0^1 (f((1-s)h_{j-1} + sh_j) - f(\gamma(t_j))) \, ds \right|$$

$$\leq |h_j - h_{j-1}| \int_0^1 |f((1-s)h_{j-1} + sh_j) - f(\gamma(t_j))|\, ds$$

$$\leq \frac{\epsilon |h_j - h_{j-1}|}{4l(\gamma)}.$$

Adding,

$$\left| \int_\beta f(z)\, dz - S_D(f;\gamma) \right| = \left| \sum_{j=1}^n \left(\int_{\sigma_j} f(z)\, dz - (h_j - h_{j-1}) f(\gamma(t_j)) \right) \right|$$

$$\leq \sum_{j=1}^n \left| \int_{\sigma_j} f(z)\, dz - (h_j - h_{j-1}) f(\gamma(t_j)) \right|$$

$$\leq \frac{\epsilon l(\beta)}{4l(\gamma)} \leq \epsilon/2.$$

Thus $| \int_\beta (f(z)\, dz - \int_\gamma (f(z)\, dz| < \epsilon.$

Finally, the function $\Gamma(s,t) = (1-s)\gamma(t) + s\beta(t)$ is a homotopy from γ to β in U. $\qquad\square$

Corollary 22.2.2   *The path β can be chosen to be a dyadic rectilinear path.*

Proof   Since the set $D = \{x + iy \in U : x, y$ dyadic rational numbers$\}$ is dense in U, we can take $H = D$; there is a polygonal path β with vertices in D which satisfies the conditions of the theorem. Let us retain the notation of the theorem. Suppose that $1 \leq j \leq k$. There is a dyadic rectilinear path $\zeta_j : [t_{j-1}, t_j] \to U$ from h_{j-1} to h_j, obtained by changing one co-ordinate at a time. Then $l(\zeta_j) \leq \sqrt{2}|h_j - h_{j-1}|$. Let

$$\zeta = \zeta_1 \vee \dots \vee \zeta_k,$$

so that $\zeta : [a, b] \to U$ is a dyadic rectilinear path with $l(\zeta) \leq \sqrt{2}l(\beta) \leq 2\sqrt{2}l(\gamma)$. Then $|\zeta(t) - \gamma(t)| < \epsilon$ for $t \in [a, b]$, and, arguing as in the theorem, if $1 \leq j \leq k$ then

$$\left| \int_{\zeta_j} f(z)\, dz - (h_j - h_{j-1}) f(\gamma(t_j)) \right| < \sqrt{2}\epsilon |h_j - h_{j-1}|/4l(\gamma),$$

from which the result follows. $\qquad\square$

The next important result illustrates the usefulness of Theorem 22.2.1.

Theorem 22.2.3 *Suppose that U is a domain, that $\gamma : [a, b] \to U$ is a rectifiable path in U, and that F is a holomorphic function on U with continuous derivative f. Then*

$$\int_\gamma f(z)\, dz = F(\gamma(b)) - F(\gamma(a)).$$

If, further, γ is closed, then $\int_\gamma f(z)\, dz = 0$.

Proof Suppose that $\epsilon > 0$. Let β be a piecewise-linear path which satisfies the conclusions of Theorem 22.2.1. Then, adopting the notation of Theorem 22.2.1, and using Example 22.1.8,

$$\int_\beta f(z)\, dz = \sum_{j=1}^n \left(\int_{\sigma_j} f(z)\, dz \right)$$

$$= \sum_{j=1}^n \left((\gamma(t_j) - \gamma(t_{j-1})) \int_0^1 f((1 - s)\gamma(t_{j-1}) + s\gamma(t_j))\, ds \right).$$

But

$$(\gamma(t_j) - \gamma(t_{j-1})) \int_0^1 f((1 - s)\gamma(t_{j-1}) + s\gamma(t_j))\, ds = F(\gamma(t_j)) - F(\gamma(t_{j-1})),$$

by Theorem 22.1.7, and so

$$\int_\beta f(z)\, dz = \sum_{j=1}^n (F(\gamma(t_j)) - F(\gamma(t_{j-1}))) = F(\gamma(b)) - F(\gamma(a)).$$

Thus $\left| \int_\gamma f(z)\, dz - F(\gamma(b)) - F(\gamma(a)) \right| < \epsilon$. Since ϵ is arbitrary, the result follows. □

We write $\int_{[a,b]} f(z)\, dz$ for $\int_{\sigma(a,b)} f(z)\, dz$.

Corollary 22.2.4 *Suppose that U is a domain, that $\gamma : [a, b] \to U$ is a closed rectifiable path in U, and that $p = a_0 + \cdots + a_n z^n$ is a polynomial function on U with continuous derivative f. Then $\int_\gamma p(z)\, dz = 0$.*

Proof Let $P(z) = \sum_{j=0}^n a_j z^{j+1}/(j + 1)$. Then P is holomorphic, and $P' = p$. □

Exercises

22.2.1 Use Theorem 22.2.3 to show that if U is a domain in $\mathbf{C}^*$ containing $\mathbf{T}$ then there does not exist a holomorphic function F on U for which $F'(z) = 1/z$, for $z \in U$.

22.2.2 *(Integration by parts)* Suppose that U is a domain, that $\gamma : [a, b] \to U$ is a rectifiable path in U, and that F and G are holomorphic functions on U with continuous derivatives f and g respectively. Show that

$$\int_\gamma f(z)G(z)\,dz = F(\gamma(b))G(\gamma(b)) - F(\gamma(a))G(\gamma(a)) - \int_\gamma F(z)g(z)\,dz,$$

and that if γ is closed, then

$$\int_\gamma f(z)G(z)\,dz = - \int_\gamma F(z)g(z)\,dz.$$

22.3 Cauchy's theorem

So far, the complex analysis that we have studied is very similar to the real analysis of Part Two. We now show that path integrals provide a very powerful tool, which we use to obtain some remarkable results of a completely different nature. We begin with Cauchy's theorem. We shall prove this in several stages, obtaining more and more general results. First we begin with a square path.

Theorem 22.3.1 (Cauchy's theorem for a square) *Suppose that f is a holomorphic function on a simply connected domain U, and that γ is a square path in U. Then $\int_\gamma f(z)\,dz = 0$.*

Proof   This theorem is the heart of Cauchy's theorem. By scaling and translation, we can suppose that $\gamma = \gamma_0$ is the dyadic rectilinear path $sq_{0,0}^{(0)}$ with vertices $(0,0), (1,0), (1,1)$ and $(0,1)$, so that $[\gamma_0]$ is the boundary of the 0-square $Q_0 = Q_{(0,0)}^{(0)}$. Since U is simply connected, $\overline{Q}_0 \subseteq U$. Suppose that

$$\int_{\gamma_0} f(z)\,dz = I_0 \neq 0.$$

$\overline{Q}_0$ is the union of four 1-squares $\overline{Q}_{0,0}^{(1)}, \overline{Q}_{1,0}^{(1)}, \overline{Q}_{0,1}^{(1)}$ and $\overline{Q}_{1,1}^{(1)}$. Let

$$\gamma_1^{(1)} = sq_{0,0}^{(1)}, \ \gamma_2^{(1)} = sq_{1,0}^{(1)}, \ \gamma_3^{(1)} = sq_{0,1}^{(1)}, \ \gamma_4^{(1)} = sq_{1,1}^{(1)}.$$

Then

$$\sum_{j=1}^{4} \int_{\gamma_j^{(1)}} f(z)\,dz = \int_{\gamma_0} f(z)\,dz,$$

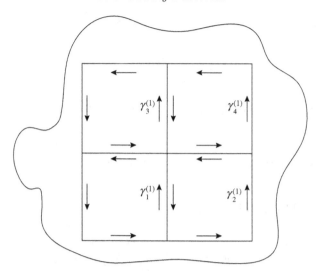

Figure 22.3.

since the contributions from edges inside $[\gamma_0]$ cancel in pairs. Consequently, there exists $1 \le j \le 4$ such that

$$\left| \int_{\gamma_j^{(1)}} f(z)\,dz \right| = |I_1| \ge |I_0|/4.$$

Set $\gamma_1 = \gamma_j^{(1)}$; then $[\gamma_1]$ is the boundary of a 1-square Q_1 contained in Q_0.

We now iterate the procedure, to obtain a sequence $(\gamma_j)_{j=0}^{\infty}$ of simple closed square paths, such that

(a) γ_j is a j-square path, and

$$\left| \int_{\gamma_j} f(z)\,dz \right| \ge |I_0|/4^j;$$

(b) $[\gamma_j] = \partial Q_j$, where Q_j is a j-square;
(c) $(\overline{Q}_j)$ is a decreasing sequence of compact sets, and diam $\overline{Q}_j = \sqrt{2}/2^j$.

Thus $\cap_{j=0}^{\infty} \overline{Q}_j$ is a singleton set, $\{z_\infty\}$, say. Then $z_\infty \in U$, and f is differentiable at z_∞. Thus there exists $\delta > 0$ such that $N_\delta(z_\infty) \subseteq U$, and such that if $|z| < \delta$ then

$$f(z_\infty + z) = f(z_\infty) + f'(z_\infty)z + r(z), \text{ where } |r(z)| \le |I_0||z|/6.$$

Now there exists j such that $\overline{Q}_j \subseteq N_\delta(z_\infty)$. Since

$$\int_{\gamma_j} (f(z_\infty) + f'(z_\infty)z)dz = 0,$$

by Corollary 22.2.4, it follows that $\int_{\gamma_j} f(z)\, dz = \int_{\gamma_j} r(z)\, dz$. But

$$\left| \int_{\gamma_j} r(z)\, dz \right| \le \sup\{|r(z)| : z \in [\gamma_j]\}.l(\gamma_j) \le \frac{|I_0|(\sqrt{2}/2^j)}{6} \cdot \frac{4}{2^j} < \frac{|I_0|}{4^j},$$

by Proposition 22.1.3; this contradicts (a).   □

Theorem 22.3.2   *Suppose that f is a continuous function on a simply connected domain U, for which $\int_\gamma f(z)\, dz = 0$ for every dyadic square path in U. Then $\int_\gamma f(z)\, dz = 0$ for every closed rectifiable path γ in U.*

Proof   First we prove the theorem for simple closed k-dyadic rectilinear paths in U. Suppose that γ is a simple closed k-dyadic rectilinear path in U. Let $n_k(\gamma)$ be the number of k-squares in $in[\gamma]$. We prove the result by induction on $n_k(\gamma)$. The result holds when $n_k(\gamma) = 1$, by Theorem 22.3.1. Suppose that the result holds for all simple closed k-dyadic rectilinear paths in U with $n_k(\gamma) < n$, and that γ is a simple closed k-dyadic rectilinear path in U with $n_k(\gamma) = n$. There exists a vertex $v_0 = (m_0 + in_0)/2^k$ in $[\gamma]$ for which m_0+n_0 is minimal, so that $((m_0+1)+in_0)/2^k$ and $(m_0+i(n_0+1))/2^k$ are the two adjacent vertices. Let γ' be the path obtained by replacing v_0 by $v_0' = ((m_0 + 1) + i(n_0 + 1))/2^k$. Then

$$\int_{\gamma'} f(z)\, dz = \int_\gamma f(z)\, dz - \int_{sq^{(k)}_{m_0,n_0}} f(z)\, dz = \int_\gamma f(z)\, dz.$$

There are now two possibilities. First, γ' is simple. Then $n_k(\gamma') = n - 1$, and so $\int_{\gamma'} f(z)\, dz = 0$. Secondly, γ' is not simple. Then $\gamma' = \delta \vee \epsilon$, where δ an ϵ are simple closed k-dyadic rectilinear path in U with $n_k(\delta) < n$ and $n_k(\epsilon) < n$. Then

$$\int_{\gamma'} f(z)\, dz = \int_\delta f(z)\, dz + \int_\epsilon f(z)\, dz = 0.$$

Thus $\int_\gamma f(z)\, dz = 0$.

Secondly, we prove the theorem for closed k-dyadic rectilinear paths in U. If γ is such a path, then $\gamma = \gamma_1 \vee \cdots \vee \gamma_n$, where each γ_j is a simple closed

k-dyadic rectilinear path in U. Then

$$\int_\gamma f(z)\,dz = \sum_{m=1}^{n}\left(\int_{\gamma_m} f(z)\,dz\right) = 0.$$

Finally, suppose that γ is a closed rectifiable path in U, and that $\eta > 0$. By Corollary 22.2.2, there is a closed dyadic rectilinear path δ in U such that

$$\left|\int_\gamma f(z)\,dz - \int_\delta f(z)\,dz\right| < \eta.$$

Since $\int_\delta f(z)\,dz = 0$, it follows that $|\int_\gamma f(z)\,dz| < \eta$. Since η is arbitrary, it follows that $\int_\gamma f(z)\,dz = 0$. $\qquad\square$

Theorem 22.3.3 *If f is a continuous function on a domain U, for which $\int_\gamma f(z)\,dz = 0$ for every closed polygonal path in U, then there exists a holomorphic function F on U such that $F' = f$.*

Proof Pick $z_0 \in U$ as a base point. Suppose that $w \in U$. Since U is path-connected, there exists a polygonal path γ_1 from z_0 to w. If γ_2 is another such path, then $\gamma_1 \vee \gamma_2^\leftarrow$ is a closed polygonal path, and

$$\int_{\gamma_1} f(z)\,dz - \int_{\gamma_2} f(z)\,dz = \int_{\gamma_1 \vee \gamma_2^\leftarrow} f(z)\,dz = 0,$$

by Theorem 22.3.2. Thus $\int_{\gamma_1} f(z)\,dz = \int_{\gamma_2} f(z)\,dz$, and so the quantity $F(w) = \int_{\gamma_1} f(z)\,dz$ does not depend upon the choice of rectilinear path from z_0 to w.

We shall show that F is holomorphic and that $F' = f$. Suppose that $w \in U$ and that $\epsilon > 0$. There exists $\delta > 0$ such that $N_\delta(w) \subseteq U$, and such that if $|\zeta| < \delta$ then $|f(w + \zeta) - f(w)| < \epsilon$. Suppose that $|\zeta| < \delta$. If γ_0 is a polygonal path from z_0 to w, then $\gamma_0 \vee \sigma(w, w + \zeta)$ is a polygonal path from z_0 to $w + \zeta$. Thus

$$F(w + \zeta) = \int_{\gamma_0} f(z)\,dz + \int_{[w,w+\zeta]} f(z)\,dz$$

$$= F(w) + \zeta \int_0^1 f(w + t\zeta)\,dt = F(w) + f(w)\zeta + r(\zeta),$$

where

$$r(\zeta) = \zeta \int_0^1 (f(w + t\zeta) - f(w))\,dt.$$

But $|\int_0^1 (f(w + t\zeta) - f(w))\,dt| < \epsilon$, so that $|r(\zeta)| \leq \epsilon|\zeta|$. Thus F is differentiable at w, with derivative $f(w)$. $\qquad\square$

Combining Theorems 22.3.1, 22.3.2 and 22.3.3 we have the following.

Theorem 22.3.4 (Cauchy's theorem for simply connected domains)   *Suppose that f is a holomorphic function on a simply connected domain U.*

(i) *If γ is a closed rectifiable path in U then $\int_\gamma f(z)\,dz = 0$.*
(ii) *There exists a holomorphic function F on U such that $F' = f$.*

Exercises

There are other ways of proving Cauchy's theorem for a simply connected domain. The following exercises provide another proof, preferable in some respects to the one given above.

22.3.1 Suppose that $a, b, c \in \mathbf{C}$. Let $a' = (b+c)/2$, $b' = (c+a)/2$, $c' = (a+b)/2$. Calculate $|b' - c'|$, $|c' - a'|$ and $|a' - b'|$.

22.3.2 Use a', b' and c' to divide the triangle abc into four triangles. Argue as in Theorem 22.3.1 to prove Cauchy's theorem for triangular paths.

22.3.3 We want to prove Cauchy's theorem for closed polygonal paths in a simply connected domain, using induction on the number of vertices. Suppose that the result holds for polygonal paths with fewer than n vertices, and that γ has n vertices. Show that the result holds if γ is not simple.

22.3.4 Now suppose that γ is a simple closed polygonal path with vertices $v_0, v_1, \ldots, v_n = v_0$. Show that it is enough to show that there is a linear path from a vertex v_j to a point in $[\gamma]$ which is inside $[\gamma]$ and divides $[\gamma]$ into two polygons, each with less than n vertices.

22.3.5 There are several ways of doing this; here is one. Maybe you can find a better one. We can suppose that $v_0 = (x_0, y_0)$, with y_0 minimal. Let $\theta_j = \arg(v_j - v_0)$, for $1 \le j \le n - 1$. Thus $0 \le \theta_j \le \pi$, for $1 \le j \le n - 1$. We can suppose that $\theta_1 < \theta_{n-1}$. Consider three possibilities:

- $\theta_2 < \theta_1$; consider the ray $\{v_1 + \lambda e^{i\theta_1} : \lambda > 0\}$.
- $\theta_{n-2} > \theta_{n-1}$; consider the ray $\{v_{n-1} + \lambda e^{i\theta_{n-1}} : \lambda > 0\}$.
- $\theta_1 < \theta_2$ and $\theta_{n-2} < \theta_{n-1}$. Show that either $\theta_1 < \theta_2 < \theta_{n-1}$ or $\theta_1 < \theta_{n-2} < \theta_{n-1}$, so that $S = \{v_j : \theta_1 < \theta_2 < \theta_{n-1}\}$ is non-empty. Consider $[v_0, v_k]$, where $v_k \in S$ and $|v_k - v_0| \le |v_j - v_0|$ for $v_j \in S$.

22.3.6 Complete the proof of Cauchy's theorem for a simply connected domain.

22.4 The Cauchy kernel

We use the function $k(z) = -1/2\pi i z$ on $\mathbf{C} \setminus \{0\}$ as a convolution kernel, and call it the *Cauchy kernel*.

Theorem 22.4.1 *Suppose that g is a continuous function on a rectifiable path γ. Let*

$$f(w) = \int_\gamma k(w - z)g(z)\, dz = \frac{1}{2\pi i} \int_\gamma \frac{g(z)}{z - w}\, dz \text{ for } w \notin [\gamma].$$

Then f is an analytic function on $\mathbf{C} \setminus [\gamma]$. If $z_0 \notin [\gamma]$ then

$$f^{(n)}(z_0) = \frac{n!}{2\pi i} \int_\gamma \frac{g(z)}{(z - z_0)^{n+1}}\, dz.$$

Proof   Let $M = \sup\{|g(z)| : z \in [\gamma]\}$, and let $d = d(z_0, [\gamma])$. Suppose that $|h| < d$. Using the formula in Proposition 20.3.8, we find that

$$\frac{1}{z - (z_0 + h)} =$$

$$\frac{1}{z - z_0} + \frac{h}{(z - z_0)^2} + \cdots + \frac{h^n}{(z - z_0)^{n+1}} + \frac{h^{n+1}}{(z - z_0)^{n+1}(z - (z_0 + h))}.$$

Multiplying by $g(z)/2\pi i$ and integrating, it follows that

$$f(z_0 + h) = \sum_{j=0}^{n} \left(\frac{1}{2\pi i} \int_\gamma \frac{g(z)}{(z - z_0)^{j+1}}\, dz \right) h^j + R_n(h),$$

where

$$R_n(h) = \frac{h^{n+1}}{2\pi i} \int_\gamma \frac{g(z)}{(z - z_0)^{n+1}(z - (z_0 + h))}\, dz.$$

Then

$$|R_n(h)| \leq \frac{|h|^{n+1} l(\gamma) M}{2\pi d^{n+1}(d - |h|)} = \left(\frac{l(\gamma) M}{2\pi (d - |h|)} \right) \left(\frac{|h|}{d} \right)^{n+1}$$

so that $R_n(h) \to 0$ as $n \to \infty$. Thus the series

$$\sum_{j=0}^{\infty} \left(\frac{1}{2\pi i} \int_\gamma \frac{g(z)}{(z - z_0)^{j+1}}\, dz \right) h^j$$

converges to the value $f(z_0 + h)$. Since this holds for all $z_0 + h \in N_d(z_0)$, the power series has radius of convergence at least d, and

$$f^{(n)}(z_0) = \frac{n!}{2\pi i} \int_\gamma \frac{g(z)}{(z - z_0)^{n+1}} \, dz.$$

$\square$

How does the analytic function f on $\mathbf{C} \setminus [\gamma]$ relate to the continuous function g on $[\gamma]$? This is something that we shall investigate in the rest of this chapter.

Exercises

22.4.1 Let $g(z) = \bar{z}^n$, for $z \in \mathbf{T}$ and $n \in \mathbf{N}$. Show that

$$\int_{\kappa_1(0)} k(w - z)g(z) \, dz = 0 \text{ for } w \in D.$$

22.4.2 Suppose that $f(z) = \sum_{n=0}^\infty a_n z^n$ is a power series with radius of convergence greater than 1. Show that $\int_{\kappa_1(0)} k(w - z)\bar{f}(z) \, dz = a_0$, for $w \in D$.

22.4.3 What are the real and imaginary parts of the Cauchy kernel?

22.4.4 The *Poisson kernel* is defined as

$$P_y(x) = \frac{y}{\pi(x^2 + y^2)}, \text{ for } x \in \mathbf{R}, \ y > 0.$$

Show that the function $(x, y) \to P_y(x)$ is harmonic.

22.5 The winding number as an integral

Recall that $\kappa_r(w)$ is the circular path $\kappa_r(w) = w + re^{it}$ for $t \in [0, 2\pi]$ and that its track is denoted by $\mathbf{T}_r(w)$. Recall also (Example 22.1.9) that if f is a continuous function on $\mathbf{T}_r(w)$ then

$$\int_{\kappa_r(w)} f(z) \, dz = ir \int_0^{2\pi} f(w + re^{it})e^{it} \, dt.$$

In particular, putting $f(z) = (z - w)^j$, where $j \in \mathbf{Z}$,

$$\int_{\kappa_r(w)} (z - w)^j \, dz = ir \int_0^{2\pi} r^j e^{i(j+1)t} \, dt = \begin{cases} 2\pi i & \text{if } j = -1 \\ 0 & \text{otherwise.} \end{cases}$$

Thus

$$n(\kappa_r(w), w) = 1 = \frac{1}{2\pi i} \int_{\kappa_r(w)} \frac{dz}{z - w};$$

the winding number of $\kappa_r(w)$ is expressed as an integral. We can extend this result to more general paths, to obtain the following fundamental theorem.

Theorem 22.5.1 *Suppose that $\gamma : [a, b] \to \mathbf{C}$ is a closed rectifiable path and that $w \notin [\gamma]$. Then*

$$n(\gamma, w) = \frac{1}{2\pi i} \int_\gamma \frac{dz}{z - w}.$$

Proof   First we consider the case where γ is piecewise smooth. Let $\gamma(a) - w = re^{i\theta}$. For $a \le s \le b$ let

$$h(s) = \int_a^s \frac{\gamma'(t)}{\gamma(t) - w} dt, \text{ and let } h(s) = j(s) + ik(s),$$

where j and k are the real and imaginary parts of h. Then

$$h(a) = 0, \quad h(b) = \int_\gamma \frac{dz}{z - w} \text{ and } h'(s) = \frac{\gamma'(s)}{\gamma(s) - w}.$$

As s varies, we unwind $\gamma(s) - w$. Let $f(s) = (\gamma(s) - w)e^{-h(s)}$. Then

$$f'(s) = (\gamma'(s) - (\gamma(s) - w)h'(s))e^{-h(s)} = 0.$$

Consequently, $f(s) = f(a)$, so that $e^{-h(s)}(\gamma(s) - w) = \gamma(a) - w$ and

$$\gamma(s) - w = e^{h(s)}(\gamma(a) - w) = re^{j(s)}e^{i(k(s)+\theta)}.$$

Thus $k(s) + \theta$ is a branch of $\mathrm{Arg}\,(\gamma(s) - w)$ on $[a, b]$, and

$$n(\gamma, w) = \frac{k(b) - k(a)}{2\pi} = \frac{k(b)}{2\pi}.$$

Now

$$(\gamma(b) - w)e^{-h(b)} = f(b) = f(a) = \gamma(a) - w = \gamma(b) - w,$$

so that $e^{-h(b)} = e^{-j(b)}e^{-ik(b)} = 1$, and $j(b) = 0$. Thus

$$2\pi i n(\gamma, w) = ik(b) = j(b) + ik(b) = h(b) = \int_\gamma \frac{dz}{z - w}.$$

Next, suppose that γ is a rectifiable path. Let $U = \mathbf{C} \setminus \{w\}$. By Corollary 22.2.2, if $\epsilon > 0$ there exists a dyadic rectilinear path $\beta : [a, b] \to U$ homotopic to γ in U such that

$$\left| \frac{1}{2\pi i} \int_\beta \frac{dz}{z - w} - \frac{1}{2\pi i} \int_\gamma \frac{dz}{z - w} \right| < \epsilon.$$

By Theorem 21.2.1, $n(\beta, w) = n(\gamma, w)$, and so

$$\left| n(\gamma, w) - \frac{1}{2\pi i} \int_\gamma \frac{dz}{z - w} \right| < \epsilon.$$

Since ϵ is arbitrary, the result follows.   □

Corollary 22.5.2   *If $f : U \to \mathbf{C}$ is holomorphic and $w \notin f([\gamma])$ then*

$$n(f \circ \gamma, w) = \frac{1}{2\pi i} \int_\gamma \frac{f'(z)}{f(z) - w} \, dz.$$

Proof   As in the theorem, it is enough to prove this when γ is piecewise smooth. Then, by the chain rule, $f \circ \gamma$ is a piecewise smooth path, with derivative

$$(f \circ \gamma)'(t) = \frac{df}{dz}(\gamma(t))\gamma'(t).$$

Thus, making a change of variables,

$$n(f \circ \gamma, w) = \frac{1}{2\pi i} \int_{f \circ \gamma} \frac{dz}{z - w} = \frac{1}{2\pi i} \int_a^b \frac{(f \circ \gamma)'(t)}{f(\gamma(t)) - w} \, dt$$

$$= \frac{1}{2\pi i} \int_a^b \frac{f'(\gamma(t))\gamma'(t)}{f(\gamma(t)) - w} \, dt = \frac{1}{2\pi i} \int_\gamma \frac{f'(z)}{f(z) - w} \, dz. \qquad \square$$

22.6 Cauchy's integral formula for circular and square paths

Recall that $\kappa_r(w)$ is the circular closed path $\kappa_r(w)(t) = w + re^{it}$, for $t \in [0, 2\pi]$, so that $M_r(w) = \overline{in}[\kappa_r(w)]$.

Theorem 22.6.1 (Cauchy's integral formula for a circular path)   *Suppose that f is a holomorphic function defined on a domain U, that $M_r(w) \subseteq U$ and that $\zeta \in N_r(w)$. Then*

$$f(\zeta) = \frac{1}{2\pi i} \int_{\kappa_r(w)} \frac{f(z)}{z - \zeta} \, dz.$$

Proof   Let $g(z) = f(z)/2\pi i(z - \zeta)$ for $z \in U \setminus \{\zeta\}$; g is holomorphic on $U \setminus \{\zeta\}$. Let $t = r - |\zeta - w|$, and suppose that $0 < s < t$. Let

$$I = \int_{\kappa_r(w)} g(z) \, dz, \text{ and let } I_s = \int_{\kappa_s(\zeta)} g(z) \, dz.$$

First we show that $I = I_s$. The set $U \setminus \{\zeta\}$ is not simply connected. We split each of the paths γ and $\kappa_r(\zeta)$ into two parts, each contained in a simply connected domain.

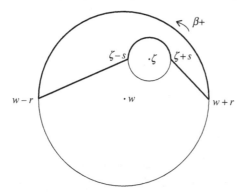

Figure 22.6.

Let

$$\kappa_r(w)_+ = \kappa_r(w)_{[0,\pi]}, \kappa_r(w)_- = \kappa_r(w)_{[\pi,2\pi]}$$
$$\kappa_s(\zeta)_+ = \kappa_s(\zeta)_{[0,\pi]}, \kappa_s(\zeta)_- = \kappa_s(\zeta)_{[\pi,2\pi]},$$

and let

$$\beta_+ = \kappa_r(w)_+ \vee \overleftarrow{\sigma}(w - r, \zeta - s) \vee \kappa_s(\zeta)_+^{\leftarrow} \vee \sigma(\zeta + s, w + r),$$
$$\beta_- = \kappa_r(w)_+^{\leftarrow} \vee \sigma(w - r, \zeta - s) \vee \kappa_s(\zeta)_- \vee \overleftarrow{\sigma}(\zeta + s, w + r).$$

The track $[\beta_+]$ is contained in the simply connected set $U \cap (\zeta + \mathbf{C}_{\pi/2})$, so that $\int_{\beta_+} g(z)\,dz = 0$, by Cauchy's theorem. Similarly, the track $[\beta_-]$ is contained in the simply connected set $U \cap (\zeta + \mathbf{C}_{-\pi/2})$, so that $\int_{\beta_-} g(z)\,dz = 0$.

Since the integrals along the linear paths cancel, it follows that

$$I - I_s = \int_{\beta_+} g(z)\,dz - \int_{\beta_-} g(z)\,dz = 0.$$

Suppose that $\epsilon > 0$. Since f is continuous at w, there exists $0 < \delta < t$ such that if $|z - \zeta| < \delta$ then $|f(z) - f(\zeta)| < \epsilon$. Since

$$\frac{1}{2\pi i} \int_{\kappa_s(\zeta)} \frac{dz}{z - \zeta} = n(\kappa_s(\zeta), \zeta) = 1,$$

it follows that

$$|I - f(\zeta)| = |I_s - f(\zeta)| = \frac{1}{2\pi} \left| \int_{\kappa_s(\zeta)} \frac{f(z) - f(\zeta)}{z - \zeta}\,dz \right|,$$

so that if $0 < s < \delta$ then

$$|I - f(\zeta)| \le \frac{1}{2\pi}\left(\frac{\epsilon}{s}\right) l(\kappa_s(\zeta)) = \epsilon.$$

Since ϵ is arbitrary, the result follows.   □

A similar result holds for square paths.

Theorem 22.6.2 (Cauchy's integral formula for a square path)   *Suppose that f is a holomorphic function defined on a domain U. Let $sq_r(w)$ be the square path with vertices $w-r-ir$, $w+r-ir$, $w+r+ir$, $w-r+ir$. Suppose that $\overline{in}[sq_r(w)] \subseteq U$ and that ζ is inside $[sq_r(w)]$. Then*

$$f(w) = \frac{1}{2\pi i}\int_{sq_r(w)}\frac{f(z)}{z-\zeta}\,dz.$$

Proof   Replace $\kappa_r(w)$ by $sq_r(w)$ in the proof of Theorem 22.6.1, and make obvious changes to the proof.   □

Corollary 22.6.3   *If $M_r(w) \subseteq U$, then*

$$f(w) = \frac{1}{2\pi}\int_{-\pi}^{\pi} f(w + re^{i\theta})\,d\theta.$$

Proof   For

$$f(w) = \frac{1}{2\pi i}\int_{\kappa_r(w)}\frac{f(z)}{z-w}\,dz = \frac{1}{2\pi i}\int_{\kappa_r(0)}\frac{f(w+z)}{z}\,dz$$
$$= \frac{1}{2\pi}\int_{-\pi}^{\pi} f(w+re^{i\theta})\,d\theta.$$
□

We can apply this to harmonic functions.

Corollary 22.6.4   *Suppose that g is a harmonic function on a domain U, and that $M_r(w) \subseteq U$. Then*

$$g(w) = \frac{1}{2\pi}\int_{-\pi}^{\pi} g(w + re^{i\theta})\,d\theta.$$

Proof   By considering real and imaginary parts, we can suppose that g is real-valued. There exists $s > r$ and a function h on $N_s(w)$, such that

$N_s(w) \subseteq U$, and such that $f = g + ih$ is holomorphic on $N_s(w)$. Then

$$f(w) = g(w) + ih(w) = \frac{1}{2\pi} \int_{-\pi}^{\pi} f(w + re^{i\theta}) \, d\theta$$

$$= \frac{1}{2\pi} \int_{-\pi}^{\pi} g(w + re^{i\theta}) \, d\theta + i \frac{1}{2\pi} \int_{-\pi}^{\pi} h(w + re^{i\theta}) \, d\theta.$$

The result now follows by considering the real part of this equation.   □

We have seen in Volume I, Example 7.1.9 that there are continuous functions on $\mathbf{R}$ with no points of differentiability, and so there are continuously differentiable functions on $\mathbf{R}$ which are not twice differentiable at any point of $\mathbf{R}$. For functions of a complex variable, the situation is completely different. The most important application of Theorem 22.6.1 is the following.

Theorem 22.6.5 (Taylor's theorem for holomorphic functions) *Suppose that f is a holomorphic function on a domain U. Then f is analytic on U. If $w \in U$ and $M_R(w) \subseteq U$ then*

$$f(w + h) = \sum_{n=0}^{\infty} a_n h^n = \sum_{n=0}^{\infty} \frac{f^{(n)}(w)}{n!} h^n, \text{ for } |h| < R,$$

where

$$a_n = \frac{1}{2\pi i} \int_{\kappa_r(w)} \frac{f(z)}{(z - w)^{n+1}} \, dz, \text{ for } r < R.$$

Proof   If $|h| < r < R$ then

$$f(w + h) = \frac{1}{2\pi i} \int_{\kappa_r(w)} \frac{f(z)}{z - (w + h)} \, dz.$$

by Theorem 22.6.1. The result now follows from Theorem 22.4.1.   □

Corollary 22.6.6   A complex-valued function on a domain U is holomorphic if and only if it is analytic.

Authors define holomorphic functions and analytic functions in various ways. This corollary shows that this is not important. We shall generally refer to 'holomorphic functions', rather than 'analytic functions'.

Corollary 22.6.7   *If f is an entire function, and $z_0 \in \mathbf{C}$, then the Taylor series expansion of f around z_0 has infinite radius of convergence.*
    Suppose that U is a domain which is a proper subset of $\mathbf{C}$. If f is a holomorphic function on U and $z_0 \in U$, then the radius of convergence of the Taylor series expansion of f around z_0 is at least $d(z_0, \partial U)$.

Proof   Let us prove the second statement. If $0 < R < d(z_0, \partial U)$ then $M_R(z_0) \subseteq U$, and so the radius of convergence is at least R. Since this holds for all $R < d(z_0, \partial U)$, the radius of convergence is at least $d(z_0, \partial U)$. The proof of the first statement is similar.   □

Example 22.6.8 (The complex binomial theorem)   If $\alpha \in \mathbf{C}$ then

$$(1+z)^\alpha = 1 + \alpha z + \sum_{n=2}^{\infty} \frac{\alpha(\alpha-1)\ldots(\alpha-n+1)}{n!} z^n,$$

the sum converging locally absolutely uniformly on **D**.

For $(1+z)^\alpha$ is holomorphic on $\mathbf{C} \setminus (-\infty, -1]$, and is therefore analytic on **D**, and if $f(z) = (1+z)^\alpha$ then $f^{(k)}(z) = \alpha(\alpha-1)\ldots(\alpha-k+1)(1+z)^{\alpha-k}$. Note how much simpler this proof is than the corresponding proof for real-valued functions (Volume I, Theorem 7.6.4).

Taylor's theorem enables us to prove a converse of Cauchy's theorem.

Theorem 22.6.9 (Morera's theorem)   *Suppose that f is a continuous function on a domain U, and that $\int_\gamma f(z)\,dz = 0$ for every dyadic square path γ for which $\overline{in}[\gamma] \subseteq U$. Then f is holomorphic on U.*

Proof   Suppose that $z_0 \in U$. Then there exists a neighbourhood $N_r(z_0)$ with $N_r(z_0) \subseteq U$. Since $N_r(z_0)$ is simply connected, it follows from Theorem 22.3.2 that $\int_\gamma f(z)\,dz = 0$ for every rectifiable closed path in $N_r(z_0)$, and it therefore follows from Theorem 22.3.3 that there exists a holomorphic function F on $N_r(z_0)$ such that $F' = f$. But F is analytic, and is therefore infinitely differentiable, and so f is differentiable at z_0.   □

Let $H(U)$ denote the vector space of holomorphic functions on a domain U. $H(U)$ is a linear subspace of the vector space $C(U)$ of continuous complex-valued functions on U. As in Volume II, Section 15.8, we give $C(U)$ a complete metric d which defines the topology of local uniform convergence.

Theorem 22.6.10   *$H(U)$ is a closed linear subspace of $(C(U), d)$.*

Proof   We use Cauchy's theorem and Morera's theorem. Suppose that $(f_n)_{n=1}^\infty$ is a sequence in $H(U)$ which converges locally uniformly to a function f in $C(U)$. Suppose that γ is a dyadic square path with $\overline{in}[\gamma] \subseteq U$. Since $[\gamma]$ is compact, $f_n \to f$ uniformly on $[\gamma]$, and so

$$\int_\gamma f(z)\,dz = \lim_{n\to\infty} \int_\gamma f_n(z)\,dz = 0.$$

Thus f is holomorphic, by Morera's theorem.   □

Thus $(H(U), d)$ is a complete metric space.

Here is a useful consequence of Morera's theorem: integrals of holomorphic functions are holomorphic.

Theorem 22.6.11 *Suppose that U is a domain and that f is a continuous complex-valued function on $U \times [a, b]$ such that, setting $f_t(z) = f(z, t)$, the function f_t is holomorphic on U for all $t \in [a, b]$. Let $F(z) = \int_a^b f(z, t)\, dt$. Then F is a holomorphic function on U.*

Proof Let γ be a dyadic square path in U with $\overline{in}[\gamma] \subseteq U$. Then

$$\int_\gamma F(z)\, dz = \int_\gamma \left(\int_a^b f(z, t)\, dt \right) dz$$

$$= \int_a^b \left(\int_\gamma f(z, t)\, dz \right) dt = 0,$$

so that F is holomorphic, by Morera's theorem. □

Corollary 22.6.12 *Suppose that U is a domain and that f is a continuous complex-valued function on $U \times [a, \infty)$ such that, setting $f_t(z) = f(z, t)$, the function f_t is holomorphic on U for all $t \in [a, \infty)$. If $\int_a^b f(z, t)\, dt$ converges locally uniformly to $\int_a^\infty f(z, t)\, dt$ as $b \to \infty$ then $\int_a^\infty f(z, t)\, dt$ is a holomorphic function on U.*

Proof Apply Theorem 22.6.10. □

Clearly, similar results also hold for improper integrals on open intervals.

Exercises

22.6.1 Suppose that γ is a convex path in $\mathbf{C}$, and that w is inside $[\gamma]$. If $\theta \in [0, 2\pi]$, let $\rho_\theta = \{w + re^{i\theta} : r \geq 0\}$. Show carefully that $\rho_\theta \cap [\gamma]$ is a singleton $\gamma(\theta)$, and that the mapping $\theta \to \gamma(\theta)$ from $[0, 2\pi]$ to $[\gamma]$ is a parametrization of $[\gamma]$.

22.6.2 Suppose that f is an entire function and that there exists $R > 0$ and $k \in \mathbf{N}$ such that $|f(z)| \leq |z|^k$ for $|z| \geq R$. Show that f is a polynomial function of degree at most k.

22.6.3 Suppose that f is a holomorphic function on a domain U and that $z_0 \in U$. Suppose that the radius of convergence r of the Taylor series expansion of f about z_0 is greater than $d(z_0, \partial U)$. Can the Taylor series be used to extend f to a holomorphic function on $U \cup N_r(z_0)$?

22.6.4 The function $f(z) = 1/(1 - z - z^2)$ is holomorphic in $\{z \in \mathbf{C} : |z| < 1/2\}$. Let its Taylor series be $\sum_{n=0}^\infty F_n z^n$. What recurrence relation does the sequence $(F_n)_{n=0}^{\infty}$ satisfy? Show that

$$F_n = \frac{g^{n+1} - (1-g)^{n+1}}{\sqrt{5}},$$

where $g = (\sqrt{5}+1)/2$ is the *golden ratio*.

22.6.5 Suppose that f is a holomorphic function on $\mathbf{D}$ taking values in $\mathbf{D}$. Show that $|f^{(n)}(0)| \le n!$, for $n \in \mathbf{N}$.

22.6.6 Suppose that $(p_n)_{n=1}^{\infty}$ is a sequence of polynomials, each of degree less than or equal to d, which converges locally uniformly on a domain U to a function f. Show that f is a polynomial, of degree at most d.

22.6.7 Use Corollary 22.6.4 to show that a non-constant harmonic function on a domain U has no local maxima.

22.7 Simply connected domains

Using Cauchy's theorem, we can give further characterizations of simply connected domains.

Theorem 22.7.1 *Suppose that U is a domain. The following are equivalent.*

(i) *U is simply connected.*

(ii) *If f is a holomorphic function on U then $\int_{\gamma} f(z)\,dz = 0$ for all polygonal closed paths γ in U.*

(iii) *If f is a holomorphic function on U then $\int_{\gamma} f(z)\,dz = 0$ for all rectifiable closed paths γ in U.*

(iv) *If f is a holomorphic function on U then there exists a holomorphic function F on U such that $F' = f$.*

(v) *If f is a holomorphic function on U such that $f(z) \ne 0$ for $z \in U$ then there exists a continuous branch of $\operatorname{Log} f$ on U.*

(vi) *If $w \notin U$ then there exists a continuous branch of $\operatorname{Arg}(z-w)$ on U.*

Proof Cauchy's theorem for simply connected domains (Theorem 22.9.1) shows that (i) implies (iii). (iii) certainly implies (ii), and (ii) implies (iv), by Theorem 22.3.3.

Let us show that (iv) implies (v). Suppose that f is a holomorphic function on U and that $f(z) \ne 0$ for $z \in U$. Then the function f'/f is holomorphic on U, and so there exists a holomorphic function G on U such that $G' = f'/f$. Let $h = e^{-G}f$. Then

$$h' = -G'e^{-G}f + e^{-G}f' = 0.$$

so that h is a constant function taking a non-zero value k. Thus $f = ke^G = e^F$, where $F = \log k + G$. F is then a continuous branch of Log f on U.

Next we show that (v) implies (vi). The function $z - w$ does not vanish on U, so that there is a continuous branch of $\text{Log}(z-w)$ on U. Since $\text{Log}(z-w) = \log|z - w| + i\text{Arg}(z - w)$, there is a continuous branch of $\text{Arg}(z - w)$ on U.

Finally we show that (vi) implies that $n(\gamma, w) = 0$ for all closed paths $\gamma \in U$ and all $w \notin U$. If α is a continuous branch of $\text{Arg}(z - w)$ on U, then $l(z) = \log|z - w| + i\alpha(z - w)$ is a continuous branch of $\text{Log}(z - w)$ of U. Since $l'(z) = 1/(z - w)$,

$$n(\gamma, w) = \frac{1}{2\pi i} \int_\gamma \frac{dz}{z - w} = \frac{1}{2\pi i} \int_\gamma l'(z)\, dz = 0,$$

by Theorem 22.2.3. This implies that U is simply connected, by Theorem 21.5.1. $\square$

Corollary 22.7.2 *Suppose that U is simply connected and that $\beta \in \mathbf{C}$. If f is a holomorphic function on U for which $f(z) \neq 0$ for $z \in U$, there exists a continuous branch of f^β on U: that is, there exists a holomorphic function g on U such that $g(z) \in \{f(z)^\beta\}$, for $z \in U$.*

Proof There exists a continuous branch l_f of Log f on U. Let $g(z) = e^{\beta l_f(z)}$, for $z \in U$. $\square$

Exercises

22.7.1 Suppose that u is a real-valued harmonic function on a domain U. A real-valued function v is called a *harmonic conjugate* of u if the complex-valued function $u + iv$ is holomorphic.

Show that if v and v' are harmonic conjugates of u then $v - v'$ is constant.

Show that if U is simply connected then a harmonic conjugate exists.

Give an example on a domain U and a real-valued harmonic function u on U which does not have a harmonic conjugate on U.

22.8 Liouville's theorem

Theorem 22.8.1 (Liouville's theorem) *A bounded entire function is constant.*

Proof   Let $M = \sup\{|f(z)| : z \in \mathbf{C}\}$. Suppose that $w \in \mathbf{C}$. We show that $f(w) = f(0)$. If $R > |w|$ then $n(\kappa_R(0), 0) = n(\kappa_R(0), w) = 1$, so that, using Cauchy's integral formula,

$$f(w) - f(0) = \frac{1}{2\pi i} \int_{\kappa_R(0)} \frac{f(z)}{z - w} dz - \frac{1}{2\pi i} \int_{\kappa_R(0)} \frac{f(z)}{z} dz$$

$$= \frac{1}{2\pi i} \int_{\kappa_R(0)} \frac{wf(z)}{z(z - w)} dz.$$

Thus

$$|f(w) - f(0)| \le \frac{l(\kappa_R(0))M|w|}{2\pi R(R - |w|)} = \frac{M|w|}{R - |w|}.$$

Since $M|w|/(R - |w|) \to 0$ as $R \to \infty$, $f(w) = f(0)$.   □

We can use Liouville's theorem to give another proof of the fundamental theorem of algebra.

Theorem 22.8.2   *If $p(z)$ is a non-constant polynomial function, there exists $z_0 \in \mathbf{C}$ such that $p(z_0) = 0$.*

Proof   If not, then $f(z) = 1/p(z)$ is an entire function. As in Corollary 20.6.3, $|p(z)| \to \infty$ as $z \to \infty$, and so $f(z) \to 0$ as $z \to \infty$. Thus there exists $R > 0$ such that $|f(z)| \le 1$ for $|z| \ge R$. But the continuous function f is bounded on the compact set $\{z : |z| \le R\}$, and so f is a bounded entire function. Thus f is constant, and so therefore is p; this gives a contradiction.   □

Exercises

22.8.1   Use Taylor's theorem to show that if f is an entire function and if $|f(z)| = O(|z|^n)$ as $|z| \to \infty$ then f is a polynomial of degree at most n. Use this to give another proof of Liouville's theorem.

22.8.2   Prove the following extension of Liouville's theorem: if f is an entire function for which $f(z)/z \to 0$ as $z \to \infty$ then f is constant.

22.8.3   Suppose that f is a non-constant entire function. Show that the image $f(\mathbf{C})$ is dense in $\mathbf{C}$.

22.9 Cauchy's theorem revisited

We now prove a more general version of Cauchy's theorem.

Theorem 22.9.1 (Cauchy's theorem) *Suppose that f is a holomorphic function on a domain U, and that β is a closed rectifiable path in U for which $n(\beta, w) = 0$ for $w \notin U$. Then $\int_\beta f(z)\, dz = 0$.*

Note that this extends Theorem 22.3.1, since if U is simply connected then $n(\beta, w) = 0$ for $w \notin U$.

Proof By Theorem 21.4.1, there exist $l \in \mathbf{Z}$ and a finite set $\{\gamma_0, \ldots, \gamma_j\}$ of simple closed l-dyadic rectilinear paths in U such that $[\beta]$ is inside $[\gamma_0]$ and outside $[\gamma_i]$ for $1 \le i \le j$. The set $\overline{in}[\gamma_0] \cap (\cap_{i=1}^{j} \overline{out}[\gamma_i])$ is the union of a finite set $F = \{\bar{Q}_1, \ldots, \bar{Q}_{u_0}\}$ of closed l-squares.

If e is an edge of two adjacent squares Q_u and Q_v then e has opposite orientations in sq_u and sq_v. Thus if g is a continuous function on $\cup_{u=1}^{u_0} \partial Q_u$ then

$$\sum_{i=0}^{j} \int_{\gamma_i} g(z)\, dz = \sum_{u=1}^{u_0} \int_{sq_u} g(z)\, dz.$$

Suppose now that w is an interior point of some Q_u. Then by Cauchy's integral formula for square paths,

$$\frac{1}{2\pi i} \int_{sq_u} \frac{f(z)}{z - w}\, dz = f(w).$$

On the other hand, if $v \ne u$, let $\delta = d(w, \bar{Q}_v)$. Then $f(z)/(z - w)$ is holomorphic on the simply connected set $N_\delta(\bar{Q}_v)$, and so

$$\frac{1}{2\pi i} \int_{sq_v} \frac{f(z)}{z - w}\, dz = 0.$$

Adding, and using the remark above,

$$\sum_{i=0}^{j} \left(\frac{1}{2\pi i} \int_{\gamma_i} \frac{f(z)}{z - w}\, dz \right) = f(w).$$

Now the expression on the left-hand side is a continuous function of w for $w \in in[\gamma_0] \cap (\cap_{i=1}^{m} out[\gamma_i])$, as is the right-hand side, and so the formula holds for all such w. In particular, it holds for all $w \in [\beta]$. Thus

$$\int_\beta f(w)\,dw = \int_\beta \left(\sum_{i=0}^j \frac{1}{2\pi i} \int_{\gamma_i} \frac{f(z)}{z-w}\,dz \right) dw$$

$$= \sum_{i=0}^j \int_{\gamma_i} f(z) \left(\frac{1}{2\pi i} \int_\beta \frac{dw}{z-w} \right) dz$$

$$= -\sum_{i=0}^j \int_{\gamma_i} f(z) n(\beta, z)\,dz,$$

the change of order being justified, since the integrands are continuous.

Now $n(\beta, z)$ is a continuous integer-valued function on the connected set $\overline{in}[\gamma_i]$, and so is constant there. Let its constant value be ν_i. Thus

$$\int_\beta f(w)\,dw = -\sum_{i=0}^j \nu_i \int_{\gamma_i} f(z)\,dz.$$

If $z \in [\gamma_0]$ then z is in the unbounded component of $\mathbf{C} \setminus [\beta]$, and so $\nu_0 = 0$. If $1 \le i \le j$, there are two possibilities. First, there exists $w \in in[\gamma_i] \setminus U$; in this case $\nu_i = 0$, by hypothesis. Secondly, $in[\gamma_i] \subseteq U$. In this case, there exists $\delta > 0$ such that $N_\delta(\overline{in}[\gamma_i])$ is a simply connected subset of U, and so $\int_{\gamma_i} f(z)\,dz = 0$, by Cauchy's theorem for simply connected domains. Thus each summand is zero, and $\int_\beta f(z)\,dz = 0$. $\square$

22.10 Cycles; Cauchy's integral formula revisited

We now prove a more general version of Cauchy's integral formula. First, we consider integrals along more general sets than closed rectifiable paths. A *cycle* Γ in a domain U is an expression of the form $\Gamma = \sum_{i=1}^j a_i\gamma_i$, where $a_i \in \mathbf{Z}$ and γ_i is a closed rectifiable path in U, for $1 \le i \le j$. We set $[\Gamma]$, the *track* of Γ, to be $[\Gamma] = \cup_{i=0}^j[\gamma_i]$. If $w \notin U$, we define the *winding number* $n(\Gamma, w)$ *of* Γ *about* w to be $n(\Gamma, w) = \sum_{i=1}^j a_i n(\gamma_i, w)$, and if f is a continuous function on $[\Gamma]$, we set

$$\int_\Gamma f(z)\,dz = \sum_{i=1}^j \left(a_i \int_{\gamma_i} f(z)\,dz \right).$$

For example, if $\gamma_0, \gamma_1, \ldots, \gamma_j$ are the paths in Theorem 21.4.1 then $\Gamma = \sum_{i=0}^j \gamma_i$ is a cycle for which $n(\Gamma, w) = 1$ for all $w \in K$.

We can deduce results about winding numbers and integrals for cycles from the corresponding results for closed paths. Suppose that $\Gamma = \sum_{i=1}^{j} a_i \gamma_i$ is a cycle in a domain U and that $w \notin [\Gamma]$, and suppose that f is a continuous function on $[\Gamma]$. Suppose that $\gamma_i : [c_i, d_i] \to U$ is a parametrization of γ_i, for $1 \leq i \leq j$. Let

$$\tilde{\gamma}_i = \begin{cases} \gamma_i \vee \ldots \vee \gamma_i & a_i \text{ times}, & \text{if } a_i > 0, \\ \text{the constant path at } \gamma_i(c_i), & \text{if } a_i = 0, \\ \gamma_i^{\leftarrow} \vee \ldots \vee \gamma_i^{\leftarrow} & |a_i| \text{ times}, & \text{if } a_i < 0. \end{cases}$$

Since $U \setminus \{w\}$ is path-connected, for $2 \leq i \leq j$ there exists a rectilinear path β_i in U from $\gamma_1(c_1)$ to $\gamma_i(c_i)$, with $w \notin [\beta_i]$. Let $\delta_i = \beta_i \vee \tilde{\gamma}_i \vee \beta_i^{\leftarrow}$, for $2 \leq i \leq j$, and let $\delta = \tilde{\gamma}_1 \vee \delta_2 \vee \ldots \vee \delta_j$. Then δ is a closed rectifiable path in U. Further, the function f can be extended to a continuous function on $[\Gamma] \cup [\delta]$.

Proposition 22.10.1 *With the notation above,*

$$n(\Gamma, w) = n(\delta, w) \text{ and } \int_\Gamma f(z)\, dz = \int_\delta f(z)\, dz.$$

Proof This follows from the facts that $n(\gamma_i^{\leftarrow}, w) = -n(\gamma_i, w)$, that $\int_{\gamma_i^{\leftarrow}} f(z)\, dz = -\int_{\gamma_i} f(z)\, dz$ and that the integrals along β_i and $\beta_i^{\leftarrow}$ cancel each other. □

Consequently, we have the following extensions of Theorems 22.9.1 and 22.5.1, and of Corollary 22.5.2.

Theorem 22.10.2 (Cauchy's theorem for cycles) *Suppose that Γ is a cycle in a domain U for which $n(\Gamma, w) = 0$ for $w \notin U$. Then*

$$\int_\Gamma f(z)\, dz = 0.$$

Theorem 22.10.3 *Suppose that $\Gamma = \sum_{j=1}^{k} a_j \gamma_j$ is a cycle and that $w \notin [\Gamma]$. Then*

$$n(\Gamma, w) = \frac{1}{2\pi i} \int_\Gamma \frac{dz}{z - w}.$$

Theorem 22.10.4 *If $f : U \to \mathbb{C}$ is holomorphic, and $w \notin f([\Gamma])$ then*

$$n(f \circ \Gamma, w) = \frac{1}{2\pi i} \int_\Gamma \frac{f'(z)}{f(z) - w}\, dz.$$

We can also establish Cauchy's integral formula for cycles.

Theorem 22.10.5 (Cauchy's integral formula for cycles)   *Suppose that Γ is a cycle in a domain U for which $n(\Gamma, w) = 0$ for $w \notin U$. Suppose that f is a holomorphic function on U and that $z_0 \in U \setminus [\Gamma]$. Then*

$$n(\Gamma, z_0) f(z_0) = \frac{1}{2\pi i} \int_\Gamma \frac{f(z)}{z - z_0} \, dz.$$

Proof $f(z)/(z - z_0)$ is holomorphic on $V = U \setminus \{z_0\}$. The result is true if $n(\Gamma, z_0) = 0$, for then $n(\Gamma, w) = 0$ for $w \notin V$, and so the result follows from Theorem 22.10.2.

Otherwise, let $\nu = n(\Gamma, z_0)$. There exists $r > 0$ such that $N_r(z_0) \subseteq U$. Let $\Gamma' = \Gamma - \nu \kappa_r(z_0)$. Then $n(\Gamma', w) = n(\Gamma, w) - \nu n(\kappa_r(z_0), w) = 0 - 0 = 0$, for $w \notin U$ and $n(\Gamma', z_0) = n(\Gamma, z_0) - \nu n(\kappa_r(z_0), z_0) = \nu - \nu = 0$. Thus $n(\Gamma', w) = 0$ for $z \notin V$, and so

$$\frac{1}{2\pi i} \int_\Gamma \frac{f(z)}{z - z_0} \, dz - \frac{\nu}{2\pi i} \int_{\kappa_r(z_0)} \frac{f(z)}{z - z_0} \, dz = 0.$$

Now

$$\frac{1}{2\pi i} \int_{\kappa_r(z_0)} \frac{f(z)}{z - z_0} \, dz = f(z_0),$$

by Theorem 22.6.1, and so the result follows.   □

22.11 Functions defined inside a contour

So far we have been concerned with holomorphic functions defined on a domain U, and with paths with tracks in U. There is another situation which is worth considering. Recall that a *contour* is a positively oriented, rectifiable, simple closed path in $\mathbf{C}$. Suppose that γ is a contour and that f is a continuous function on the closed set $\overline{in}[\gamma]$ which is holomorphic on the open set $in[\gamma]$. Do Cauchy's theorem and the Cauchy integral formula hold?

In general, the answer is 'yes', but the proofs are difficult. There is, however, one situation where the proof is quite easy, and which is quite sufficient for most needs. We need a definition. Suppose that K is a subset of $\mathbf{C}$, and that k_0 is an interior point of K. We say that K is *star-shaped about k_0* if for each $z \in K$ the open line segment $(k_0, z) = \{(1 - \lambda)k_0 + \lambda z : 0 < \lambda < 1\}$ is contained in the interior K° of K. As an important example, if K is a convex body, then K is star-shaped about each point of K°.

Theorem 22.11.1   *Suppose that $\gamma : [a, b] \to \mathbf{C}$ is a contour, that $in[\gamma]$ is star-shaped about k_0, and that $0 < r < 1$. Let $\gamma_r(t) = (1 - r)k_0 + r\gamma(t)$, for $t \in [a, b]$. Then γ_r is a contour inside $[\gamma]$.*

Suppose that $0 < r_0 < 1$ and that g is a continuous function on $\overline{in}[\gamma] \cap \overline{out}[\gamma_{r_0}]$. Then $\int_{\gamma_r} g(z)\,dz \to \int_{\gamma} g(z)\,dz$ as $r \nearrow 1$.

Proof It follows from the definition of 'star-shaped' that γ_r is a simple closed path inside $[\gamma]$. Since $|\gamma_r(t) - \gamma_r(t')| = r|\gamma(t) - \gamma(t')|$, it also follows that γ_r is a contour.

If $r_0 < r < 1$ and if $z \in [\gamma]$, let $g_r(z) = rg((1-r)k_0 + rz)$. Then $g_r(z)$ converges uniformly to $g(z)$ on $[\gamma]$ as $r \nearrow 1$, and $\int_{\gamma_r} g(z)\,dz = \int_{\gamma} g_r(z)\,dz$. Consequently, $\int_{\gamma_r} g(z)\,dz \to \int_{\gamma} g(z)\,dz$ as $r \nearrow 1$. □

Corollary 22.11.2 *If f is continuous on $\overline{in}[\gamma]$ and holomorphic on $in[\gamma]$ then $\int_{\gamma} f(z)\,dz = 0$, and*

$$f(w) = \frac{1}{2\pi i} \int_{\gamma} \frac{f(z)}{z-w}\,dz \text{ for } w \in in[\gamma].$$

Proof Since $\int_{\gamma_r} f(z)\,dz = 0$, the first equation follows immediately. For the second, there exists $0 < r_0 < 1$ such that $w \in in[\gamma_{r_0}]$. Then the function $g(z) = f(z)/2\pi i(z-w)$ is continuous on $\overline{in}[\gamma] \cap \overline{out}[\gamma_{r_0}]$, and if $r_0 < r < 1$ then

$$f(w) = \frac{1}{2\pi i} \int_{\gamma_r} \frac{f(z)}{z-w}\,dz = \int_{\gamma_r} g(z)\,dz,$$

so that

$$f(w) = \lim_{r \nearrow 1} \int_{\gamma_r} g(z)\,dz = \int_{\gamma} g(z)\,dz = \frac{1}{2\pi i} \int_{\gamma} \frac{f(z)}{z-w}\,dz.$$

□

22.12 The Schwarz reflection principle

We use Morera's theorem to establish the Schwarz reflection principle.

Theorem 22.12.1 (The Schwarz reflection principle) *Suppose that*

(i) *U is a domain in $H^+ = \{z = x + iy \in \mathbf{C} : y > 0\}$;*
(ii) *∂U contains an open interval (a,b) in $\mathbf{R}$;*
(iii) *f is a continuous function on $U \cup (a,b)$ which is holomorphic on U, and*
(iv) *f is real-valued on (a,b).*

Let $U^* = \{z : \bar{z} \in U\}$, and let $V = U \cup (a,b) \cup U^*$. If $z \in U^*$ let $f(z) = \overline{f(\bar{z})}$. Then f is holomorphic on V.

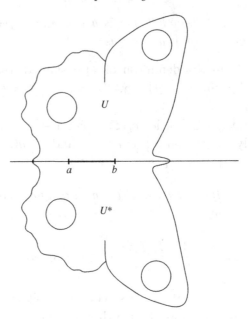

Figure 22.12.

Proof It follows from (iv) that f is a continuous function on V, and it follows from the definition of differentiability, or from the Cauchy–Riemann equations, that f is holomorphic on U^*. But we need to establish differentiability at points of (a, b). We therefore use Morera's theorem and Corollary 22.11.2. Suppose that γ is a square path, with $\overline{in}[\gamma] \subseteq V$. If $[\gamma] \subseteq \overline{U}$ or $[\gamma] \subseteq \overline{U^*}$, then $\int_\gamma f(z)dz = 0$. Otherwise, suppose that $[\gamma] \cap (a, b) = \{w_1, w_2\}$. There exist rectangular dyadic paths γ_1 in $\overline{U}$ and γ_2 in $\overline{U^*}$ such that $[\gamma_1] \cap (a, b) = [\gamma_2] \cap (a, b) = [w_1, w_2]$ and $[\gamma_1] \cup [\gamma_2] = [\gamma] \cup [w_1, w_2]$. Then, with suitable orientation, $\int_\gamma f(z)dz = \int_{\gamma_1} f(z)dz + \int_{\gamma_2} f(z)dz = 0$. Thus f is holomorphic on V, by Morera's theorem. □

Exercises

22.12.1 Suppose that
  (i) U is a domain in $\mathbf{D}$;
  (ii) ∂U contains an open circular arc $A_{\alpha,\beta} = \{e^{it} : \alpha < t < \beta\}$ in $\mathbf{T}$;
  (iii) f is a continuous function on $U \cup A_{\alpha,\beta}$ which is holomorphic on U, and
  (iv) f is real-valued on $A_{\alpha,\beta}$.

Let $U^* = \{z : 1/\bar{z} \in U\}$, and let $V = U \cup (a,b) \cup U^*$. If $z \in U^*$ let $f(z) = \overline{f(1/\bar{z})}$. Show that f is holomorphic on V.

22.12.2 Suppose that condition (iv) of the previous exercise is replaced by

(iv') $|f(z)| = 1$ for $z \in A_{\alpha,\beta}$.

Let $f(z) = 1/\overline{f(1/\bar{z})}$, for $z \in U^*$. Show that f is holomorphic on V.

23

Zeros and singularities

23.1 Zeros

Suppose that f is a holomorphic function on a domain U. We denote the zero set $\{z \in U : f(z) = 0\}$ by Z_f. What can we say about Z_f? It is certainly closed, since f is continuous.

Theorem 23.1.1 *Suppose that f is a non-constant holomorphic function on a domain U, and that $f(z_0) = 0$. Then there exists a least $k \in \mathbf{N}$ such that $f^{(k)}(z_0) \neq 0$, and there exists $s > 0$ such that $N_s(z_0) \subseteq U$ and $f(z) \neq 0$ for z in the punctured neighbourhood $N_s^*(z_0)$.*

Proof By Proposition 20.3.9, there exists a least $k \in \mathbf{N}$ such that $f^{(k)}(z_0) \neq 0$, so that there exists $r > 0$ such that $N_r(z_0) \subseteq U$ and

$$f(z) = \sum_{n=k}^{\infty} \frac{f^{(n)}(z_0)}{n!} (z - z_0)^n = (z - z_0)^k h(z),$$

$$\text{where } h(z) = \sum_{n=0}^{\infty} \frac{f^{(n+k)}(z_0)}{(n+k)!} (z - z_0)^n,$$

for $z \in N_r(z_0)$. Then $h(z_0) = f^{(k)}(z_0)/k! \neq 0$. Since $h(z) \to h(z_0)$ as $z \to z_0$, there exists $0 < s \le r$ such that $h(z) \neq 0$ for $z \in N_s(z_0)$, and so $f(z) \neq 0$ for $z \in N_s^*(z_0)$. $\square$

Thus the points of Z_f are isolated points of Z_f; the subspace topology on Z_f is the discrete topology. In general, a subspace S of a topological space (X, τ) is a *discrete subspace* if the subspace topology on S is the discrete topology, so that each of its points is an isolated point in X. If $z_0 \in Z_f$, f is said to have a *zero of order k*, or *multiplicity k*, at z_0 if k is the least integer for which $f^{(k)}(z_0) \neq 0$. Then $f(z) = (z - z_0)^k h(z)$, where h is a holomorphic function for which $h(z_0) \neq 0$.

It is important that we only consider points of the domain U. For example, let $U = \mathbf{C}^* = \mathbf{C} \setminus \{0\}$, and let $f(z) = \exp(2\pi i/z) - 1$. Then f is holomorphic on U, and $Z_f = \{\pm 1/k : k \in \mathbf{N}\}$. The point 0 is a closure point of Z_f in $\mathbf{C}$, but it is not in U.

A closed subset S of a domain for which every point of S is an isolated point of S can be infinite, but it must be locally finite.

Proposition 23.1.2 *Suppose that S is a discrete subspace of a Hausdorff topological space (X, τ). If K is a compact subset of U then $K \cap S$ is a finite set.*

Proof For $K \cap S$ is a compact Hausdorff space with the discrete topology, and so must be a finite set.

$\square$

Corollary 23.1.3 *If S is a closed discrete subspace of a domain U then S is countable.*

Proof For U is σ-compact; it is the union of countably many compact sets.

$\square$

In particular, if f is a non-constant holomorphic function on U then the zero set Z_f is countable.

Proposition 23.1.4 *If S is a closed discrete subspace of a domain U, then $V = U \setminus S$ is a domain.*

Proof Since S is closed in U, V is an open subset of $\mathbf{C}$. We must show that it is connected; we shall show that it is path-connected. Suppose that $v, v' \in V$. Since U is path-connected, there is a simple path $\gamma : [0, 1] \to U$ from v to v'. Since its track $[\gamma]$ is compact, $[\gamma] \cap S$ is finite. Thus there exist $0 < t_1 < \cdots < t_k < 1$ such that $[\gamma] \cap S = \{\gamma(t_j) : 1 \le j \le k\}$. It is now easy to perturb the path so that it avoids S. For each $1 \le j \le k$ there exists $\epsilon_j > 0$ such that $N_{\epsilon_j}(\gamma(t_j)) \subseteq U$ and $N_{\epsilon_j}(\gamma(t_j)) \cap S = \{\gamma(t_j)\}$. Since γ is continuous, for each j there exist l_j and r_j with $l_j < t_j < r_j$ such that $\gamma([l_j, r_j]) \subseteq N_{\epsilon_j}(\gamma(t_j))$. We can suppose that $r_j < l_{j+1}$ for $1 \le j < k$. Since each punctured neighbourhood $N^*_{\epsilon_j}(\gamma(t_j))$ is path-connected, we can replace $\gamma_{[l_j, r_j]}$ by a path in $N^*_{\epsilon_j}(\gamma(t_j))$ from $\gamma(l_j)$ to $\gamma(r_j)$. In this way, we obtain a path in V from v to v'.

$\square$

Exercises

23.1.1 Where are the zeros of the function $\sin((1+z)/(1-z))$? Show that they have an accumulation point in $\mathbf{C}$. Why does this not contradict Proposition 23.1.2?

23.1.2 Suppose that f and g are holomorphic functions on a domain U and that $|f(z)| = |g(z)|$ for $z \in U$. Show that there exists $\alpha \in \mathbf{C}$ such that $f = e^{i\alpha}g$.

23.1.3 Give an example of a connected Hausdorff topological space (X, τ) for which $X \setminus \{x\}$ is not connected, for each $x \in X$.

23.2 Laurent series

Suppose that f is a non-constant holomorphic function on a domain U, with zero set Z_f. Then $V = U \setminus Z_f$ is a domain, and the function $1/f$ is holomorphic on V. If $z_0 \in Z_f$ then there exists $r > 0$ such that the punctured neighbourhood $N_r^*(z_0)$ is contained in V. The function $1/f$ is holomorphic on $N_r^*(z_0)$, and $|1/f(z)| \to \infty$ as $z \to z_0$. We are therefore led to study holomorphic functions on such punctured neighbourhoods. In fact, we consider holomorphic functions on rather more general sets. Suppose that $z_0 \in \mathbf{C}$ and that $0 \leq r < R \leq \infty$. The *(open) annulus* $A_{r,R}(z_0)$ is defined to be the set

$$A_{r,R}(z_0) = \{z \in \mathbf{C} : r < |z - z_0| < R\}.$$

Thus $N_R^*(z_0) = A_{0,R}(z_0)$. An annulus $A_{r,R}(z_0)$ is an open connected subset of $\mathbf{C}$, but it is not simply connected; its complement has two connected components, namely $\{z \in \mathbf{C} : |z - z_0| \leq r\}$ and $\{z \in \mathbf{C} : |z - z_0| \geq R\}$. If $r < s < R$ then $\kappa_s(z_0)$ is a closed path in $A_{r,R}(z_0)$ which is not homotopic to a constant path. If f is a holomorphic function on $A_{r,R}(z_0)$, we cannot always represent f by a Taylor series, as the example $1/(z - z_0)$ shows. Instead, we represent it by a doubly infinite series.

Theorem 23.2.1 *Suppose that f is a holomorphic function on the annulus $A_{r,R}(z_0)$ and that $n \in \mathbf{Z}$. If $r < s < R$, the quantity*

$$a_n = \frac{1}{2\pi i} \int_{\kappa_s(z_0)} \frac{f(z)}{(z - z_0)^{n+1}} \, dz$$

does not depend upon s.

  Suppose that $r < s \leq t < R$. The series $\sum_{n=0}^{\infty} a_n (z - z_0)^n$ converges absolutely uniformly on $M_t(z_0)$, and the series $\sum_{n=1}^{\infty} a_{-n}(z-z_0)^{-n}$ converges absolutely uniformly on the set $\{z : |z - z_0| \geq s\}$. Thus the doubly infinite series $\sum_{n=-\infty}^{\infty} a_n(z-z_0)^n$ *converges absolutely uniformly on the set* $\{z : s \leq |z-z_0| \leq t\}$: *its sum is* $f(z)$.

Proof Let Γ be the cycle $\kappa_t(z_0) - \kappa_s(z_0)$. The $n(\Gamma, w) = 0$ for $w \notin A_{r,R}(z_0)$, and the function $f(z)/(z-z_0)^{n+1}$ is holomorphic on the annulus. By Cauchy's theorem for cycles,

$$\int_\Gamma \frac{f(z)}{(z-z_0)^{n+1}} dz = \int_{\kappa_t(z_0)} \frac{f(z)}{(z-z_0)^{n+1}} dz - \int_{\kappa_s(z_0)} \frac{f(z)}{(z-z_0)^{n+1}} dz = 0.$$

Thus a_n does not depend upon s.

Suppose now that $z_0 + h \in A_{r,R}(z_0)$. Choose $r < s < |h| < t < R$. Then $n(\Gamma, z_0 + h) = 1$, so that

$$f(z_0 + h) = \frac{1}{2\pi i} \int_\Gamma \frac{f(z)}{z - (z_0 + h)} dz$$

$$= \frac{1}{2\pi i} \int_{\kappa_t(z_0)} \frac{f(z)}{z - (z_0 + h)} dz - \frac{1}{2\pi i} \int_{\kappa_s(z_0)} \frac{f(z)}{z - (z_0 + h)} dz.$$

Arguing exactly as in the proof of Theorem 22.4.1,

$$\frac{1}{2\pi i} \int_{\kappa_t(z_0)} \frac{f(z)}{z - (z_0 + h)} dz = \sum_{n=0}^{\infty} a_n h^n,$$

and the series has radius of convergence at least R. It therefore converges absolutely uniformly on $M_t(z_0)$.

We use an argument similar to the one used in the proof of Theorem 22.4.1 to deal with the remaining terms. Choose $r < s' < s$. If $z \in T_{s'}(z_0)$ then $|z - z_0| = s' < |h|$. Since

$$\frac{-1}{z - (z_0 + h)}$$

$$= \frac{1}{h} \left(\frac{1}{1 - (z-z_0)/h} \right) = \frac{1}{h} + \cdots + \frac{(z-z_0)^n}{h^{n+1}} - \frac{(z-z_0)^{n+1}}{h^{n+1}(z-(z_0+h))},$$

it follows that

$$-\frac{1}{2\pi i} \int_{\kappa_{s'}(z_0)} \frac{f(z)}{z - (z_0 + h)} dz$$

$$= \frac{1}{2\pi i} \int_{\kappa_{s'}(z_0)} f(z) \left(\frac{1}{h} + \frac{z - z_0}{h^2} + \cdots + \frac{(z-z_0)^n}{h^{n+1}} - \frac{(z-z_0)^{n+1}}{h^{n+1}(z-(z_0+h))} \right) dz$$

$$= \sum_{j=1}^{n} \frac{a_{-j}}{h^j} - R_n(h),$$

where
$$R_n(h) = \frac{1}{2\pi i h^{n+1}} \int_{\kappa_{s'}(z_0)} f(z) \frac{(z - z_0)^{n+1}}{z - (z_0 + h)} dz.$$

Let $M_{s'} = \sup\{|f(z)| : z \in \mathbf{T}_{s'}(z_0)\}$. Then

$$|R_n(h)| \le \frac{2\pi s'}{2\pi |h|^{n+1}} \cdot \frac{(s')^{n+1} M_{s'}}{|h| - s'} \le \frac{s' M_{s'}}{s - s'} \cdot \left(\frac{s'}{s}\right)^{n+1},$$

so that $R_n(h) \to 0$ as $n \to \infty$. Thus

$$-\frac{1}{2\pi i} \int_{\kappa_{s'}(z_0)} \frac{f(z)}{z - (z_0 + h)} dz = \sum_{j=1}^{\infty} \frac{a_{-j}}{h^j}.$$

Consequently the series $\sum_{j=1}^{\infty} a_{-j} z^j$ has radius of convergence at least $1/s'$. Since $1/s' > 1/s$, the series $\sum_{j=1}^{\infty} a_{-j}/h^j$ converges absolutely uniformly on the set $\{h : |h - z_0| \ge s\}$.

Adding the two infinite series, we obtain the result.   □

The doubly infinite series $\sum_{n=-\infty}^{\infty} a_n (z - z_0)^n$ is called the *Laurent series* for f. The function $f_p(w) = \sum_{n=1}^{\infty} a_{-n}/(w - z_0)^n$, defined for $|w - z_0| > r$, is called the *principal part* of f: if $r < s < |w - z_0|$ then

$$f_p(w) = \frac{-1}{2\pi i} \int_{\kappa_s(z_0)} \frac{f(z)}{z - w} dz.$$

Let us show that the Laurent series is unique.

Theorem 23.2.2   *Suppose that f is a holomorphic function on the annulus $A_{r,R}(z_0)$, and that $f(z_0 + h) = \sum_{n=-\infty}^{\infty} b_n h^n$ for $z_0 + h \in A_{r,R}(z_0)$. Then*

$$b_n = \frac{1}{2\pi i} \int_{\kappa_s(z_0)} \frac{f(z)}{(z - z_0)^{n+1}} dz$$

where $r < s < R$.

Proof   As in Theorem 23.2.1, the series $\sum_{n=0}^{\infty} b_n h^n$ and $\sum_{n=1}^{\infty} b_{-n} h^{-n}$ converge uniformly on $[\kappa_s(z_0)]$. Since

$$\frac{1}{2\pi i} \int_{\kappa_s(z_0)} \frac{(z - z_0)^j}{(z - z_0)^{n+1}} dz = \begin{cases} 1 & \text{if } j = n \\ 0 & \text{otherwise,} \end{cases}$$

it follows that

$$\frac{1}{2\pi i} \int_{\kappa_s(z_0)} \frac{\sum_{j=-M}^{N} b_j (z - z_0)^j}{(z - z_0)^{n+1}} dz = b_n, \quad \text{for } -M \le n \le N.$$

Thus

$$b_n = \lim_{M,N\to\infty} \frac{1}{2\pi i} \int_{\kappa_s(z_0)} \frac{\sum_{j=-M}^{N} b_j(z-z_0)^j}{(z-z_0)^{n+1}} \, dz$$

$$= \frac{1}{2\pi i} \int_{\kappa_s(z_0)} \frac{f(z)}{(z-z_0)^{n+1}} \, dz.$$

□

Exercises

23.2.1 Find the Laurent series of the function $f(z) = 1/z(z-1)(z-2)$, defined on $\mathbf{C} \setminus \{0, 1, 2\}$, in each of the annuli $A_{0,1}(0)$, $A_{1,2}(0)$, $A_{2,\infty}(0)$ and $A_{0,1}(1)$.

23.2.2 Let $\sum_{n=-\infty}^{\infty} a_n z^n$ be the Laurent series for the function $e^{z+1/z}$ defined on $\mathbf{C}^*$. Show that

$$a_n = a_{-n} = \sum_{j=0}^{\infty} \frac{1}{j!(n-j)!} = \frac{1}{\pi} \int_0^{\pi} e^{2\cos t} \cos nt \, dt.$$

23.2.3 Let $\sum_{n=-\infty}^{\infty} b_n z^n$ be the Laurent series for the function $e^{z-1/z}$ defined on $\mathbf{C}^*$. Show that if $n \in \mathbf{N}$ then

$$b_n = (-1)^n b_{-n} = \sum_{j=0}^{\infty} \frac{(-1)^j}{j!(n-j)!} = \frac{1}{\pi} \int_0^{\pi} \cos(2\sin t - nt) \, dt.$$

23.2.4 Find the coefficients in the Laurent series for the functions $\cos(z+1/z)$ and $\sin(z+1/z)$ defined on $\mathbf{C}^*$.

23.3 Isolated singularities

Suppose that f is a holomorphic function on a domain U with zero set Z_f. We can then define the function $1/f$ on the domain $V = U \setminus Z_f$. The points of Z_f are isolated points of $\mathbf{C} \setminus V$. If $z_0 \in Z_f$ then $|1/f(z)| \to \infty$ as $z \to z_0$.

This leads to the following definition. If f is a holomorphic function on a domain U and z_0 is an isolated point of $\mathbf{C} \setminus U$ then z_0 is an *isolated singularity* of f. There then exists $r > 0$ such that $N_r^*(z_0) \subseteq U$, and the restriction of f to $N_r^*(z_0)$ has a Laurent series

$$f(z) = \sum_{n=-\infty}^{\infty} a_n(z-z_0)^n \text{ for } z \in N_r^*(z_0).$$

Further, if $0 < s < r$ then

$$a_n = \frac{1}{2\pi i} \int_{\kappa_s(z_0)} \frac{f(z)}{(z-w)^{n+1}} \, dz.$$

As we shall see, the coefficient $a_{-1} = (1/2\pi i) \int_{\kappa_s(z_0)} f(z) \, dz$ is particularly important; it is called the *residue* of f at z_0, and is denoted by res $_f(z_0)$.

We use the coefficients in the Laurent series to classify the singularity. We define the *spectrum* $\sigma_f(z_0)$ of f at z_0 to be $\sigma_f(z_0) = \{n \in \mathbf{Z} : a_n \neq 0\}$. If $f \neq 0$ then $\sigma_f(z_0)$ is not empty. We then classify the singularity in the following way.

- If $f = 0$ or $\sigma_f(z_0) \subseteq \mathbf{Z}^+$ then f has a *removable singularity* at z_0.
- If $\sigma_f(z_0)$ is bounded below, but $\inf(\sigma_f(z_0)) = -k < 0$, then f has a *pole* at z_0, of *order* k. If $k = 1$ then z_0 is a *simple pole* of f.
- If $\sigma_f(z_0)$ is not bounded below, then f has an *essential isolated singularity* at z_0.

In the next three theorems, we characterize each of these possibilities.

Theorem 23.3.1 *Suppose that f is a non-zero holomorphic function on a domain U, that z_0 is an isolated singularity of f and that $N_r^*(z_0) \subseteq U$. The following are equivalent:*

(i) *f has a removable singularity at z_0;*
(ii) *f can be extended to an analytic function on $N_r(z_0)$;*
(iii) *there exists $l \in \mathbf{C}$ such that $f(z) \to l$ as $z \to z_0$;*
(iv) *$(z - z_0)f(z) \to 0$ as $z \to z_0$;*
(v) *if $0 < t < r$ then f is bounded on the closed punctured neighbourhood $M_t^*(z_0)$.*

Proof If f has a removable singularity at z_0, the series $\sum_{n=0}^{\infty} a_n(z - z_0)^n$ defines an analytic function on $N_r(z_0)$, with $f(z_0) = a_0$, which agrees with f on $N_r^*(z_0)$. Thus (i) implies (ii). Then (ii) implies (iii), (iii) implies (iv) and (iv) implies (v). Suppose that (v) holds, and that $0 < t < r$; let $K_t = \sup\{|f(z)| : z \in M_t^*(z_0)\}$. If $0 < s \leq t$ and $n \in \mathbf{N}$ then

$$|a_{-n}| = \left| \frac{1}{2\pi i} \int_{\kappa_s(z_0)} f(z)(z - z_0)^{n-1} \, dz \right| \leq \frac{2\pi s}{2\pi}.K_t|s|^{n-1} = K_t s^n.$$

Since s can be taken to be arbitrarily small, $a_{-n} = 0$, and so (i) holds. □

Thus the singularity can be removed, by setting $f(z_0) = a_0$.

Theorem 23.3.2   *Suppose that f is a non-zero holomorphic function on a domain U, that z_0 is an isolated singularity of f and that $N_r^*(z_0) \subseteq U$. The following are equivalent:*

(i) *f has a pole at z_0;*

(ii) *There exists a holomorphic function g on $U \cup \{z_0\}$ and $k \in \mathbf{N}$ such that $g(z_0) \neq 0$ and $f(z) = g(z)/(z - z_0)^k$ for $z \in U$;*

(iii) *$|f(z)| \to \infty$ as $z \to z_0$;*

(iv) *there exists $0 < s \leq r$ such that $f(z) \neq 0$ on $N_s^*(z_0)$ (so that $1/f$ is defined on $N_s^*(z_0)$), and $1/f(z) \to 0$ as $z \to z_0$.*

Proof   Suppose that f has a pole of order k at z_0 and that

$$f(z) = \sum_{n=-k}^{\infty} a_n (z - z_0)^n, \text{ for } z \in N_r^*(z_0).$$

Let $g(z) = (z - z_0)^k f(z)$ for $z \in U$. Then $g(z) = \sum_{n=0}^{\infty} a_{n-k}(z - z_0)^n$ for $z \in N_r^*(z_0)$, so that g has a removable singularity at z_0, and can therefore be extended to a holomorphic function on $U \cup \{z_0\}$. Thus (i) implies (ii). Clearly (ii) implies (iii) and (iii) implies (iv).

Suppose that (iv) holds. Then $1/f$ has a removable singularity at z_0, and it can be extended to a holomorphic function on $N_s(z_0)$ by setting $1/f(z_0) = 0$. Thus this extension has a zero at z_0, of order k, say. There therefore exists a holomorphic function g on $N_s(z_0)$, with $g(z_0) \neq 0$, such that $1/f(z) = (z - z_0)^k g(z)$ for $z \in N_s(z_0)$. Then $g(z) \neq 0$ for $z \in N_s(z_0)$. Let $h(z) = 1/g(z)$, for $z \in N_s(z_0)$. The function h is holomorphic on $N_s(z_0)$, and so has a Taylor series expansion $h(z) = \sum_{n=0}^{\infty} h_n (z - z_0)^n$, for $z \in N_s(z_0)$, with $h_0 = 1/g(z_0) \neq 0$. Then

$$f(z) = (z - z_0)^{-k} h(z) = \sum_{n=-k}^{\infty} h_{n+k}(z - z_0)^n \text{ for } z \in N_s^*(z_0).$$

Since $h_0 \neq 0$, it follows that f has a pole of order k at z_0.   □

Finally we characterize essential isolated singularities.

Theorem 23.3.3 (Weierstrass' theorem)   *Suppose that f is a non-zero holomorphic function on a domain U, that z_0 is an isolated singularity of f and that $N_r^*(z_0) \subseteq U$. The following are equivalent:*

(i) *f has an essential isolated singularity at z_0;*

(ii) *for each $0 < s < r$ the set $f(N_s^*(z_0))$ is dense in $\mathbf{C}$ – that is, if $w \in \mathbf{C}$, $\delta > 0$ and $0 < s < r$ there exists $z \in N_s^*(z_0)$ such that $|f(z) - w| < \delta$;*

(iii) if $w \in \mathbf{C}$ there exists a sequence $(z_k)_{k=1}^{\infty}$ in U such that $z_k \to z_0$ and $f(z_k) \to w$ as $k \to \infty$.

Proof   It is an easy exercise to show that (ii) and (iii) are equivalent, and it follows from Theorems 23.3.1 and 23.3.2 that either implies (i). It remains to show that (i) implies (ii). Suppose that (ii) does not hold, so that there exist $w \in \mathbf{C}$, $\delta > 0$ and $0 < s < r$ for which $|f(z) - w| > \delta$ for $z \in N_s^*(z_0)$. Then the function $g(z) = 1/(f(z) - w)$ is a bounded holomorphic function on $N_s^*(z_0)$, and by Theorem 23.3.1, it has a removable singularity at z_0. It can therefore be extended to a holomorphic function g on $N_s(z_0)$. If $g(z_0) = 0$, then $f(z) - w$ has a pole at z_0, and so therefore does f; if $g(z_0) \neq 0$, then $f(z) - w$ has a removable singularity at z_0, and so therefore does f. In either case, f does not have an essential isolated singularity at z_0.   □

Corollary 23.3.4   *Suppose that f is a holomorphic function on the domain $\{z \in \mathbf{C} : |z| > R\}$ with Laurent series $f(z) = \sum_{n=-\infty}^{\infty} a_n z^n$, and that its spectrum $\{n \in \mathbf{Z} : a_n \neq 0\}$ is not bounded above. If $w \in \mathbf{C}$, $S \geq R$ and $\epsilon > 0$ then there exists $z \in \mathbf{C}$ with $|z| > S$ such that $|f(z) - w| < \epsilon$; that is, $f(\{z \in \mathbf{C} : |z| > S\})$ is dense in $\mathbf{C}$.*

Proof   Let $h(z) = f(1/z)$, for $0 < |z| < 1/R$. Then h has Laurent series $\sum_{n=-\infty}^{\infty} a_{-n} z^n$, so that h has an isolated essential singularity at 0. Thus there exists ζ with $0 < |\zeta| < 1/S$ such that $|h(\zeta) - w| < \epsilon$. If $z = 1/\zeta$ then $|z| > S$ and $|f(z) - w| < \epsilon$.   □

As an example, the function $f(z) = e^{-1/z^2}$ on $\mathbf{C} \setminus \{0\}$ has an essential isolated singularity at 0, since its Laurent series is

$$f(z) = \sum_{n=0}^{\infty} \frac{(-1)^n}{n!} \left(\frac{1}{z}\right)^{2n}.$$

If we consider its restriction to $\mathbf{R} \setminus \{0\}$, and define $f(0) = 0$, then f is an infinitely differentiable function all of whose derivatives vanish at 0 (see Volume I, Section 7.6). On the other hand, $f(it) = e^{1/t^2}$, so that, as we approach 0 along the imaginary axis, $f(it) \to \infty$ as $t \to 0$.

Weierstrass' theorem shows that functions behave badly near an essential isolated singularity. In fact, more can be said.

Theorem 23.3.5 (Picard's theorem)   *Suppose that f is a non-zero holomorphic function on a domain U, that z_0 is an essential isolated singularity of f and that $N_r^*(z_0) \subseteq U$. Then $\mathbf{C} \setminus f(N_r^*(z_0))$ consists of at most one point.*

Thus f takes all values, except perhaps one, arbitrarily close to z_0. We cannot do better: the function $f(z) = e^{-1/z^2}$ on $\mathbf{C} \setminus \{0\}$ fails to take the value 0. The proof of this theorem is beyond the scope of this book[1].

Exercises

23.3.1 Where are the singularities of the following functions? If a singularity is isolated, determine whether it is removable, or a pole, or an isolated singularity. If the function has a removable singularity at z, determine the value that the function should take at z to make it continuous at z.

    (i) $(\sin z)/z$
    (ii) $(1 - \cos z)/z^2$
    (iii) $\tan z$
    (iv) $e^{1/z}$
    (v) $(\log z^n)/(1 - z)^n$ (defined in $N_1(1)$)
    (vi) $1/(e^z - 1)$
    (vii) $z^n \cos(1/z)$
    (viii) $z^n \tan(1/z)$

23.3.2 Suppose that f is an entire function and that $|f(z)| \to \infty$ as $|z| \to \infty$. Show that f is a polynomial function.

23.3.3 If $0 < |a| < 1$, let b_a be the rational function $b_a(z) = (z - a)/(1 - \bar{a}z)$ defined on the set $N_{1/|a|}(0)$, and let $b_0(z) = z$. Show that if $|z| = 1$ then $|b_a(z)| = 1$.

23.3.4 Suppose that f is a non-constant holomorphic function on a domain U, that $M_1(0) \subseteq U$, and that $|f(z)| = 1$ for $|z| = 1$. Let $a_1, \ldots, a_n$ be the zeros of f in $N_1(0)$, with multiplicities $k_1, \ldots, k_n$ respectively. Show that there exists $\theta \in (-\pi, \pi]$ such that f is the rational function

$$f = e^{i\theta} b_{a_1}^{k_1} \ldots b_{a_n}^{k_n},$$

where b_a is the function defined in the previous exercise.

23.3.5 Suppose that f is a holomorphic function on a domain U and that f has a singularity at z_0. Show that if z_0 is a non-removable singularity then the function e^f has an isolated essential singularity at z_0. Deduce that if $\Re f$ is bounded in a neighbourhood of z_0 then z_0 is a removable singularity.

---

[1] See John B. Conway *Functions of one complex variable*, Springer-Verlag 1978.

23.4 Meromorphic functions and the complex sphere

Theorem 23.3.2 shows that zeros and poles are closely related. This leads to the following definition. A *meromorphic* function f on a domain U is a pair (f, S_f), where S_f (the *singular set*) is a discrete closed subset of U, together with a holomorphic function f on $U \setminus S_f$ which has a pole at each point of S_f. If f is a non-zero meromorphic function on U with singular set S_f and zero set Z_f, then $1/f$ is also a meromorphic function on U, with singular set Z_f and zero set S_f. It follows from this that the meromorphic functions on U form a field. Rational functions are examples of meromorphic functions on $\mathbf{C}$; another example is the function $\cot z$, with singular set $\{n\pi : n \in \mathbf{Z}\}$ and zero set $\{(n + \frac{1}{2})\pi : n \in \mathbf{Z}\}$.

A meromorphic function f on a domain U concerns a function defined on a subset $U \setminus S_f$ of U. This appears to be a misuse of the word 'function', but is excusable since 'mero' means 'part'. But we can remedy this misuse by enlarging the range of f. If $z_0 \in S_f$ then $|f(z)| \to \infty$ as $z \to z_0$, and so we need to adjoin a 'point at infinity' in a suitable way. The following construction provides a concrete way of doing so.

The complex plane $\mathbf{C}$ is isomorphic as a real vector space to $\mathbf{R}^2$, and we can identify $\mathbf{C}$ with a linear subspace of $\mathbf{R}^3$ by identifying $z = x + iy$ with the point $(x, y, 0)$. If we give $\mathbf{R}^3$ the Euclidean metric d defined by the norm $\|(x, y, t)\| = (x^2 + y^2 + t^2)^{\frac{1}{2}}$, then distances are preserved: $d(z, w) = |z - w|$. We now consider the unit sphere $\mathbf{S} = \{(x, y, t) : x^2 + y^2 + t^2 = 1\}$ and the point $N = (0, 0, 1)$: N is the *north pole* of $\mathbf{S}$. If $z \in \mathbf{C}$ then the straight line l_z through N and z meets $\mathbf{S}$ in two points: one is N, and we denote the other by $\phi(z)$. The mapping $z \to \phi(z)$ is a bijection of $\mathbf{C}$ onto $\mathbf{S} \setminus \{N\}$, called the *stereographic projection*.

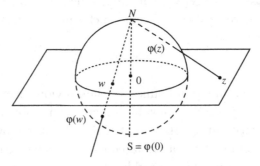

Figure 23.4.

Let us calculate $\phi(z)$. If $z = x + iy = re^{i\theta}$, the straight line l_z is the set

$$\{(1 - \lambda)N + \lambda z : \lambda \in \mathbf{R}\} = \{(\lambda r \cos \theta, \lambda r \sin \theta, 1 - \lambda) : \lambda \in \mathbf{R}\}.$$

This meets S where $\lambda^2 r^2 + (1-\lambda)^2 = 1$; that is, where $\lambda = 0$ (the point N) and $\lambda = 2/(1+r^2)$ (the point $\phi(z)$). Thus

$$\phi(z) = \left(\frac{2x}{1+r^2}, \frac{2y}{1+r^2}, \frac{r^2-1}{1+r^2} \right).$$

We now adjoin an extra point ∞ to $\mathbf{C}$, and denote $\mathbf{C} \cup \{\infty\}$ by $\mathbf{C}_\infty$. We extend ϕ to $\mathbf{C}_\infty$ by setting $\phi(\infty) = N$, so that ϕ is a bijection of $\mathbf{C}_\infty$ onto $\mathbf{S}$. We define a metric ρ on $\mathbf{C}_\infty$ by setting $\rho(z,w) = d(\phi(z), \phi(w)) = \|\phi(z) - \phi(w)\|$. Then $(\mathbf{C}_\infty, \rho)$ is a compact metric space, isometrically homeomorphic to $(\mathbf{S}, d)$; for this reason, $\mathbf{C}_\infty$ is called the *complex sphere*. The inclusion mapping $(\mathbf{C}, d) \rightarrow (\mathbf{C}_\infty, \rho)$ is then a homeomorphism of $(\mathbf{C}, d)$ onto the dense subset $\mathbf{C}_\infty \setminus \{\infty\}$ of $(\mathbf{C}_\infty, \rho)$: $(\mathbf{C}_\infty, \rho)$ is a *one-point compactification* of $(\mathbf{C}, d)$.

If $a \neq 0$, we set $a.\infty = \infty.a = \infty$, and if $b \in \mathbf{C}$ we set $\infty + b = b + \infty = \infty$. The quantities $0.\infty$, $0/0$, ∞/∞ and $\infty + \infty$ are not defined.

If now f is a meromorphic function on a domain U, with singular set S_f, we can extend f to a continuous function from U to $\mathbf{C}_\infty$ by setting $\phi(z) = \infty$ for $z \in S_f$. As an example, the function $J(z) = 1/z$ is meromorphic on $\mathbf{C}$, with a simple pole at 0, and so we define $J(0) = 1/0 = \infty$. Since $J(z) \rightarrow 0$ as $z \rightarrow \infty$, we set $J(\infty) = 1/\infty = 0$. Then J is a homeomorphism (*inversion*) of $\mathbf{C}_\infty$ onto itself.

An open connected subset of $\mathbf{C}_\infty$ is called a *domain in $\mathbf{C}_\infty$*. If U is a domain in $\mathbf{C}_\infty$ then either $\infty \notin U$, in which case U is a domain in $\mathbf{C}$, or $\infty \in U$, in which case $U \cap \mathbf{C}$ is a domain in $\mathbf{C}$ with the property that there exists $R \geq 0$ such that $\{z \in \mathbf{C} : |z| > R\} \subseteq U$.

Suppose that U is a domain in $\mathbf{C}_\infty$, that $\infty \in U$ and that f is a meromorphic function on $U \cap \mathbf{C}$ with singular set S_f. We consider the function $f \circ J$ on $J(U \cap \mathbf{C})$ (so that $f \circ J(z) = f(1/z)$ for $z \in J(U \cap \mathbf{C})$). It is a meromorphic function on $J(U \cap \mathbf{C})$, which is a subset of $\mathbf{C}^* = \mathbf{C} \setminus \{0\}$. If S_f is unbounded, then 0 is an accumulation point of the singular set of $f \circ J$, but if S_f is bounded, then 0 is an isolated singularity of $f \circ J$. There are then three possibilities. First, 0 is a removable singularity of $f \circ J$, in which case $f(z)$ tends to a finite limit l as $z \rightarrow \infty$, and we set $f(\infty) = l$. Secondly, 0 is a pole of of $f \circ J$, in which case $f(z) \rightarrow \infty$ as $z \rightarrow \infty$, and we set $f(\infty) = \infty$. Thirdly, 0 is an essential isolated singularity, in which case $f(\infty)$ is not defined. If the first or second possibility holds, we call the mapping $f : U \rightarrow \mathbf{C}_\infty$ a *meromorphic function* on U.

Exercises

23.4.1 Verify that the mapping $\phi : (\mathbf{C}, d) \to (\phi(\mathbf{C}), d)$ is a homeomorphism.

23.4.2 Suppose that $\phi(z) = (u, v, w)$. Show that $z = (u + iv)/(1 - w)$.

23.4.3 Suppose that $\phi(z) = (u, v, w)$ and that $\phi(z') = (u', v', w')$. Show that $\rho(z, z')^2 = 2 - 2(uu' + vv' + ww')$. Deduce that

$$\rho(z, z') = \frac{2|z - z'|}{(1 + |z|^2)^{\frac{1}{2}}(1 + |z'|^2)^{\frac{1}{2}}}.$$

Show that

$$\rho(z, \infty) = \frac{2}{(1 + |z|^2)^{\frac{1}{2}}}.$$

23.4.4 Suppose that f is a meromorphic function on $\mathbf{C}$ and that there exist $R > 0$ and $k \in \mathbf{N}$ such that $|f(z)| \le |z|^k$ for $|z| \ge R$. Show that f is a rational function: there exist polynomials p and q such that $f(z) = p(z)/q(z)$ for $z \in \mathbf{C} \setminus S_f$.

23.4.5 Suppose that f is a meromorphic function on $\mathbf{C}$ and that $f \circ J$ is also meromorphic on $\mathbf{C}$. Show that the singular set S_f is finite. Show that f is a rational function.

23.4.6 Let R be the rotation of $\mathbf{R}^3$ by π about the first axis, so that $R(x, y, t) = (x, -y, -t)$. If $z \in \mathbf{C}$, what is $\phi^{-1} R\phi(z)$?

23.5 The residue theorem

We now apply Cauchy's theorem to a meromorphic function f. This involves the residues at the poles of f.

Proposition 23.5.1 *Suppose that f is a meromorphic function on a domain U with zero set Z_f and singular set S_f. Suppose that Γ is a cycle in $U \setminus S_f$ such that $n(\Gamma, w) = 0$ for $w \notin U$. Let*

$$U_\Gamma = \{z \in U \setminus [\Gamma] : n(\Gamma, z) \ne 0\} \text{ and let } K_\Gamma = [\Gamma] \cup U_\Gamma.$$

Then K_Γ is a compact subset of U.

Proof $\mathbf{C} \setminus K_\Gamma = \{w \in \mathbf{C} \setminus [\Gamma] : n(\Gamma, w) = 0\}$ is the union of some of the connected components of the open set $\mathbf{C} \setminus [\Gamma]$, including the unbounded one, so that K_Γ is a compact subset of $\mathbf{C}$. Further, $K_\Gamma \subseteq U$, since $n(\Gamma, w) = 0$ for $w \notin U$. □

Corollary 23.5.2   *The sets*

$$S_f(\Gamma) = K_\Gamma \cap S_f = \{z \in S_f : n(\Gamma, f) \neq 0\}$$
$$\text{and } Z_f(\Gamma) = K_\Gamma \cap Z_f = \{z \in Z_f : n(\Gamma, f) \neq 0\}$$

are finite.

Proof   This follows from Proposition 23.1.2.   □

Theorem 23.5.3 (The residue theorem)   *Suppose that f is a meromorphic function on a domain U with singular set S_f, and suppose that Γ is a cycle in $V = U \setminus S_f$ such that $n(\Gamma, w) = 0$ for $w \notin U$. Then*

$$\frac{1}{2\pi i} \int_\Gamma f(z)\, dz = \sum \{n(\Gamma, s) \text{res}_f(s) : s \in S_f(\Gamma)\}.$$

Proof   By Corollary 23.5.2, the set $S_f(\Gamma)$ is a finite subset of U, and so the sum is finite. The restriction of f to V is a holomorphic function, but we cannot immediately apply Cauchy's theorem, since if $S_f(\Gamma) \neq \emptyset$ then $n(\Gamma, s)$ may be non-zero for $s \in S_f(\Gamma) \subseteq \mathbf{C} \setminus V$. There exists $r > 0$ such that $M_r(s) \subseteq U_\Gamma$, for each $s \in S_f(\Gamma)$, and such that $M_r(s) \cap M_r(s') = \emptyset$, for s, s' distinct elements of $S_f(\Gamma)$. Let

$$\Gamma' = \Gamma - \sum_{s \in S_f(\Gamma)} n(\Gamma, s)\kappa_r(s).$$

Then Γ' is a cycle for which $n(\Gamma', s) = 0$ for $s \in S_f(\Gamma)$, and also $n(\Gamma', s) = 0$ for $s \in S_f \setminus S_f(\Gamma)$. Thus $n(\Gamma', w) = 0$ for $w \notin V$. We can apply Cauchy's theorem for cycles: $\int_{\Gamma'} f(z)\, dz = 0$.
    Since

$$\frac{1}{2\pi i} \int_{\Gamma'} f(z)\, dz = \frac{1}{2\pi i} \int_\Gamma f(z)\, dz - \sum_{s \in S_f(\Gamma)} n(\Gamma, s) \left(\frac{1}{2\pi i} \int_{\kappa_r(s)} f(z)\, dz \right)$$

$$= \frac{1}{2\pi i} \int_\Gamma f(z)\, dz - \sum_{s \in S_f(\Gamma)} n(\Gamma, s) \text{res}_f(s),$$

the result follows.   □

    This theorem can be used to calculate certain definite integrals; we shall give examples in the next chapter.
    The rest of this section can be omitted on a first reading: the results are used in Chapter 26. Suppose that f is a meromorphic function on a domain U and that $\zeta \in U \setminus S_f$. Then, as in Cauchy's integral formula, we may

consider the meromorphic function $f(z)/(z - \zeta)$. Then this has a simple pole at ζ, with residue $f(\zeta)$. The next proposition gives information about the residues at the other poles.

Proposition 23.5.4 *Suppose that f is a meromorphic function on a domain U with singular set S_f, that $s \in S_f$ is a pole of order k, that*

$$f(z) = \sum_{j=-k}^{\infty} a_j(z - s)^j$$

is the Laurent expansion of f in a neighbourhood of s and that $\zeta \neq s$. Then $g(z) = f(z)/(z - \zeta)$ has a pole of order k at s, and

$$\operatorname{res}_g(s) = -\left(\frac{a_{-k}}{(\zeta - s)^k} + \frac{a_{-k+1}}{(\zeta - s)^{k-1}} + \cdots + \frac{a_{-1}}{\zeta - s} \right).$$

Proof   The function g certainly has a pole of order k at s. Let $g(z) = \sum_{j=-k}^{\infty} b_j(z - s)^j$ be its Laurent expansion in a punctured neighbourhood of s. If $|z - s| < |s - \zeta|$, then

$$\frac{1}{z - \zeta} = \frac{1}{(z - s) - (\zeta - s)} = -\frac{1}{\zeta - s} \left(\frac{1}{1 - \frac{z-s}{\zeta-s}} \right)$$

$$= -\frac{1}{\zeta - s} \left(1 + \frac{z - s}{\zeta - s} + \left(\frac{z - s}{\zeta - s} \right)^2 + \cdots \right).$$

Multiplying the Laurent series for f by this power series, we see that the coefficient b_{-1} of $1/(z - s)$ is equal to

$$-\left(\frac{a_{-k}}{(\zeta - s)^k} + \frac{a_{-k+1}}{(\zeta - s)^{k-1}} + \cdots + \frac{a_{-1}}{\zeta - s} \right). \qquad \square$$

The residue theorem has the following corollary.

Corollary 23.5.5   *If $s \in S_f(\Gamma)$, let $\sum_{j=-k_s}^{\infty} a_j^{(s)}(z - s)^j$ be the Laurent expansion of f in a punctured neighbourhood of s. If $\zeta \in U \setminus S_f$ then*

$$n(\Gamma, \zeta)f(\zeta) = \frac{1}{2\pi i} \int_{\Gamma} \frac{f(z)}{z - \zeta} \, dz + \sum_{s \in S_f(\Gamma)} n(\Gamma, s) \sum_{j=1}^{k_s} \frac{a_{-j}^{(s)}}{(\zeta - s)^j}.$$

This has the following consequence for certain meromorphic functions defined on **C**.

Theorem 23.5.6   *Suppose that f is a meromorphic function on **C** with the following property: there exists a sequence $(r_n)_{n=1}^{\infty}$ of real numbers* *increasing to* ∞, *such that* $\mathbf{T}_{r_n} \cap S_f = \emptyset$ *for each* $n \in \mathbf{N}$, *and such that* $M_n = \sup\{|f(z)| : z \in \mathbf{T}_{r_n}\} \to 0$ *as* $n \to \infty$. *If* $s \in S_f$, *let* $\sum_{j=-k_s}^{\infty} a_j^{(s)}(z-s)^j$ *be the Laurent expansion of* f *in a punctured neighbourhood of* s. *Suppose that* $\zeta \in \mathbf{C} \setminus S_f$. *Then*

$$f(\zeta) = \lim_{n \to \infty} \sum_{s \in S_f, |s| < r_n} \left(\sum_{j=1}^{k_s} \frac{a_{-j}^{(s)}}{(\zeta - s)^j} \right)$$

and the limit exists locally uniformly on $\mathbf{C} \setminus S_f$.

Proof Suppose that K is a compact subset of $\mathbf{C} \setminus S_f$. Let $L = \sup_{\zeta \in K} |\zeta|$. If $n > L$ and $\zeta \in K$, then

$$\left| \int_{\kappa_{r_n}} \frac{f(z)}{\zeta - z} dz \right| \leq (2\pi r_n) \cdot \frac{M_n}{r_n - L} \to 0,$$

uniformly on K. The result therefore follows from Corollary 23.5.5. $\square$

Thus we have a 'partial fractions' expansion of f.

One important case occurs when all the singularities of f are simple: here we impose weaker conditions on f.

Theorem 23.5.7 *Suppose that* f *is a meromorphic function on* $\mathbf{C}$, *all of whose singularities are simple poles, and suppose that* $0 \notin S_f$. *Suppose also that there exists a sequence* $(r_n)_{n=1}^{\infty}$ *of real numbers increasing to* ∞, *such that* $\mathbf{T}_{r_n} \cap S_f = \emptyset$ *for* $n \in \mathbf{N}$ *and such that if* $M_n = \sup\{f(z) : z \in \mathbf{T}_{r_n}\}$ *then* $(M_n)_{n=1}^{\infty}$ *is bounded. If* $s \in S_f$, *let* b_s *be the residue of* f *at* s. *If* $\zeta \notin S_f$ *then*

$$f(\zeta) = f(0) + \lim_{n \to \infty} \sum_{s \in S_f, |s| < r_n} b_s \left(\frac{1}{\zeta - s} + \frac{1}{s} \right).$$

Proof The meromorphic function $h(z) = (f(z) - f(0))/z$ satisfies the conditions of Theorem 23.5.6. The residue of $h(z)$ at $s \in S_f$ is $b(s)/s$, so that, applying Theorem 23.5.6,

$$f(\zeta) = f(0) + \zeta h(\zeta) = f(0) + \lim_{n \to \infty} \sum_{s \in S_f, |s| < r_n} \frac{\zeta b_s}{s(\zeta - s)}$$

$$= f(0) + \lim_{n \to \infty} \sum_{s \in S_f, |s| < r_n} b_s \left(\frac{1}{\zeta - s} + \frac{1}{s} \right).$$

$\square$

Exercise

23.5.1 What is the corresponding result if, in Theorem 23.5.7, f has a simple pole at 0?

23.6 The principle of the argument

If f is a non-zero meromorphic function on a domain U, then so are f' and f'/f. Where are the poles of f'/f, and what are their residues?

Proposition 23.6.1 *Suppose that f is a non-constant meromorphic function on a domain U with singular set S_f and zero set Z_f. The function f'/f is a meromorphic function on U with singular set $S_f \cup Z_f$. If f has a zero of order k at $z_0 \in Z_f$ then f'/f has a simple pole at z_0 with residue k. If f has a pole of order k at $z_0 \in S_f$ then f'/f has a simple pole at z_0 with residue $-k$.*

Proof The function f'/f is defined on $U \setminus (S_f \cup Z_f)$ and is holomorphic there. If f has a zero of order k at $z_0 \in Z_f$, then $f(z) = (z - z_0)^k g(z)$, where g is a holomorphic function on $U \setminus S_f$ and $g(z_0) \neq 0$. Then

$$f'(z) = k(z - z_0)^{k-1} g(z) + (z - z_0)^k g'(z).$$

Thus

$$\frac{f'(z)}{f(z)} = \frac{k}{z - z_0} + \frac{g'(z)}{g(z)} \quad \text{for } z \in U \setminus (S_f \cup Z_f).$$

Since g'/g is holomorphic in a neighbourhood of z_0, it follows that f'/f has a simple pole at z_0 with residue k.

The argument for a pole is very similar. If f has a pole of order k at $z_0 \in Z_f$, then $f(z) = (z - z_0)^{-k} h(z)$, where h is a holomorphic function on $U \setminus S_f$ for which $h(z_0) \neq 0$. Then

$$f'(z) = -k(z - z_0)^{-k-1} h(z) + (z - z_0)^{-k} h'(z).$$

Thus

$$\frac{f'(z)}{f(z)} = \frac{-k}{z - z_0} + \frac{h'(z)}{h(z)} \quad \text{for } z \in U \setminus (S_f \cup Z_f).$$

Since h'/h is holomorphic in a neighbourhood of z_0, it follows that f'/f has a simple pole at z_0 with residue $-k$.

If $z_0 \in U \setminus (S_f \cup Z_f)$ then $f(z_0) \neq 0$, and f'/f is holomorphic in a neighbourhood of z_0.   □

We use this to prove the principle of the argument.

Theorem 23.6.2 (The principle of the argument) *Suppose that f is a meromorphic function on a domain U with zero set Z_f and singular set S_f, and suppose that Γ is a cycle in $U \setminus (Z_f \cup S_f)$ such that $n(\Gamma, w) = 0$ for $w \notin U$. Then*

$$n(f \circ \Gamma, 0) = \sum_{\zeta \in Z_f} n(\Gamma, \zeta) l_f(\zeta) - \sum_{s \in S_f} n(\Gamma, s) k_f(s),$$

where $l_f(\zeta)$ is the order of the zero of f at ζ and $k_f(s)$ is the order of the pole of f at s.

Proof   By Corollary 22.10.4,

$$n(f \circ \Gamma, 0) = \frac{1}{2\pi i} \int_\Gamma \frac{f'(z)}{f(z)}\, dz,$$

and

$$\frac{1}{2\pi i} \int_\Gamma \frac{f'(z)}{f(z)}\, dz = \sum_{\zeta \in Z_f(\Gamma)} n(\Gamma, \zeta) \operatorname{res}_{f'/f}(\zeta) + \sum_{s \in S_f(\Gamma)} n(\Gamma, s) \operatorname{res}_{f'/f}(s)$$

$$= \sum_{\zeta \in Z_f(\Gamma)} n(\Gamma, \zeta) l_f(\zeta) - \sum_{s \in S_f(\Gamma)} n(\Gamma, s) k_f(s).$$

$\square$

Let us apply Proposition 21.1.8.

Corollary 23.6.3 (Rouché's theorem) *Suppose that g is a meromorphic function on U for which $S_g \cap [\Gamma] = \emptyset$ and*

$$|f(w) - g(w)| < |f(w)| + |g(w)| \text{ for } w \in [\Gamma].$$

Then

$$\sum_{\zeta \in Z_f(\Gamma)} n(\Gamma, \zeta) l_f(\zeta) - \sum_{s \in S_f(\Gamma)} n(\Gamma, s) k_f(s) =$$

$$\sum_{\zeta \in Z_g(\Gamma)} n(\Gamma, \zeta) l_g(\zeta) - \sum_{s \in S_g(\Gamma)} n(\Gamma, s) k_g(s),$$

where $l_g(\zeta)$ is the order of the zero of g at ζ and $k_g(s)$ is the order of the pole of g at s.

Proof   The conditions imply that f and g have no zeros and no poles in $[\Gamma]$. Applying Proposition 21.1.8 to each of the paths in Γ, we see that $n(f \circ \Gamma, 0) = n(g \circ \Gamma, 0)$.

$\square$

Rouché's theorem is usually stated (and used) with the stronger condition that $0 < |f(w) - g(w)| < |f(w)|$, for $w \in [\Gamma]$.

These results can be used to locate the zeros of holomorphic functions. We shall give some examples in Section 23.7.

We use the principle of the argument to consider the behaviour of a holomorphic function at a point where the derivative may be 0.

Theorem 23.6.4 *Suppose that f is a non-constant holomorphic function on a domain U and that $z_0 \in U$. Let d be the least positive integer such that $f^{(d)}(z_0) \neq 0$. Then there exist $\rho > 0$ and $r > 0$ such that for each $w \in N_\rho^*(f(z_0))$ there exist exactly d points in $N_r^*(z_0)$ satisfying $f(z) = w$. These points are simple zeros of the function $f - w$.*

Proof   Since the zeros of $f - f(z_0)$ and f' are isolated, there exists $r > 0$ such that $M_r(z_0) \subseteq U$ and such that $f(z) - f(z_0) \neq 0$ and $f'(z) \neq 0$ for $z \in M_r^*(z_0)$. Then $f(z) \neq f(z_0)$ for $z \in [\kappa_r(z_0)]$, and $n(\kappa_r(z_0), z) = 1$ for $z \in N_r(z_0)$. By the principle of the argument, $n(f \circ \kappa_r(z_0), f(z_0)) = d$. Let V be the connected component of $\mathbf{C} \setminus f([\kappa_r(z_0)])$ to which $f(z_0)$ belongs. Since V is open, there exists $\rho > 0$ such that $N_\rho(f(z_0)) \subseteq V$. Since the winding number $n(f \circ \kappa_r(z_0), w)$ is constant on V, $n(f \circ \kappa_r(z_0), w) = d$, for $w \in N_\rho(f(z_0))$, and so $f - w$ has d zeros, counted according to multiplicity, in $N_r^*(z_0)$. Since $f'(z) \neq 0$ for $z \in N_r^*(z_0)$, each of these zeros is a simple zero, and so there are d distinct solutions to the equation $f(z) = w$ in $N_r^*(z_0)$. $\square$

We use this to describe the behaviour of a meromorphic function near a pole.

Corollary 23.6.5 *Suppose that f is a meromorphic function on a domain U, with a pole of order k at z_0. Then there exist $R > 0$ and $r > 0$ such that if $|\zeta| > R$ there exist exactly d points in $N_r^*(z)$ satisfying $f(z) = \zeta$. These points are simple zeros of the function $f - \zeta$.*

Proof   There exists $\delta > 0$ such that $f(z) = g(z)/(z - z_0)^k$ for $z \in N_\delta^*(z_0)$, where g is a holomorphic function on $N_\delta(z_0)$ with no zeros in $N_\delta(z_0)$. Let $h(z) = (z - z_0)^k/g(z)$ for $z \in N_\delta(z_0)$. Then h has a zero of order k at z_0, and so there exist $\rho > 0$ and $0 < r \leq \delta$ such that the conclusions of the theorem hold (with h in place of f). Let $R = 1/\rho$. If $|\zeta| > R$, then $1/\zeta \in N_\rho^*(0)$, and there exist exactly d points in $N_r^*(z_0)$ satisfying $h(z) = 1/\zeta$, and these points are simple zeros of the function $h - 1/\zeta$. Since $h(z) = 1/f(z)$ for $z \in N_r^*(z_0)$, the result follows. $\square$

This gives another proof of the open mapping theorem.

Corollary 23.6.6 (The open mapping theorem)   *Suppose that f is a non-constant holomorphic function on a domain U. If V is an open subset of U then $f(V)$ is an open subset of $\mathbf{C}$.*

Proof   Suppose that $z_0 \in V$. Let W be the connected component of V to which z_0 belongs, and apply the theorem to the restriction of f to W. If $w \in N_\rho(f(z_0))$ then the equation $f(z) = w$ has at least one solution in $N_r(z_0)$, so that $N_\rho(f(z_0)) \subseteq f(N_r(z_0)) \subseteq f(V)$. Thus $f(V)$ is open.   □

It also provides another proof of the maximum modulus principle.

Corollary 23.6.7   *Suppose that f is a non-constant holomorphic function on a domain U. Then $|f|$ has no local maxima on U, and the only local minima are the zeros of f.*

Proof   If $N_r(z_0)$ is a neighbourhood of z_0 contained in U, then $f(z_0)$ is an interior point of the open set $f(N_r(z_0))$, and so $|f(z_0)|$ is not the supremum of $|f|$ on $N_r(z_0)$, and is the infimum only if $f(z_0) = 0$.   □

We now give an improved version of Theorem 20.2.3.

Theorem 23.6.8   *Suppose that f is a univalent function on a domain U. Then $f(U)$ is a domain, f is a homeomorphism of U onto $f(U)$, $f'(z) \neq 0$ for $z \in U$, $f^{-1} : f(U) \to U$ is holomorphic and if $f(z) = w$ then $(f^{-1})'(w) = 1/f'(z)$.*

Proof   If $f'(z_0) = 0$ for some $z_0 \in U$ then the equation $f(z) = w$ has more than one solution in U for values of w close to $f(z_0)$, contradicting the fact that f is univalent. Thus $f'(z) \neq 0$ for $z \in U$. The derivative f' is holomorphic, and is therefore continuous. The result therefore follows from Theorem 20.2.3.   □

Exercises

23.6.1 Suppose that f is a non-constant continuous complex-valued function on $\overline{\mathbf{D}}$ whose restriction to $\mathbf{D}$ is holomorphic. Show that if $f(\overline{\mathbf{D}}) \subseteq \mathbf{D}$ then f has exactly one fixed point.

23.6.2 Suppose that $|a| < 1$. Show that the function

$$z^m \left(\frac{z - a}{1 - \bar{a}z} \right)^n - a$$

has $m + n$ zeros in $\mathbf{D}$.

23.6.3 Suppose that $p(z) = a_0 + \cdots + a_n z^n$ is a non-constant polynomial of degree n. Show that there exists $R > 0$ such that $|p(z) - a_n z^n| < |a_n z^n|$ for $|z| \geq R$. Use Rouché's theorem to give another proof of the fundamental theorem of algebra.

23.6.4 How many zeros does the function $z \sin z - 1$ have in the disc $N_{(n+\frac{1}{2})\pi}(0)$? Use this to show that all the solutions of the equation $z \sin z = 1$ are real.

23.6.5 *(The inverse mapping theorem.)* Suppose that f is a non-constant holomorphic function on a domain U and that $z_0 \in U$. What is the residue of the meromorphic function $z f'(z)/(f(z) - f(z_0))$ at z_0? Suppose that f is univalent, that γ is a contour in U and that $V = in[\gamma]$. If $w \in f(V)$ let

$$g(w) = \frac{1}{2\pi i} \int_\gamma \frac{z f'(z)}{f(z) - w}\, dz.$$

Show that g is the restriction of the inverse mapping f^{-1} to $f(V)$.

23.6.6 Suppose that f is a meromorphic function on a domain U with the property that the residue at every pole is an integer. Suppose that $z_0 \in U \setminus S_f$. If $z \in U \setminus S_f$ and γ is a rectifiable path in $U \setminus S_f$ from z_0 to z, let $F_\gamma(z) = \int_\gamma f(z)\, dz$. Show that $e^{F_\gamma(z)}$ does not depend upon the choice of γ. Show that there exists a holomorphic function g on $U \setminus S_f$ such that $f = g'/g$. Show further that g is meromorphic on U.

23.6.7 This exercise extends the results of Theorem 23.6.4.

(i) Suppose that U, f, z_0 and d satisfy the conditions of Theorem 23.6.4 and that r and ρ satisfy its conclusions. Suppose that $z_0 = 0$ and that $f(z_0) = 0$. Show that there exists a holomorphic function h on U such that $f(z) = z^d h(z)$ for $z \in U$, and that $h(0) \neq 0$.

(ii) Show that there exist $0 < r_1 < r$ and a univalent function k on $N_{r_1}(0)$ such that $h(z) = k(z)^d$ for $z \in N_{r_1}(0)$.

(iii) Let $l(z) = z k(z)$ for $z \in N_{r_1}(0)$. Observe that $f(z) = l(z)^d$, for $z \in N_{r_1}(0)$. Show that there exists $0 < r_2 \leq r_1$ such that l is univalent on $N_{r_2}(0)$.

(iv) Let $0 < s < r_2$. Let $\gamma_0(t) = s e^{it}$ for $0 \leq t \leq 2\pi/d$. Let δ_0 be the simple closed path $\sigma(0, s) \vee \gamma_0 \vee \sigma(s e^{2\pi i/d}, 0)$. Let $\epsilon_0 = l^{-1} \circ \delta_0$, and let $V_0 = in[\epsilon_0]$. Show that the restriction of f to V_0 is a univalent mapping of V_0 onto the cut disc $N_s(0) \setminus (-s, 0]$.

(v) Carry out similar constructions for the paths γ_j and δ_j, for $1 \le j < d$, where $\gamma_t(t) = se^{it}$ for $2\pi j/d \le t \le 2\pi(j+1)/d$, and δ_j is the simple closed path $\sigma(0, se^{2\pi ij/d}) \vee \gamma_j \vee \sigma(se^{2\pi i(j+1)/d}, 0)$.

(vi) Draw a sketch to illustrate these constructions.

(vii) Show that there is no loss of generality in taking $z_0 = 0$ and $f(z_0) = 0$.

23.6.8 Let $P_n = \{a = (a_0, \ldots, a_n) \in \mathbf{C}^{n+1} : a_n \ne 0\}$. If $a \in P_n$, let $r(a) = \{z \in \mathbf{C} : a_0 + a_1 z + \cdots + a_n z^n = 0\}$ be the set of roots of the polynomial $p_a(z) = a_0 + a_1 z + \cdots + a_n z^n$, counted according to multiplicity. Explain why $r(a)$ can be considered as an element of the weighted configuration space $W_n(\mathbf{C})$ defined in the exercises of Volume II, Section 15.6.

Suppose that $\epsilon > 0$. Show that there exists a finite set Γ of disjoint circular paths in $\mathbf{C} \setminus r(a)$, each of radius less than ϵ, with centres the elements of $r(a)$.

Let $m = \inf\{|p_a(z)| : z \in [\Gamma]\}$. Show that $m > 0$.

Show that there exists $\delta > 0$ such that if $b \in P_n$ and $\|a - b\|_\infty < \delta$ then $|p_a(z) - p_b(z)| < m$ for $z \in [\Gamma]$.

Use Rouché's theorem to show that the mapping r from P_n to $(W_n(\mathbf{C}), d_W)$ is continuous. (The roots of a polynomial depend continuously on the coefficients.)

23.6.9 Let $\mathbf{Z}(i) = \{m + in : m, n \in \mathbf{Z}\}$ be the set of *Gaussian integers*. If $z \in \mathbf{C} \setminus \mathbf{Z}(i)$, let

$$f(z) = \frac{1}{z^2} + \sum_{w \in \mathbf{Z}(i)} \left(\frac{1}{(z-w)^2} - \frac{1}{w^2} \right).$$

Prove carefully that the sum converges locally uniformly to a meromorphic function f on $\mathbf{C}$. Show that $f(z + w) = f(z)$ for $w \in \mathbf{Z}(i)$. (f is *doubly periodic*.) Suppose that $z_0 \in \mathbf{C} \setminus \mathbf{Z}(i)$ and that f has no zeros on the sides of the square with vertices z_0, $z_0 + 1$, $z_0 + 1 + i$ and $z_0 + i$. Show that f has two zeros (counted according to multiplicity) inside the square.

23.6.10 Suppose that f is a meromorphic function on $\mathbf{C}$ for which $f(z) = f(z + w)$ for $w \in \mathbf{Z}(i)$. Show that if f is holomorphic, then f is constant. Suppose that f is not constant, and that f does not have a pole on the edges of the square with vertices z_0, $z_0 + 1$, $z_0 + 1 + i$ and $z_0 + i$. Show that the sum of the residues of the poles within the square is zero. Show that the number of zeros within the square (counted according to multiplicity) is equal to the number of poles (counted according to multiplicity), and that the number is at least 2.

23.7 Locating zeros

We now give examples to show how the principle of the argument and Rouché's theorem can be used to provide information about the location of zeros of polynomials and of other holomorphic functions.

Example 23.7.1   The polynomial $p(z) = z^4 - z^3 - z + 5$ has four simple roots in the annulus $A_{1,2}(0) = \{z : 1 < |z| < 2\}$, and has one in each of the four quadrants of $\mathbf{C}$.

If $|z| = 2$ then

$$16 = |z^4| > 15 = |z|^3 + |z| + 5 \geq |z^3 + z - 5|,$$

so that by Rouché's theorem all the zeros of p lie in $N_2(0)$. If $|z| \leq 1$ then $|z^4| + |z^3| + |z| \leq 3$, so that $|p(z)| \geq 2$. Thus the zeros of p lie in the annulus $A_{1,2}(0)$. Further, $p(x) = (x^4 - x^3) + (5 - x) \geq 5$ for $1 \leq x \leq 2$, and $p(x) > 5$ for $x < 0$, and so p has no real roots. Similarly $p(iy) = (y^4 + 5) + i(y^3 - y) \neq 0$, so that there are no purely imaginary roots.

We show that p has one root in the quadrant $\{x + iy : x > 0, y > 0\}$. Let $\gamma : [0, 3] \to \mathbf{C}$ be the path defined as

$$\gamma(t) = \begin{cases} 2t & \text{for } 0 \leq t \leq 1, \\ 2e^{i(t-1)\pi/2} & \text{for } 1 \leq t \leq 2, \\ 2i(3 - t) & \text{for } 2 \leq t \leq 3, \end{cases}$$

and let $\theta : [0, 3] \to \mathbf{R}$ be a continuous branch of $\mathrm{Arg}\,(p \circ \gamma)$ with $\theta(0) = 0$. First, since p is real and positive on $[0, 2]$, $\theta(t) = 0$ for $t \in [0, 1]$. Next, suppose that $1 \leq t \leq 2$. Then $\arg((\gamma(t))^4) = 2(t - 1)\pi$. Since $|p(\gamma(t)) - (\gamma(t))^4| < |(\gamma(t))^4|$ it follows that $|\theta(t) - 2(t - 1)\pi| < \pi$. In particular $|\theta(2) - 2\pi| < \pi$, so that

$$\theta(2) = \arg(p(\gamma(2))) + 2\pi = \arg(21 + 6i) + 2\pi \in (2\pi, 2\tfrac{1}{2}\pi).$$

Finally, if $2 \leq t \leq 3$ and $\gamma(t) = ir$ then $p(t) = (r^4 + 5) + i(r^3 - r)$ lies in the right-hand half-plane $H = \{z = x + iy : x > 0\}$, so that $|\theta(t) - \theta(2)| < \pi$, and $\pi < \theta(3) < 3\tfrac{1}{2}\pi$. But $\theta(3)$ is an integer multiple of 2π, and so $\theta(3) = 2\pi$. By the principle of the argument, there is therefore exactly one zero of p inside $[\gamma]$.

We can carry out similar calculations for the other three quadrants, but in this case it is easier to argue differently. Since p has real coefficients, z is a zero of p if and only if $\bar{z}$ is, and so there is one zero in the quadrant $\{x + iy : x > 0, y < 0\}$. There remain two more zeros to account for. For the same reason, there must be one root in each of the other quadrants.

As a second, harder, example, let us consider the entire function $f(z) = e^z - z$. The function e^z has no zeros, and $f(x) \geq 1$ for $x \in \mathbf{R}$. Does f have any zeros?

Example 23.7.2 The function $f(z) = e^z - z$ has one simple zero in each of the semi-infinite open strips

$$A_n = \{z = x + iy : x > 0, 2n\pi < y < 2(n+1)\pi\},$$

and has no other zeros.

Proof In this proof, certain details are left for the reader to verify. Draw a diagram!

First we show that if $x \leq 0$ then $f(x + iy) \neq 0$. Suppose not. Since $f(x + iy) = (e^x \cos y - x) + i(e^x \sin y - y)$, it follows that

$$|\sin y| = e^{-x}|y| \geq |y|.$$

Thus $y = 0$. But then $x = e^x$, which is not possible.

Next, if $z = x + 2n\pi i$, with $n \in \mathbf{Z}$, then $f(z) = (e^x - x) - 2n\pi i \neq 0$. Suppose now that $n \in \mathbf{Z}$. Choose $x_n > 0$ such that

$$e^{x_n} > 2(x_n + 2(|n| + 1)\pi),$$

and consider the closed rectangular path $\gamma : [0, 4] \to \mathbf{C}$ with

$$\gamma(0) = 2n\pi i, \ \gamma(1) = x_n + 2n\pi i, \ \gamma(2) = x_n + 2(n+1)\pi i, \ \gamma(3) = 2(n+1)\pi i.$$

Let θ be a continuous branch of $\mathrm{Arg}\,(f \circ \gamma)$ on $[0, 4]$, with

$$\theta(0) = \arg\,(f(2n\pi i)) = \arg\,(1 - 2n\pi i).$$

If $0 \leq t \leq 1$ and $\gamma(t) = s + 2n\pi i$ then $\Re(f(\gamma(t))) = e^s - s > 0$. Thus $f([\gamma_{[0,1]}])$ is contained in the right-hand half-plane $\{z = x + iy : x > 0\}$, from which it follows that $\theta(t) = \arg\,(f(\gamma(t)))$; in particular, $\theta(1) = \arg\,(f(\gamma(1)) = \arg\,(e^{x_n} - x_n - 2n\pi i)$, and $|\theta(1)| < \pi/6$.

If $1 \leq t \leq 2$ we can suppose that $\gamma(t) = x_n + 2\pi(n + (t - 1))i$. Then

$$|f(\gamma(t)) - e^{x_n}e^{2\pi it}| = |x_n + 2\pi(n + (t - 1))i| < \tfrac{1}{2}e^{x_n} = \tfrac{1}{2}|e^{x_n}e^{2\pi it}|.$$

Since $|\theta(1)| < \pi/6$, it follows that $|\theta(t) - 2\pi(t - 1)| < \pi/3$. Consequently $|\theta(2) - 2\pi| < \pi/3$, so that $\theta(2) = \arg f(\gamma(2)) + 2\pi$.

Arguing as for the interval $[0, 1]$, $f([\gamma_{[2,3]}])$ is contained in the right-hand half-plane $\{z = x + iy : x > 0\}$, and it follows from this that $\theta(3) = \arg f(\gamma(3)) + 2\pi$.

Finally, since $f(iy) = \cos y + (\sin y - y)i$, the imaginary part of $f(iy)$ is negative when $y > 0$ and positive when $y < 0$. From this it follows that if $3 \leq t \leq 4$ then $|\theta(t) - \theta(3)| < \pi$ and $|\arg f(\gamma(t)) - \arg f(\gamma(3))| < \pi$, so that $|\theta(t) - (\arg f(\gamma(t)) + 2\pi)| < 2\pi$. Thus $\theta(t) = \arg(f(\gamma(t)) + 2\pi$, and in particular $\theta(4) = \arg(f(\gamma(4)) + 2\pi = \theta(0) + 2\pi$.

It therefore follows from the principle of the argument that there is one simple zero of f inside $[\gamma]$. Since x_n can be chosen to be arbitrarily large, there is just one simple zero of f in A_n.   □

Exercises

23.7.1 Let $p(z) = z^3 + ikz - 1$, where $0 < k < 1$. Show that the roots of p lie in the annulus $\{z : \frac{1}{2} < z < 2\}$ and lie in different quadrants. Which quadrant does not contain a root of p?

23.7.2 Show that the track $[\gamma]$ of Example 23.7.2 meets $(-\infty, 0]$ in just one point. Use this to give another proof of the result.

24

The calculus of residues

24.1 Calculating residues

In this chapter, we show how the residue theorem (Theorem 23.5.3) can be used to calculate certain definite integrals. First, we see how to calculate residues.

Theorem 24.1.1 *Suppose that f is a meromorphic function on a domain U, with a pole at z_0.*

(i) *If z_0 is a simple pole then* $\operatorname{res}_f(z_0) = \lim_{z \to z_0}(z - z_0)f(z)$.

(ii) *Suppose that z_0 is a simple pole and that g and h are holomorphic functions in a neighbourhood $N_r(z_0)$, with $g(z_0) \neq 0$ and $h(z_0) = 0$, such that $f(z) = g(z)/h(z)$ for $z \in N_r^*(z_0)$. Then* $\operatorname{res}_f(z_0) = g(z_0)/h'(z_0)$.

(iii) *If z_0 is a pole of order k, with $k > 1$, then $(z - z_0)^k f(z)$ extends to a holomorphic function g in a neighhbourhood $N_r(z_0)$, and* $\operatorname{res}_f(z_0) = g^{(k-1)}(z_0)/(k - 1)!$.

Proof

(i) We can write

$$f(z) = \frac{\operatorname{res}_f(z_0)}{z - z_0} + j(z),$$

where j is holomorphic in a neighbourhood of z_0. Thus

$$(z - z_0)f(z) = \operatorname{res}_f(z_0) + (z - z_0)j(z) \to \operatorname{res}_f(z_0) \text{ as } z \to z_0.$$

(ii) If $f(z) = g(z)/h(z)$ for $z \in N_r^*(z_0)$ then

$$\operatorname{res}_f(z_0) = \lim_{z \to z_0} \frac{(z - z_0)g(z)}{h(z)} = \lim_{z \to z_0} g(z) \left(\frac{z - z_0}{h(z) - h(z_0)} \right) = \frac{g(z_0)}{h'(z_0)}.$$

(iii) Suppose that

$$f(z) = \frac{a_{-k}}{(z-z_0)^k} + \cdots + \frac{a_{-1}}{z-z_0} + \sum_{j=0}^{\infty} a_j(z-z_0)^j$$

is the Laurent series for f in a punctured neighbourhood $N_r^*(z_0)$ of z_0. Let $g(z) = (z-z_0)^k f(z) = \sum_{j=0}^{\infty} a_{j-k}(z-z_0)^j$, for $z \in N_r^*(z_0)$. Then g has a removable singularity at z_0. If we set $g(z_0) = a_{-k}$ then g is analytic on $N_r(z_0)$, and

$$g(z) = \sum_{j=0}^{\infty} a_{j-k}(z-z_0)^j = a_{-k} + \sum_{j=1}^{\infty} \frac{g^{(j)}(z_0)}{j!}(z-z_0)^j.$$

Equating the coefficient of $(z-z_0)^{k-1}$, we obtain the result.   □

24.2 Integrals of the form $\int_0^{2\pi} f(\cos t, \sin t)\, dt$

First we consider integrals of the form $\int_0^{2\pi} f(\cos t, \sin t)\, dt$.
    Since

$$\cos t = \frac{e^{it} + e^{-it}}{2} \quad \text{and} \quad \sin t = \frac{e^{it} - e^{-it}}{2i},$$

we consider the function

$$g(z) = f\left(\frac{1}{2}\left(z + \frac{1}{z}\right), \frac{1}{2i}\left(z - \frac{1}{z}\right)\right) \quad \text{on } \mathbf{T}.$$

Then $g(e^{it}) = f(\cos t, \sin t)$. Suppose that there exists a function g, meromorphic in a domain U containing the closed unit disc $M_1(0)$, with no poles on the unit circle $\mathbf{T} = \{z : |z| = 1\}$, and for which $f(\cos t, \sin t) = g(e^{it})$. We consider the circular path $\kappa = \kappa_1(0)$. Then $\kappa'(t) = ie^{it}$, so that

$$\int_\kappa \frac{g(z)}{z}\, dz = i \int_0^{2\pi} g(e^{it})\, dt = i \int_0^{2\pi} f(\cos t, \sin t)\, dt.$$

Let $h(z) = g(z)/z$, so that h is meromorphic on U. Let S_h be the set of poles of h in the open unit disc $N_1(0)$. Then, applying the theorem of residues,

$$\int_0^{2\pi} f(\cos t, \sin t)\, dt = -i \int_\kappa h(z)\, dz = 2\pi \sum_{w \in S_h} \operatorname{res} h(w).$$

Example 24.2.1   If $2k$ is an even integer then

$$I_{2k} = \int_0^{2\pi} (\cos t)^{2k}\, dt = \frac{2\pi(2k)!}{2^{2k}(k!)^2} = \frac{2\pi}{2^{2k}}\binom{2k}{k}.$$

By the binomial theorem,

$$h(z) = \frac{1}{z} \cdot \left(\tfrac{1}{2}\left(z + \tfrac{1}{z}\right) \right)^{2k} = \frac{1}{2^{2k}} \sum_{j=0}^{2k} \binom{2k}{j} z^{2j-2k-1}.$$

Then h has a pole of order $2k + 1$ at 0, and the residue is $(2k)!/2^{2k}(k!)^2$. Compare this calculation with the calculation of I_{2k} in Volume I, Section 10.3.

Example 24.2.2 If a is real and $m \in \mathbf{Z}^+$ then

$$\int_0^{2\pi} \frac{\cos mt}{1 + a^2 - 2a\cos t}\, dt = \begin{cases} 2\pi a^m/(1 - a^2) & \text{for } |a| < 1, \\[2mm] 2\pi/a^m(a^2 - 1) & \text{for } |a| > 1. \end{cases}$$

The integrand is real, and is the real part of $e^{imt}/(1 + a^2 - 2a\cos t)$. We therefore consider the function $g(z) = z^m/(1 + a^2 - a(z + 1/z))$. Then

$$h(z) = \frac{g(z)}{z} = \frac{-z^m}{az^2 - (1+a^2)z + a} = \frac{1}{1-a^2}\left(\frac{z^m}{z-a} - \frac{z^m}{z - 1/a} \right).$$

If $a \neq \pm 1$, then h has a pole at a with residue $a^m/(1 - a^2)$ and a pole at $1/a$ with residue $1/a^m(a^2 - 1)$. If $|a| < 1$ then the pole at a is inside $[\gamma]$ and the pole at $1/a$ is outside $[\gamma]$, giving the first equality. If $|a| > 1$ then the pole at a is outside $[\gamma]$ and the pole at $1/a$ is inside $[\gamma]$, giving the second equality.

Exercises

24.2.1 Show that if $a > 1$ then

$$\int_0^\pi \frac{dt}{a + \cos t} = \frac{\pi}{\sqrt{a^2 - 1}}.$$

24.2.2 Show that if $b > 0$ then

$$\int_0^\pi \frac{dt}{b + \sin^2 t} = \frac{\pi}{\sqrt{b^2 + b}},$$

first by using the calculus of residues, and secondly by making a change of variables, and using the result of the previous example.

24.2.3 What is the value of

$$\int_0^{2\pi} \frac{\cos mt}{1 + a^2 - 2a\cos t}\, dt$$

when $a = 1$ or -1?

24.2.4 Show that if a is real and $|a| < 1$ then

$$\int_0^{2\pi} \frac{\sin^2 t}{1 + a^2 - 2a \cos t} \, dt = \pi.$$

What is the value if $|a| > 1$? Does the integral exist if $a = \pm 1$?

24.2.5 Show that if $a > 0$ then

$$\int_0^{2\pi} \frac{dt}{a^2 + (\tan t)^2} = \frac{2\pi}{a(a+1)}.$$

24.2.6 Suppose that $0 < a < b$. Show that

$$\int_0^{2\pi} \frac{dt}{a^2 \cos^2 t + b^2 \sin^2 t} = \frac{2\pi}{ab}$$

by considering $\int_\gamma f(z) \, dz$, where f is a suitable meromorphic function and γ is the contour with track $[\gamma] = \{z = x+iy : x^2/a^2 + y^2/b^2 = 1\}$.

24.2.7 By considering the function e^z/z^{n+1}, show that

$$\int_0^{2\pi} e^{\cos t} \cos(nt - \sin t) \, dt = 2\pi/n! \text{ and } \int_0^{2\pi} e^{\cos t} \sin(nt - \sin t) \, dt = 0.$$

24.3 Integrals of the form $\int_{-\infty}^{\infty} f(x) \, dx$

Suppose that f is a meromorphic function on the open upper half-plane $H_+ = \{x + iy : y > 0\}$ with a finite singular set S_f, and that f has a continuous extension (also denoted by f) to the closed upper half-plane $\overline{H}_+$. Suppose that $-R < 0 < S$. We consider a contour of the form $\gamma = \sigma(-R, S) \vee \delta(S, -R)$, where $\delta(S, -R)$ is either a semicircular path in $\overline{H}_+$ from S to $-R$, or a rectilinear path from S to $-R$ with vertices S, $S + i(R + S)$, $-R + i(R + S)$ and $-R$. For large enough R and S, S_f is inside $[\gamma]$, so that by the residue theorem

$$\int_{-R}^{S} f(x) \, dx = 2\pi i \sum_{s \in S_f} \text{res}_f(s) - \int_{\delta(S, -R)} f(z) \, dz.$$

If $\int_{\delta(S, -R)} f(z) \, dz \to 0$ as $R, S \to \infty$ then

$$\int_{-\infty}^{\infty} f(x) \, dx = 2\pi i \sum_{s \in S_f} \text{res}_f(s).$$

If f is an even function, then $\int_{-\infty}^{\infty} f(x) \, dx = 2 \int_0^{\infty} f(x) \, dx$ and it is sufficient to consider paths $\delta(R, -R)$.

Example 24.3.1

$$\int_{-\infty}^{\infty} \frac{dx}{1+x^4} = \frac{\pi}{\sqrt{2}}.$$

Let $\eta = e^{i\pi/4} = (1+i)/\sqrt{2}$. Then the meromorphic function

$$f(z) = \frac{1}{1+z^4} = \frac{1}{(z-\eta)(z-\eta^3)(z-\eta^5)(z-\eta^7)}$$

has simple poles at η, η^3, η^5 and η^7, but only the first two of these are in H_+. Then

$$\operatorname{res}_f(\eta) = \frac{1}{(\eta-\eta^3)(\eta-\eta^5)(\eta-\eta^7)} = \frac{1}{\sqrt{2}(\sqrt{2}(1+i))(i\sqrt{2})} = \frac{-(1+i)}{4\sqrt{2}},$$

$$\operatorname{res}_f(\eta^3) = \frac{1}{(\eta^3-\eta)(\eta^3-\eta^5)(\eta^3-\eta^7)} = \frac{1}{(-\sqrt{2})(i\sqrt{2})(\sqrt{2}(1-i))} = \frac{1-i}{4\sqrt{2}}.$$

If $R > 1$ then $l(\delta(R,-R)) \le 6R$ and $|f(z)| \le 1/(R^4-1)$ for $z \in [\delta(-R,R)]$, so that

$$\int_{\delta(R,-R)} f(z)\,dz \le \frac{6R}{R^4-1}, \quad \text{and} \quad \int_{\delta(R,-R)} f(z)\,dz \to 0 \text{ as } R \to \infty.$$

Thus

$$\int_{-\infty}^{\infty} \frac{dx}{1+x^4} = 2\pi i\left(\frac{-(1+i)}{4\sqrt{2}} + \frac{1-i}{4\sqrt{2}}\right) = \frac{\pi}{\sqrt{2}}.$$

Example 24.3.2 If $k \in \mathbf{Z}^+$ then

$$\int_{-\infty}^{\infty} \frac{dx}{(1+x^2)^{k+1}} = \frac{\pi(2k)!}{2^{2k}(k!)^2} = \frac{\pi}{2^{2k}}\binom{2k}{k}.$$

The function $f(z) = 1/(1+z^2)^{k+1}$ has poles of order $k+1$ at i and $-i$, but only the former is in H_+. Now $f(z) = g(z)/(z-i)^{k+1}$, where $g(z) = 1/(z+i)^{k+1}$, and

$$g^{(k)}(z) = \frac{(-1)^k(k+1)(k+2)\dots(2k)}{(z+i)^{2k+1}} = \frac{(-1)^k(2k)!}{k!(z+i)^{2k+1}},$$

so that

$$\operatorname{res}_f(i) = \frac{(-1)^k(2k)!}{(k!)^2(2i)^{2k+1}} = \frac{(2k)!}{(k!)^2 2^{2k+1}i}.$$

Again, it is easy to see that $\int_{\delta(R,-R)} f(z)\,dz \to 0$ as $R \to \infty$. Thus

$$\int_{-\infty}^{\infty} \frac{dx}{(1+x^2)^{k+1}} = (2\pi i)\frac{(2k)!}{(k!)^2 2^{2k+1}i} = \frac{\pi}{2^{2k}}\binom{2k}{k}.$$

We can also use the calculus of residues to calculate the Fourier transform of certain functions. If f is a Riemann integrable function on $\mathbf{R}$ for which the function $e^{-itx}f(x)$ is Riemann integrable for all $t \in \mathbf{R}$ then the *Fourier transform $\hat{f}$ of f* is defined as

$$\hat{f}(t) = \int_{-\infty}^{\infty} e^{-itx} f(x)\, dx.$$

(This definition can be greatly extended; it is also often defined in a slightly different form, including various constants.)

Example 24.3.3 If $f(x) = e^{-x^2/2}/\sqrt{2\pi}$ then $\hat{f}(t) = e^{-t^2/2}$.

The constant $1/\sqrt{2\pi}$ is included to ensure that $\int_{-\infty}^{\infty} f(x)\, dx = 1$. (The function f is then the density function of the *standard normal probability distribution*. In this setting, the function $t \to \hat{f}(-t)$ is, rather unfortunately, called the *characteristic function* of the probability distribution.) In fact, in this case we only need Cauchy's theorem to calculate the Fourier transform of f. The function $f(z) = e^{-z^2/2}/\sqrt{2\pi}$ is an entire function, so that if γ is the rectangular closed simple path with vertices $-R, S, S+it$ and $-R+it$ then $\int_\gamma f(z)\, dz = 0$.

If $z = S + iu \in [S, S+it]$ then

$$f(z) = \frac{1}{\sqrt{2\pi}} e^{-S^2/2 - iuS + u^2/2}, \quad \text{so that} \quad |f(z)| \le \frac{e^{t^2/2}}{\sqrt{2\pi}} e^{-S^2/2},$$

and so $\int_S^{S+it} f(z)\, dz \to 0$ as $S \to \infty$. Similarly, $\int_{-R+it}^{-R} f(z)\, dz \to 0$ as $R \to \infty$. Thus

$$\int_{-R}^{S} f(x)\, dx - \int_{-R+it}^{S+it} f(z)\, dz \to 0 \quad \text{as } R, S \to \infty.$$

Since $\int_{-R}^{S} f(x)\, dx \to 1$ as $R, S \to \infty$, it follows that $\int_{-R+it}^{S+it} f(z)\, dz \to 1$ as $R, S \to \infty$. But

$$\int_{-R+it}^{S+it} f(z)\, dz = \frac{1}{\sqrt{2\pi}} \int_{-R}^{S} e^{-(x+it)^2/2}\, dx$$

$$= \frac{e^{t^2/2}}{\sqrt{2\pi}} \int_{-R}^{S} e^{-x^2/2} e^{-ixt}\, dx \to e^{t^2/2} \hat{f}(t)$$

as $R, S \to \infty$. Thus $\hat{f}(t) = e^{-t^2/2}$.

When we consider integrands with a factor e^{imt}, with $m > 0$, the following result helps deal with the integral along δ.

Proposition 24.3.4 (Jordan's lemma)   *Suppose that f is a meromorphic function on the upper half-plane H_+ with a finite singular set S_f, which has a continuous extension (also denoted by f) to $\overline{H}_+$. Suppose that $|f(re^{it})| \to 0$ uniformly on $[0, \pi]$ as $r \to \infty$, and suppose that $m > 0$. If $-R < 0 < S$, let $\delta(S, -R)$ be either the semicircular path from S to $-R$ in $\overline{H}_+$, or the rectilinear path from S to $-R$ with vertices S, $S + i(R + S)$, $-R + i(R + S)$ and $-R$. Then $\int_{\delta(S,-R)} e^{imz} f(z)\,dz \to 0$ as $R, S \to \infty$.*

Proof   We consider the rectilinear path. Suppose that $-R < 0 < S$ and let $M(S, -R) = \sup\{|f(z)| : z \in \delta(S, -R)\}$. Then

$$\left| \int_{[S,S+i(R+S)]} e^{imz} f(z)\,dz \right| = \left| \int_0^{R+S} e^{imS} e^{-mt} f(S + it) i\,dt \right|$$

$$\leq \int_0^{R+S} M(S, -R) e^{-mt}\,dt \leq M(S, -R)/m,$$

and similarly $|\int_{[-R+i(R+S),-R]} e^{imz} f(z)\,dz| \leq M(S, -R)/m$. Also

$$\left| \int_{[S+i(R+S),-R+i(R+S)]} e^{imz} f(z)\,dz \right|$$

$$= |-\int_{-R}^{S} e^{imt} e^{-m(R+S)} f(t + i(R + S))\,dt|$$

$$\leq e^{-m(R+S)} \int_{-R}^{S} |f(t + i(R + S))|\,dt$$

$$\leq M(S, -R)(R + S) e^{-m(R+S)}.$$

All three terms tend to 0 as $R, S \to \infty$, and so the result follows.

    Provided that S_f is contained inside the contour with the semicircular path, the integrals along the rectangular path and the semicircular path are the same, by Cauchy's theorem, and so the result follows for the contours with semicircular paths. It is also easy to give a direct proof in this case; see Exercise 24.3.2.   □

    Inspection of the proof shows that it is essential that $m > 0$. If $m < 0$, we should consider paths in the lower half space H_-.

Example 24.3.5   If $f(x) = 1/\pi(1 + x^2)$ then $\hat{f}(t) = e^{-|t|}$.

    Once again, the numerical factor $1/\pi$ is included so that $\int_{-\infty}^{\infty} f(x)\,dx = 1$; here f is the density function of the *Cauchy distribution*. Suppose that $t > 0$. The function $e^{itz} f(z)$ has simple poles at i and $-i$, but only the former is in

H_+. The residue at i is $e^{-t}/2\pi i$. Since f satisfies the condition's of Jordan's Lemma,

$$\int_{-\infty}^{\infty} e^{itx} f(x)\, dx = e^{-t},$$

and so $\hat{f}(t) = e^{-|t|}$ for $t < 0$. Note that if we wish to calculate $\hat{f}(t)$ for $t > 0$ then Jordan's Lemma does not apply, since the exponential term grows in magnitude in the upper half-plane H_+. We could consider paths in the lower half plane H_-, but it is easier simply to consider complex conjugates: if $t > 0$ then $\int_{-\infty}^{\infty} e^{-itx} f(x)\, dx$ is the complex conjugate of $\int_{-\infty}^{\infty} e^{itx} f(x)\, dx$, and so is equal to e^{-t}. Thus $\hat{f}(t) = e^{-|t|}$ for all $t \in \mathbf{R}$.

We can also consider functions f with a finite number of simple poles on $\mathbf{R}$. In this case we need to consider the Cauchy principal value of the integral. Thus if f has one simple pole at x_0, we calculate

$$(PV) \int_{-\infty}^{\infty} f(x)\, dx = \lim_{r \to 0} \left(\int_{-\infty}^{x_0 - r} f(x)\, dx + \int_{x_0 + r}^{\infty} f(x)\, dx \right),$$

with similar conventions if there are several poles on $\mathbf{R}$. In order to do this, we indent the contour, and make use of the following proposition.

Proposition 24.3.6 *Suppose that f is holomorphic in the punctured neighbourhood $N_s^*(w)$ of w, and that f has a simple pole at w. Let $\gamma_r(t) = w + re^{it}$, for $0 < r < s$ and $t \in [\alpha, \beta]$. Then*

$$\int_{\gamma_r} f(z)\, dz \to i(\beta - \alpha)\mathrm{res}\, f(w) \quad as \ r \searrow 0.$$

Proof   We can write $f(z) = \mathrm{res}\, f(w)/(z - w) + h(z)$, where h is a holomorphic function on $N_s(w)$. Suppose that $0 < s' < s$. Then h is bounded on the closed neighbourhood $M_{s'}(w)$: let $M = \sup\{|h(z)| : z \in M_{s'}(w)\}$. If $0 < r < s'$ then

$$\left| \int_{\gamma_r} h(z)\, dz \right| \leq Ml(\gamma_r) = Mr(\beta - \alpha),$$

and so $\int_{\gamma_r} h(z)\, dz \to 0$ as $r \searrow 0$. On the other hand,

$$\int_{\gamma_r} \frac{dz}{z - w} = \int_{\alpha}^{\beta} \frac{rie^{it}}{re^{it}}\, dt = i(\beta - \alpha),$$

and so

$$\int_{\gamma_r} f(z)\, dz \to i(\beta - \alpha)\mathrm{res}\, f(w) \quad as \ r \searrow 0.$$

$\square$

Example 24.3.7 Let $f(x) = \operatorname{sinc} x = \sin x / x$.

Then

$$\hat{f}(t) = \widehat{\operatorname{sinc}}\,(t) = \begin{cases} 0 & \text{if } t < -1, \\ \pi & \text{if } -1 < t < 1, \\ 0 & \text{if } t > 1. \end{cases}$$

The function $\operatorname{sinc} z = \sin z / z$ has a removable singularity at 0, and if we set $\operatorname{sinc} 0 = 0$ then sinc is an entire function on $\mathbf{C}$. But we cannot apply Jordan's lemma to $f(z) = e^{-itz} \operatorname{sinc} z$, and so we must proceed in a different way. First we show that if $k > 0$ then

$$(PV)\int_{-\infty}^{\infty} \frac{e^{ikx}}{x}\,dx = i\pi.$$

Suppose that $-R < 0 < S$ and that $0 < r < \min(R, S)$. We consider the contour $\gamma = \sigma(-R, -r) \vee \epsilon_r \vee \sigma(r, S) \vee \delta(S, -R)$, where ϵ_r is the semicircular path in H_+ from $-r$ to r and $\delta(S, -R)$ is the semicircular path in H_+ from S to $-R$. Let $g(z) = e^{ikz}/z$. Then g has no poles inside γ, and so $\int_\gamma g(z)\,dz = 0$. Since g has a simple pole at 0 with residue 1, it follows from the proposition above that $\int_{\epsilon_r} g(z)\,dz \to -i\pi$ as $r \searrow 0$, and it follows from Jordan's lemma that $\int_\delta g(z)\,dz \to 0$ as $R, S \to \infty$. Thus

$$(PV)\int_{-\infty}^{\infty} \frac{e^{ikx}}{x}\,dx = i\pi.$$

(Equating imaginary parts, we see that $\int_{-\infty}^{\infty} \operatorname{sinc} x\,dx = \pi$; but note that this is an improper integral, since $\int_{-\infty}^{\infty} |\operatorname{sinc} x|\,dx = \infty$.) Considering complex conjugates, we see that if $k < 0$ then

$$(PV)\int_{-\infty}^{\infty} \frac{e^{-ikx}}{x}\,dx = -i\pi.$$

If we now use the equation

$$e^{-itx}\operatorname{sinc} x = \frac{1}{2i}\left(\frac{e^{i(1-t)x}}{x} - \frac{e^{-i(1+t)x}}{x}\right),$$

we find that

$$\widehat{\operatorname{sinc}}\,(t) = \begin{cases} 0 & \text{if } t < -1, \\ \pi & \text{if } -1 < t < 1, \\ 0 & \text{if } t > 1. \end{cases}$$

Exercises

24.3.1 Use the calculus of residues to calculate

$$\int_{-\infty}^{\infty} \frac{1}{(1+ax^2)(1+bx^2)}\, dx, \text{ where } 0 < a < b.$$

Verify your answer by expressing the integrand in terms of partial fractions.

24.3.2 Calculate

$$\int_{-\infty}^{\infty} \frac{\cos \pi x}{1 + x + x^2}\, dx \text{ and } \int_{-\infty}^{\infty} \frac{\sin \pi x}{1 + x + x^2}\, dx.$$

24.3.3 Show by making a change of variables that

$$\int_0^{2\pi} (\cos t)^{2k}\, dt = 2 \int_{-\infty}^{\infty} \frac{dx}{(1+x^2)^{k+1}}.$$

24.3.4 Show that if $0 \le t \le \pi/2$ then $t \le (\pi/2)\sin t$.

24.3.5 In the setting of Jordan's lemma, let $U = (S-R)/2$, $V = (R+S)/2$, and let $\epsilon(t) = U + Ve^{it}$ for $0 \le t \le \pi/2$. Show that

$$\int_{\epsilon} f(z)e^{imz}\, dz = \int_0^{\pi/2} f(\epsilon(t))e^{im(U+V\cos t)}e^{-mV\sin t}iVe^{it}\, dt.$$

Use this, and the preceding exercise, to obtain an upper bound for $|\int_{\epsilon} f(z)e^{imz}\, dz|$, and give a direct proof of Jordan's lemma for contours with semicircular paths.

24.3.6 Calculate $\widehat{\mathrm{sinc}}\,(t)$ for $t = \pm 1$ as a Cauchy principal value integral.

24.3.7 Calculate the Fourier transforms of the functions g and h defined by

$$g(x) = 1 \text{ and } h(x) = 1 - |x| \text{ if } |x| \le 1, \text{ and } g(x) = h(x) = 0 \text{ otherwise.}$$

24.4 Integrals of the form $\int_0^{\infty} x^\alpha f(x)\, dx$

Suppose that α is a real number which is not an integer. The function z^α has a *branch point* at 0; we can only define z^α as a holomorphic function on a cut plane. As we shall see, this works to our advantage. We consider the cut plane $C_\pi = \mathbf{C} \setminus [0, \infty)$, and the holomorphic function $z \to z^\alpha_{(\pi)}$ on it. We cannot extend $z^\alpha_{(\pi)}$ continuously to $\mathbf{C}$ since if $x > 0$ then $(x+iy)^\alpha_{(\pi)} \to x^\alpha$ as $y \searrow 0$, while $(x+iy)^\alpha_{(\pi)} \to e^{2\pi i \alpha}x^\alpha$ as $y \nearrow 0$.

Suppose that f is a meromorphic function on $\mathbf{C}$ with finite singular set S_f disjoint from $[0, \infty)$. Let $g(z) = z^\alpha_{(\pi)}f(z)$, for $z \in C_\pi$. For $0 < r < R < \infty$

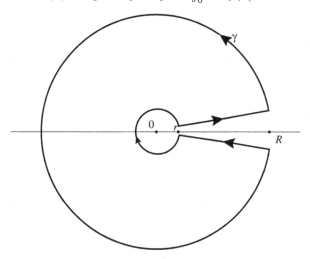

Figure 24.4.

and $0 < \delta < \pi$ we consider the contour

$$\gamma = \sigma(re^{i\delta}, Re^{i\delta}) \vee \kappa_{R,\delta} \vee \sigma(Re^{-i\delta}, re^{-i\delta}) \vee \overleftarrow{\kappa_{r,\delta}},$$

where $\kappa_{s,\delta}(t) = se^{it}$ for $t \in [\delta, 2\pi - \delta]$.

If r and δ are small enough and R is large enough, then $S_g = S_f \subseteq in[\gamma]$. By the residue theorem,

$$\int_\gamma g(z)\,dz = 2\pi i \sum_{s \in S_f} \text{res}_g(s).$$

As $\delta \searrow 0$,

$$\int_{\sigma(re^{i\delta}, Re^{i\delta})} g(z)\,dz \to \int_r^R x^\alpha f(x)\,dx,$$

$$\int_{\kappa_{R,\delta}} g(z)\,dz \to \int_{\kappa_R(0)} g(z)\,dz,$$

$$\int_{\sigma(Re^{-i\delta}, re^{-i\delta})} g(z)\,dz \to -e^{2\pi i \alpha} \int_r^R x^\alpha f(x)\,dx,$$

$$\text{and} \int_{\overleftarrow{\kappa_{r,\delta}}} g(z)\,dz \to -\int_{\kappa_r(0)} g(z)\,dz.$$

Consequently

$$(1 - e^{2\pi i \alpha}) \int_r^R x^\alpha f(x)\, dx + \int_{\kappa_R(0)} g(z)\, dz - \int_{\kappa_r(0)} g(z)\, dz = 2\pi i \sum_{s \in S_f} \mathrm{res}_g(s).$$

Thus if $\int_{\kappa_r(0)} z_{(\pi)}^\alpha f(z)\, dz \to 0$ as $r \to 0$ and $\int_{\kappa_R(0)} z_{(\pi)}^\alpha f(z)\, dz \to 0$ as $R \to \infty$ then

$$(1 - e^{2\pi i \alpha}) \int_0^\infty x^\alpha f(x)\, dx = 2\pi i \sum_{s \in S_f} \mathrm{res}_g(s).$$

If s is a simple pole of f, then $\mathrm{res}\,_g(s) = s^\alpha \mathrm{res}\,_f(s)$. Thus if all the poles of f are simple then

$$(1 - e^{2\pi i \alpha}) \int_0^\infty x^\alpha f(x)\, dx = 2\pi i \sum_{s \in S_f} s^\alpha \mathrm{res}_f(s).$$

Example 24.4.1   If $\mu, \nu \in \mathbf{R}$ and $0 < \mu + 1 < \nu$ then

$$\int_0^\infty \frac{x^\mu}{1 + x^\nu}\, dx = \frac{\pi}{\nu \sin((\mu + 1)\pi/\nu)}.$$

As always, it is a good idea to see if the problem can be simplified by a change of variables. Let $u = x^\nu$. Then

$$\int_0^\infty \frac{x^\mu}{1 + x^\nu}\, dx = \frac{1}{\nu} \int_0^\infty \frac{u^{\lambda - 1}}{1 + u}\, du, \text{ where } \lambda = \frac{\mu + 1}{\nu}.$$

Thus $0 < \lambda < 1$. It is therefore enough to show that if $0 < \lambda < 1$ then

$$\int_0^\infty \frac{u^{\lambda - 1}}{1 + u}\, du = \frac{\pi}{\sin \lambda \pi}.$$

The meromorphic function $f(z) = 1/(1 + z)$ has a simple pole at -1, with residue 1, and $(-1)^{\lambda - 1} = e^{\pi(\lambda - 1)i} = -e^{\pi \lambda i}$. Let $g(z) = z^{\lambda - 1} f(z)$. Since $0 < \lambda < 1$, $\int_{\kappa_r(0)} g(z)\, dz \to 0$ as $r \to 0$ and $\int_{\kappa_R(0)} g(z)\, dz \to 0$ as $R \to \infty$. Thus

$$(1 - e^{2\pi \lambda i}) \left(\int_0^\infty \frac{u^{\lambda - 1}}{1 + u}\, du \right) = (1 - e^{2\pi(\lambda - 1)i}) \left(\int_0^\infty \frac{u^{\lambda - 1}}{1 + u}\, du \right)$$

$$= -2\pi i e^{\pi \lambda i},$$

so that

$$\int_0^\infty \frac{u^{\lambda - 1}}{1 + u}\, du = 2\pi i \frac{e^{\pi \lambda i}}{e^{2\pi(\lambda - 1)i} - 1} = 2\pi i \frac{e^{\pi \lambda i}}{e^{2\pi \lambda i} - 1} = \frac{\pi}{\sin \lambda \pi}.$$

Exercises

We can also evaluate integrals of the form $\int_0^\infty t^{\alpha-1} f(t)\,dt$ by making the substitution $t = e^x$.

24.4.1 Show that under suitable conditions

$$\int_0^\infty t^{\alpha-1} f(t)\,dt = \int_{-\infty}^\infty e^{\alpha x} f(e^x)\,dx.$$

24.4.2 Where are the singularities of the function $f(z) = e^{\lambda z}/(1 + e^z)$?

24.4.3 Show that there is exactly one singularity of f within the rectangular contour with vertices $-S$, R, $R + 2\pi i$ and $-S + 2\pi i$.

24.4.4 Show that it is a simple pole, and calculate its residue.

24.4.5 Use this to show that if $0 < \lambda < 1$ then

$$\int_0^\infty \frac{u^{\lambda-1}}{1+u}\,du = \frac{\pi}{\sin \lambda \pi}.$$

24.4.6 By using a contour which includes part of the real axis and part of the line $\{z \in \mathbf{C} : \arg z = 2\pi/\mu\}$, evaluate

$$\int_0^\infty \frac{dx}{1 + x^\mu}, \quad \text{for } \mu > 1.$$

By making a change of variables, verify that your answer agrees with the result of the previous question.

24.5 Integrals of the form $\int_0^\infty f(x)\,dx$

If f is an even function, then $\int_0^\infty f(x)\,dx = \frac{1}{2}\int_{-\infty}^\infty f(x)\,dx$, and we can try to apply the techniques of Section 21.2. Sometimes an astute change of variables or choice of contour can be used.

Example 24.5.1 If $a > -1$ then

$$\int_0^\infty \frac{\log x}{1 + 2ax + x^2}\,dx = 0.$$

Set $u = 1/x$. Then

$$\int_0^\infty \frac{\log x}{1 + 2ax + x^2}\,dx = -\int_0^\infty \frac{\log u}{1 + 2au + u^2}\,du,$$

so that $\int_0^\infty \log x/(1 + 2ax + x^2)\,dx = 0$.

Otherwise, we can use the idea of the previous section by introducing a logarithmic factor. The function $\log z$ has a branch point at 0; we define $\log z$ on the cut plane $C_\pi = \mathbf{C} \setminus [0, \infty)$ by setting $\log(re^{it}) = \log r + it$ for $0 < t < 2\pi$. If f is a meromorphic function on $\mathbf{C}$ with finitely many poles, none of which is in $[0, \infty)$, and if $g(z) = f(z) \log z$ on C_π, then by considering the contour γ defined in the previous section, and letting δ tend to 0, we see that

$$\int_r^R f(x) \log x \, dx + \int_{\kappa_R(0)} g(z)$$

$$+ \int_R^r f(x)(\log x + 2\pi i) \, dx + \int_{\kappa_r(0)\leftarrow} g(z) \, dz$$

$$= 2\pi i \sum_{s \in S_f} \mathrm{res}_g(s),$$

for small enough r and large enough R. Thus if $\int_{\kappa_r(0)} f(z) \log z \, dz \to 0$ as $r \to 0$ and $\int_{\kappa_R(0)} f(z) \log z \, dz \to 0$ as $R \to \infty$, then

$$\int_0^\infty f(x) \, dx = - \sum_{s \in S_f} \mathrm{res}_g(s),$$

and if all the poles of f are simple then

$$\int_0^\infty f(x) \, dx = - \sum_{s \in S_f} \log s . \mathrm{res}_f(s).$$

Example 24.5.2   If $a > 0$ and $0 < t < \pi$ then

$$\int_0^\infty \frac{dx}{x^2 + 2ax \cos t + a^2} = \frac{t}{a \sin t}.$$

The rational function

$$f(z) = \frac{1}{z^2 + 2az \cos t + a^2} = \frac{1}{(z + ae^{it})(z + ae^{-it})}$$

has simple poles at $-ae^{-it} = ae^{i(\pi - t)}$ and $-ae^{it} = ae^{i(\pi + t)}$, and the residues of $f(z) \log z$ are

$$\frac{\log a + i(\pi - t)}{2ai \sin t} \quad \text{and} \quad - \frac{\log a + i(\pi + t)}{2ai \sin t},$$

respectively. Since $\int_{\kappa_r(0)} f(z) \log z \, dz \to 0$ as $r \to 0$ and $\int_{\kappa_R(0)} f(z) \log z \, dz \to 0$ as $R \to \infty$, the result follows.

We can also use this idea when the integrand has a logarithmic factor.

Example 24.5.3

$$\int_0^\infty \left(\frac{\log x}{x+1}\right)^2 dx = \frac{\pi^2}{3}.$$

Consider the meromorphic function $h(z) = (\log z)^3/(z+1)^2$ on C_π. This has a pole of order 2 at -1, with residue $3(\log(-1))^2/(-1) = 3\pi^2$. Since $\int_{\kappa_r(0)} h(z)\,dz \to 0$ as $r \to 0$ and $\int_{\kappa_R(0)} h(z)\,dz \to 0$ as $R \to \infty$,

$$\int_0^\infty \frac{(\log x)^3}{(x+1)^2}dx - \int_0^\infty \frac{(\log x + 2\pi i)^3}{(x+1)^2}dx = 2\pi i(3\pi^2) = 6\pi^3 i.$$

Expanding the integrand, and equating imaginary parts, we see that

$$\int_0^\infty \frac{-6\pi(\log x)^2 + 8\pi^3}{(x+1)^2}dx = 6\pi^3.$$

Since $\int_0^\infty dx/(x+1)^2 = 1$, it follows that

$$\int_0^\infty \left(\frac{\log x}{x+1}\right)^2 dx = \frac{\pi^2}{3}.$$

Equating real parts, we see again that $\int_0^\infty \log x/(x+1)^2\,dx = 0$.

Exercises

24.5.1 Suppose that $a > 0$. Use Example 24.5.2 to calculate

$$\int_0^\infty \frac{dx}{x^2 - 2ax \cos t + a^2}.$$

Verify your result by calculating

$$\int_{-\infty}^\infty \frac{dx}{x^2 + 2ax \cos t + a^2},$$

using the methods of the previous section.

24.5.2 Show that

$$\int_0^\infty \frac{\log x}{(1+x^2)^2}dx = -\pi/4.$$

24.5.3 Calculate
$$\int_0^\infty \frac{dx}{1+x+x^2} \quad \text{and} \quad \int_0^\infty \frac{dx}{1-x+x^2}$$
by the calculus of residues, and check your answers by calculating

$$\int_{-\infty}^\infty \frac{dx}{1+x+x^2}.$$

25

Conformal transformations

25.1 Introduction

Recall that a *univalent* function f on a domain U is a holomorphic function which takes each value at most once, and that if f is univalent on U then $f(U)$ is a domain, f^{-1} is a univalent mapping of $f(U)$ onto U, and $f'(z) \neq 0$ for all $z \in U$ (Theorem 23.6.8). In this chapter, we consider two related problems. First, if U and V are domains, is there a univalent function f mapping U onto V? If so, f is called a *conformal transformation* of U onto V, and U and V are said to be *conformally equivalent*. Secondly, what are the univalent functions mapping U onto itself? Such functions are called *conformal automorphisms* of U. Since the composition of two holomorphic functions is holomorphic, it follows that the set of conformal automorphisms of a domain U forms a group, under composition.

Why are these mappings called 'conformal'? Suppose that f is a conformal transformation of U onto V, and that $z_0 \in U$. Then $f'(z_0) \neq 0$; let $f'(z_0) = re^{i\phi}$. Since U is open, there exists $\delta > 0$ such that $N_\delta(z_0) \subseteq U$. Let $l_\theta(t) = z_0 + e^{i\theta}t$ for $t \in (-\delta, \delta)$, for $-\pi < \theta \leq \pi$ so that $l'_\theta(t) = e^{i\theta}$. Then $(f \circ l_\theta)'(0) = f'(z_0)e^{i\theta} = re^{i(\theta+\phi)}$. Thus

$$f(z_0 + e^{i\theta}t) = f(z_0) + re^{i(\theta+\phi)}t + o(t),$$

and the line $\lambda_\theta(t) = f(z_0) + re^{i(\theta+\phi)}t$ is tangent to $f \circ l_\theta$ at $f(z_0)$. If $\theta_1, \theta_2 \in \mathbf{T}$ then $\theta_1 - \theta_2$ is the oriented angle between l_{θ_1} and l_{θ_2}. But $\theta_1 - \theta_2 = (\theta_1 + \phi) - (\theta_2 + \phi)$ is also the oriented angle between λ_{θ_1} and λ_{θ_2}. Thus conformal transformations preserve oriented angles; locally, f provides a rotation through an angle $\phi = \arg f'(z_0)$ and a scaling by a factor $r = |f'(z_0)|$.

25.2 Univalent functions on C

If $\lambda, \mu \in \mathbf{C}$ and $\lambda \neq 0$, let $a_{\lambda,\mu}(z) = \lambda z + \mu$. The mapping $a_{\lambda,\mu}$ is certainly a conformal automorphism of $\mathbf{C}$, with inverse $a_{1/\lambda, -\mu/\lambda}$. Are there any more conformal automorphisms of $\mathbf{C}$?

Theorem 25.2.1 *If f is a univalent function on $\mathbf{C}$ then $f = a_{\lambda,\mu}$ for some $\lambda, \mu \in \mathbf{C}$ with $\lambda \neq 0$.*

Proof The function f is an entire function, and is therefore analytic: we can write $f(z) = \sum_{n=0}^{\infty} a_n z^n$, for all $z \in \mathbf{C}$. First we show that only finitely many coefficients are non-zero. Suppose not. By the open mapping theorem $f(\mathbf{D})$ is an open subset of $\mathbf{C}$. By Corollary 23.3.4, there exists z with $|z| > 1$ such that $f(z) \in f(\mathbf{D})$; this contradicts the univalence of f.

Thus f is a polynomial, and so therefore is f'. But $f'(z) \neq 0$ for all z, so that f' is a non-zero constant function. Thus $f = a_{\lambda,\mu}$, where $\lambda = a_1$ and $\mu = a_0$. $\square$

Corollary 25.2.2 *If U is a domain which is a proper subset of $\mathbf{C}$ then U is not conformally equivalent to $\mathbf{C}$.*

This is a remarkable result: there are very few univalent functions on $\mathbf{C}$, and if f is a univalent function on a domain U which is a proper subset of $\mathbf{C}$ then there exists w such that the equation $f(z) = w$ has no solutions.

Exercises

25.2.1 Show that the group of conformal automorphisms of $\mathbf{C}$ is isomorphic to the group of two-by-two matrices

$$\left\{ \begin{bmatrix} \lambda & \mu \\ 0 & 1 \end{bmatrix} : \lambda, \mu \in \mathbf{C} : \lambda \neq 0 \right\}.$$

25.3 Univalent functions on the punctured plane C*

Let $J(z) = 1/z$ for $z \in \mathbf{C}^*$. J, the *inversion* mapping, is a conformal automorphism of the punctured plane $\mathbf{C}^* = \mathbf{C} \setminus \{0\}$, and has a simple pole at 0 with residue 1.

Theorem 25.3.1 *If $\lambda \in \mathbf{C}^*$ then $a_{\lambda,0}$ and λJ are conformal automorphisms of $\mathbf{C}^*$.*

Conversely, if f is a univalent function on $\mathbf{C}^*$ then either $f = \lambda J + \mu$ or $f = a_{\lambda,\mu}$ for some $\lambda, \mu \in \mathbf{C}$ with $\lambda \neq 0$. In either case, f is a conformal transformation of $\mathbf{C}^*$ onto $\mathbf{C} \setminus \{\mu\}$. If f is a conformal automorphism of $\mathbf{C}^*$ then $\mu = 0$.

Proof   The first statement is obvious. Suppose conversely that f is a univalent function on $\mathbf{C}^*$. Then f has an isolated singularity at 0.

Suppose first that f has a removable singularity at 0. Then there exists ν such that $f(z) \to \nu$ as $z \to 0$. We shall show that $\nu \notin f(\mathbf{C}^*)$. Suppose not, and suppose that $f(z_0) = \nu$ for some $z_0 \in \mathbf{C}^*$. Let $\epsilon = |z_0|/2$. By the open mapping theorem, $f(N_\epsilon(z_0))$ is an open subset of $\mathbf{C}$ containing ν. But $f(z) \to \nu$ as $z \to 0$, and so there exists $w \in N_\epsilon^*(0)$ with $f(w) \in f(N_\epsilon(z_0))$. Since $N_\epsilon^*(0) \cap N_\epsilon(z_0) = \emptyset$, this contradicts the univalence of f on $\mathbf{C}^*$. Thus if we set $f(0) = \nu$ then f is an entire univalent function on $\mathbf{C}$. By Theorem 25.2.1, $f = a_{\lambda,\mu}$, for some $\lambda, \mu \in \mathbf{C}$ with $\lambda \neq 0$.

Secondly we show that f cannot have an isolated essential singularity at 0. For if it did, by Weierstrass' theorem there would be $w \in N_{1/2}^*(0)$ with $f(w)$ in the open set $f(N_{1/2}(1))$, contradicting the univalence of f.

Finally we consider the case where f has a pole at 0. By Corollary 23.6.5, this must be a simple pole. Thus f has a Laurent series expansion $\sum_{n=-1}^{\infty} a_n z^n$. By Corollary 23.3.4, there are only finitely many non-zero coefficients a_n. Thus $f(z) = p(z)/z$, where p is a polynomial with non-zero constant term a_{-1}. Since $\mathbf{C}^*$ is not conformally equivalent to $\mathbf{C}$, there exists $\mu \in \mathbf{C}$ which is not in $f(\mathbf{C}^*)$. Let $q(z) = p(z) - \mu z$. Then $q(0) = p(0) = a_{-1} \neq 0$, and $q(z) = (f(z) - \mu)z \neq 0$ for $z \neq 0$. Thus the polynomial q has no zeros, and so must be the non-zero constant function a_{-1}. Thus $f(z) = a_{-1}/z + \mu$. The final two statements now follow immediately.   □

Exercises

25.3.1   Suppose that S is a discrete closed subset of $\mathbf{C}$ and that f is a univalent function on $\mathbf{C} \setminus S$ which has a non-removable singularity at each point of S. Show that S has at most one element, and that if f has a singularity then it must be a simple pole.

25.4 The Möbius group

We next consider univalent meromorphic functions on the unit sphere $\mathbf{C}_\infty$. Recall that a function $f : \mathbf{C}_\infty \to \mathbf{C}_\infty$ is meromorphic if the restrictions of f and $f \circ J$ to $\mathbf{C}$ are meromorphic functions on $\mathbf{C}$.

Theorem 25.4.1   *If f is a univalent meromorphic function on $\mathbf{C}_\infty$ then either $f(z) = a_{\lambda,\mu}(z) = \lambda z + \mu$ or $f(z) = \lambda/(z - z_0) + \mu$ for some λ, μ and z_0 in $\mathbf{C}$, with $\lambda \neq 0$. In either case, f is a homeomorphism of $\mathbf{C}_\infty$ onto itself.*

Proof   If the restriction of f to $\mathbf{C}$ is holomorphic, then it is a univalent function on $\mathbf{C}$, and so $f = a_{\lambda,\mu}$, by Theorem 25.2.1. Otherwise, it has one simple pole, at z_0 say. Let $g(z) = f(z + z_0)$, for $z \in \mathbf{C}^*$. Then g is a univalent function on $\mathbf{C}^*$ with a pole at 0, so that $g = \lambda J + \mu$, by Theorem 25.3.1, and $f(z) = \lambda/(z - z_0) + \mu$. In either case, f is a homeomorphism of $\mathbf{C}_\infty$ onto itself.   □

A univalent meromorphism of $\mathbf{C}_\infty$ onto itself is called a *Möbius transformation* of $\mathbf{C}_\infty$. If f is a Möbius transformation, we can write

$$f(z) = \frac{az + b}{cz + d}:$$

in the former case

$$a = \lambda, \ b = \mu, \ c = 0 \text{ and } d = 1,$$

and in the latter case

$$a = \mu, \ b = \lambda - \mu z_0, \ c = 1 \text{ and } d = -z_0.$$

Note that in either case $ad - bc = \lambda \neq 0$.

Conversely, suppose that $ad - bc \neq 0$, and consider the meromorphic function $f(z) = (az + b)/(cz + d)$. If $c = 0$ then $d \neq 0$ and $f(z) = \lambda z + \mu$, where $\lambda = a/d \neq 0$ and $\mu = b/d$. If $c \neq 0$, let $z_0 = -d/c$. Then f has a simple pole at z_0, and

$$f(z) = \frac{\lambda}{z - z_0} + \mu, \text{ where } \lambda = -\frac{ad - bc}{c^2} \neq 0 \text{ and } \mu = \frac{a}{c}.$$

Thus f is a Möbius transformation.

Theorem 25.4.2   *The set $\mathcal{M}$ of Möbius transformations is a group under composition. If*

$$A = \begin{bmatrix} a & b \\ c & d \end{bmatrix} \in GL_2(\mathbf{C}), \text{ the group of invertible two-by-two matrices,}$$

let $m(A)(z) = (az + b)/cz + d)$. Then m is a homomorphism of $GL_2(\mathbf{C})$ onto $\mathcal{M}$, with kernel $\{aI : a \neq 0\}$. $(m(A))^{-1} = m(B)$, where

$$B = \begin{bmatrix} d & -b \\ -c & a \end{bmatrix}.$$

Proof   Since $A \in GL_2(\mathbf{C})$ is invertible if and only if $\det A = ad - bc \neq 0$. m maps $GL_2(\mathbf{C})$ onto $\mathcal{M}$, and $m(I)(z) = z$. Suppose that

$$A_1 = \begin{bmatrix} a_1 & b_1 \\ c_1 & d_1 \end{bmatrix} \text{ and } A_2 = \begin{bmatrix} a_2 & b_2 \\ c_2 & d_2 \end{bmatrix}$$

are in $GL_2(\mathbf{C})$. Then

$$m(A_1)m(A_2)(z) = m(A_1)\left(\frac{a_2 z + b_2}{c_2 z + d_2}\right)$$

$$= \frac{a_1\left(\frac{a_2 z + b_2}{c_2 z + d_2}\right) + b_1}{c_1\left(\frac{a_2 z + b_2}{c_2 z + d_2}\right) + d_1}$$

$$= \frac{a_1(a_2 z + b_2) + b_1(c_2 z + d_2)}{c_1(a_2 z + b_2) + d_1(c_2 z + d_2)}$$

$$= \frac{(a_1 a_2 + b_1 c_2)z + a_1 b_2 + b_1 d_2}{(c_1 a_2 + d_1 c_2)z + c_1 b_2 + d_1 d_2}$$

$$= m(A_1 A_2)(z).$$

Consequently $m(A_1)m(A_2) \in \mathcal{M}$. Since

$$m(A^{-1})m(A) = m(A)m(A^{-1}) = m(I)$$

and $m(I)$ is the identity mapping on $\mathbf{C}_\infty$, it follows that $\mathcal{M}$ is a group under composition, and that m is a homomorphism of $GL_2(\mathbf{C})$ onto $\mathcal{M}$. Clearly $m(A)(z) = z$ for all $z \in \mathbf{C}_\infty$ if and only if $a = d \neq 0$ and $b = c = 0$. Since $AB = BA = (ad - bc)I$, $m(B)$ is the inverse of $m(A)$. This last statement can also be verified directly; if $w = (az + b)/(cz + d)$, we can consider this as an equation in z, and solve it to find that $z = (dw - b)/(-cw + a)$.   □

A Möbius transformation is determined by its action on three distinct points.

Proposition 25.4.3   *Suppose that $\{z_1, z_2, z_3\}$ and $\{w_1, w_2, w_3\}$ are sets of distinct points of $\mathbf{C}_\infty$. Then there exists a unique Möbius transformation m for which $m(z_1) = w_1$, $m(z_2) = w_2$ and $m(z_3) = w_3$.*

Proof   First we show that if w_1, w_2 and w_3 are distinct points of $\mathbf{C}_\infty$ then there exists a unique Möbius transformation $m = m_{w_1, w_2, w_3}$ for which $m(w_1) = 0$, $m(w_2) = 1$ and $m(w_3) = \infty$. If $w_1, w_2, w_3 \in \mathbf{C}$ we can take

$$m(z) = \left(\frac{w_2 - w_3}{w_2 - w_1}\right)\left(\frac{z - w_1}{z - w_3}\right)$$

and otherwise we can take

$$m(z) = \frac{w_2 - w_3}{z - w_3} \text{ if } w_1 = \infty,$$

$$= \frac{z - w_1}{z - w_3} \text{ if } w_2 = \infty,$$

$$= \frac{z - w_1}{w_2 - w_1} \text{ if } w_3 = \infty.$$

Suppose that $n \in \mathcal{M}$ is another Möbius transformation for which $n(w_1) = 0$, $n(w_2) = 1$ and $n(w_3) = \infty$. Let $k = m \circ n^{-1}$. Then $k(0) = 0$, $k(1) = 1$ and $k(\infty) = \infty$. Suppose that $k(z) = (az + b)/(cz + d)$. Since $k(0) = 0$, $b = 0$. If $c \neq 0$ then $k(-d/c) = \infty$, giving a contradiction. Thus $c = 0$. Finally $k(1) = a/d = 1$, so that $a = d$ and $k(z) = z$. k is the identity mapping, and so $n = m$; m is unique.

The Möbius transformation $m_{z_1,z_2,z_3}^{-1} \circ m_{w_1,w_2,w_3}$ then has the required properties. It is unique, for if n is another Möbius transformation for which $m(z_1) = w_1$, $m(z_2) = w_2$ and $m(z_3) = w_3$, then $n \circ m_{z_1,z_2,z_3} = m_{w_1,w_2,w_3}$, so that $n = m_{w_1,w_2,w_3} \circ m_{z_1,z_2,z_3}^{-1}$. $\qquad\square$

The following Möbius transformations are called *elementary* Möbius transformations:

- $T_b(z) = z + b$ (translation);
- $D_r(z) = rz$ for r real and positive (dilation);
- $R_\theta(z) = e^{i\theta}z$ for $\theta \in \mathbf{R}$ (rotation);
- $J(z) = 1/z$ (inversion).

If $\lambda = re^{i\theta}$ with $r > 0$ then

$$\lambda z + \mu = T_\mu \circ R_\theta \circ D_r \text{ and } \frac{\lambda}{z - z_0} + \mu = (T_\mu \circ R_\theta \circ D_r \circ J \circ T_{-z_0})(z),$$

so that these elementary transformations generate $\mathcal{M}$. This is very useful in establishing properties of general Möbius transformations, as Theorem 25.4.5 will show. Our next aim is to show that a Möbius transformation 'maps circles and straight lines into circles or straight lines'.

First we must describe straight lines and circles in $\mathbf{C}$ and $\mathbf{C}_\infty$ in terms of the complex structure. A straight line L in $\mathbf{C}$ can be written as

$$L = \{z = x + iy \in \mathbf{C} : ax + by + c = 0\},$$

where a, b and c are real, and a and b are not both zero. Substituting $x = (z + \bar{z})/2$ and $y = (z - \bar{z})/2i$, we find that

$$L = \{z \in \mathbf{C} : \bar{\lambda}z + \lambda\bar{z} + \mu = 0\},$$

where $\lambda = a + ib \neq 0$ and $\mu = 2c$ is real. L is a closed unbounded subset of $\mathbf{C}$. In $\mathbf{C}_\infty$, we define a straight line L_∞ to be $L \cup \{\infty\}$, where L is a straight line in $\mathbf{C}$. Thus L_∞ is the closure of L in $\mathbf{C}_\infty$, and so is closed in $\mathbf{C}_\infty$.

A circle C in $\mathbf{C}$ can be written as

$$C = \{z \in \mathbf{C} : |z - c| = r\} = \{z \in \mathbf{C} : (z - c)\overline{(z - c)} = r^2\},$$

where $c \in \mathbf{C}$ and $r > 0$. Then $C = \{z \in \mathbf{C} : z\bar{z} - \bar{c}z - c\bar{z} = r^2 - c\bar{c}\}$, where $c \in \mathbf{C}$ and $r > 0$. C is a bounded closed subset of $\mathbf{C}$, and so is closed in $\mathbf{C}_\infty$.

It is clear that translation, dilation and rotation map straight lines to straight lines and circles to circles. What about inversion?

Proposition 25.4.4 *Suppose that L_∞ is a straight line in $\mathbf{C}_\infty$ and that C is a circle in $\mathbf{C}_\infty$.*

(i) *If $0 \in L_\infty$ then $J(L_\infty)$ is a straight line in $\mathbf{C}_\infty$ and $0 \in J(L_\infty)$.*
(ii) *If $0 \notin L_\infty$ then $J(L_\infty)$ is a circle in $\mathbf{C}_\infty$ and $0 \in J(L_\infty)$.*
(iii) *If $0 \in C$ then $J(C)$ is a straight line in $\mathbf{C}_\infty$ and $0 \notin J(C)$.*
(iv) *If $0 \notin C$ then $J(C)$ is a circle in $\mathbf{C}_\infty$ and $0 \notin J(C)$.*

Proof   This is a matter of straightforward verification.

(i) If $0 \in L_\infty$, then

$$L_\infty = \{z \in \mathbf{C} : \bar{\lambda}z + \lambda\bar{z} = 0\} \cup \{\infty\},$$

so that

$$J(L_\infty) = J^{-1}(L_\infty) = \{\infty\} \cup \{z \in \mathbf{C} \setminus \{0\} : \frac{\bar{\lambda}}{z} + \frac{\lambda}{\bar{z}} = 0\} \cup \{0\}$$
$$= \{z \in \mathbf{C} : \lambda z + \bar{\lambda}\bar{z} = 0\} \cup \{\infty\},$$

which shows that $J(L_\infty)$ is a straight line in $\mathbf{C}_\infty$ and that $0 \in J(L_\infty)$.
(ii) If $0 \notin L_\infty$, then

$$L_\infty = \{z \in \mathbf{C} : \bar{\lambda}z + \lambda\bar{z} + \mu = 0\} \cup \{\infty\},$$

with $\mu \neq 0$. Arguing as above,

$$J(L_\infty) = J^{-1}(L_\infty) = \{0\} \cup \{z \in \mathbf{C} \setminus \{0\} : \frac{\bar{\lambda}}{z} + \frac{\lambda}{\bar{z}} + \mu = 0\}$$
$$= \{z \in \mathbf{C} : z\bar{z} + \frac{\lambda z}{\mu} + \frac{\bar{\lambda}\bar{z}}{\mu} = 0\}$$
$$= \{z \in \mathbf{C} : z\bar{z} - \bar{c}z - c\bar{z} = r^2 - c\bar{c}\}$$

where $c = -\bar{\lambda}/\mu$ and $r^2 = c\bar{c}$. This shows that $J(L_\infty)$ is a circle in $\mathbf{C}_\infty$ and that $0 \in J(L_\infty)$.

(iii) If $0 \in C$, then

$$C = \{z \in \mathbf{C} : z\bar{z} - \bar{c}z - c\bar{z} = 0\},$$

so that

$$J(C) = J^{-1}(C) = \{z \in \mathbf{C} \setminus \{0\} : \frac{1}{z\bar{z}} - \frac{\bar{c}}{z} - \frac{c}{\bar{z}} = 0\} \cup \{\infty\}$$

$$= \{z \in \mathbf{C} : cz + \bar{c}\bar{z} = 1\} \cup \{\infty\},$$

which shows that $J(C)$ is a straight line in $\mathbf{C}_\infty$ and that $0 \notin J(C)$.

(iv) If $0 \notin C$, then

$$C = \{z \in \mathbf{C} : z\bar{z} - \bar{c}z - c\bar{z} = d\},$$

with $d = r^2 - c\bar{c} \neq 0$, and so

$$J(C) = J^{-1}(C) = \{z \in \mathbf{C} \setminus \{0\} : \frac{1}{z\bar{z}} - \frac{\bar{c}}{z} - \frac{c}{\bar{z}} = d\}$$

$$= \{z \in \mathbf{C} : z\bar{z} + \frac{cz}{d} + \frac{\bar{c}\bar{z}}{d} = \frac{1}{d}\},$$

which shows that $J(C)$ is a circle in $\mathbf{C}_\infty$ and that $0 \notin J(C)$. $\square$

Theorem 25.4.5   *Suppose that m is a Möbius transformation, that L_∞ is a straight line in $\mathbf{C}_\infty$ and that C is a circle in $\mathbf{C}_\infty$.*

(i) *If $\infty \in m(L_\infty)$ then $m(L_\infty)$ is a straight line in $\mathbf{C}_\infty$.*
(ii) *If $\infty \notin m(L_\infty)$ then $m(L_\infty)$ is a circle in $\mathbf{C}_\infty$.*
(iii) *If $\infty \in m(C)$ then $m(C)$ is a straight line in $\mathbf{C}_\infty$.*
(iv) *If $\infty \notin m(C)$ then $m(C)$ is a circle in $\mathbf{C}_\infty$.*

Proof   Translations, dilations and rotations map straight lines to straight lines and circles to circles. Since m is a product of elementary transformations, it follows from Proposition 25.4.4 that $m(L_\infty)$ is either a straight line or a circle. $m(L_\infty)$ is a straight line if $\infty \in m(L_\infty)$ and is a circle if not, and $m(C)$ is a straight line if $\infty \in m(C)$ and is a circle if not. $\square$

This result suggests that we may think of a straight line in $\mathbf{C}_\infty$ as an unbounded circle, or as a circle of infinite radius.

The complement of a straight line in $\mathbf{C}_\infty$ has two connected components, as does the complement of a circle (the inside, and the union of the outside and $\{\infty\}$). Since a Möbius transformation m is a homeomorphism of $\mathbf{C}_\infty$, if S is a circle or straight line and U and V are the connected components of $\mathbf{C}_\infty \setminus S$ then $m(U)$ and $m(V)$ are the connected components of $\mathbf{C}_\infty \setminus m(S)$.

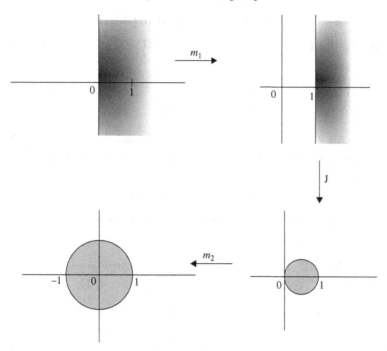

Figure 25.4.

As an example, the Möbius transformation $m(z) = (-z+1)/(z+1)$ maps the extended y-axis $Y_\infty = \{z = iy : y \in \mathbf{R}\} \cup \{\infty\}$ onto the unit circle $\mathbf{T}$. Although this can be verified directly, it is more informative to construct the mapping in several steps. First, the mapping $m_1(z) = z + 1$ maps Y_∞ onto the extended line $L_\infty = \{z = 1 + iy : y \in \mathbf{R}\} \cup \{\infty\}$. Secondly, since $0 \notin L_\infty$, J maps L_∞ onto a circle C passing through 0 and 1. J maps the extended x-axis onto itself. Since L_∞ is orthogonal to the x-axis, and since Möbius transformations are conformal, the tangent to $J(L_\infty)$ at 1 is orthogonal to the x-axis. Thus C is the circle with centre $1/2$ and radius $1/2$. The mapping $m_2(z) = 2z - 1$ then maps C to a circle passing through 1 and -1, with centre 0, so that $m_2(C) = \mathbf{T}$. Then $m = m_2 \circ J \circ m_1$. Since $m(1) = 0$, m maps the right-hand half-plane onto the unit disc D and the left-hand half-plane onto $\{z : |z| > 1\} \cup \{\infty\}$.

Exercises

25.4.1 Show that the mapping $m(z) = (z - i)/(z + i)$ defines a conformal transformation of the upper half-plane H_+ onto the open unit disc $\mathbf{D}$, and maps i to 0.

25.5 The conformal automorphisms of D

What are the conformal automorphisms of $\mathbf{D}$? To answer this, we need Schwarz' lemma.

Proposition 25.5.1 (Schwarz' lemma)   *If f is a holomorphic mapping of $\mathbf{D}$ into $\mathbf{D}$ and $f(0) = 0$ then $|f(z)| \leq |z|$ for $z \in \mathbf{D}$.*

Proof   We can write $f(z) = zg(z)$, where g is holomorphic on $\mathbf{D}$. Suppose that $z \in \mathbf{D}$ and suppose that $|z| < r < 1$. If $|w| = r$ then $|g(w)| = |f(w)|/r < 1/r$, so that

$$|g(z)| \leq \sup\{|g(w)| : |w| = r\} < 1/r,$$

by the maximum modulus principle. Since this holds for all r with $|z| < r < 1$, $|g(z)| \leq 1$ and so $|f(z)| \leq |z|$.   □

Proposition 25.5.2   *If f is a conformal automorphism of $\mathbf{D}$ and $f(0) = 0$ then there exists $e^{i\theta} \in \mathbf{T}$ such that $f(z) = e^{i\theta}z$ for $z \in \mathbf{D}$.*

Proof   If $z \in \mathbf{D}$ then $|f(z)| \leq |z|$, by Schwarz' lemma. But $|z| = |f^{-1}f(z)| \leq |f(z)|$ as well, so that $|f(z)| = |z|$. Thus if $f(z) = zg(z)$, as in Schwarz' lemma, then $|g(z)| = 1$ for $z \in \mathbf{D}$. It therefore follows from the maximum modulus principle that g is constant; there exists $e^{i\theta} \in \mathbf{T}$ such that $g(z) = e^{i\theta}$ for $z \in \mathbf{D}$. Hence $f(z) = e^{i\theta}z$ for $z \in \mathbf{D}$.   □

Theorem 25.5.3   *If $|\alpha| < 1$ the Möbius transformation*

$$m_\alpha(z) = \frac{z + \alpha}{\bar{\alpha}z + 1}$$

is a conformal automorphism of $\mathbf{D}$ with $m_\alpha(0) = \alpha$ and $m_\alpha(-\alpha) = 0$, and with inverse $m_{-\alpha}$. The transformation m_α is a homeomorphism of $\overline{\mathbf{D}}$ onto itself.

   If m is a conformal automorphism of $\mathbf{D}$ with $m(0) = \alpha$ then there exists $e^{i\theta} \in \mathbf{T}$ such that $m(z) = m_\alpha(e^{i\theta}z)$ for $z \in \mathbf{D}$.

Proof   Clearly $m_\alpha(0) = \alpha$ and $m_\alpha(-\alpha) = 0$. It follows from Theorem 25.4.2, or by direct calculation, that $(m_\alpha)^{-1} = m_{-\alpha}$. If $|z| = 1$ then

$$|m_\alpha(z)|^2 = m_\alpha(z)\overline{m_\alpha(z)} = \left(\frac{z + \alpha}{\bar{\alpha}z + 1}\right)\left(\frac{\bar{z} + \bar{\alpha}}{\alpha\bar{z} + 1}\right)$$

$$= \frac{1 + \bar{\alpha}z + \alpha\bar{z} + \alpha\bar{\alpha}}{\alpha\bar{\alpha} + \bar{\alpha}z + \alpha\bar{z} + 1} = 1.$$

It therefore follows from the maximum modulus principle that $|m_\alpha(z)| < 1$ for $z \in \mathbf{D}$. Since m_α is univalent on $\mathbf{C} \setminus \{-1/\bar{\alpha}\}$ it follows that m_α maps $\mathbf{D}$

conformally onto $m_\alpha(\mathbf{D})$ and is a homeomorphism of $\overline{\mathbf{D}}$ onto $m_\alpha(\overline{\mathbf{D}})$, and $m_\alpha(\overline{\mathbf{D}}) \subseteq \overline{\mathbf{D}}$. By the same token $m_\alpha^{-1}(\overline{\mathbf{D}}) = m_{-\alpha}(\mathbf{D}) \subseteq \overline{\mathbf{D}}$, and so m_α is a homeomorphism of $\overline{\mathbf{D}}$ onto itself, and m_α is a conformal mapping of $\mathbf{D}$ onto itself.

The mapping $m_\alpha^{-1} \circ m$ is a conformal automorphism of $\mathbf{D}$, and $m_\alpha^{-1} m(0) = 0$, so that by Proposition 25.5.2, there exists $e^{i\theta} \in \mathbf{T}$ such that $m_\alpha^{-1} m(w) = e^{i\theta} w$ for $w \in \mathbf{D}$. Thus if $z \in \mathbf{D}$ then

$$m(z) = m_\alpha m_\alpha^{-1} m(z) = m_\alpha(e^{i\theta} z) = \frac{e^{i\theta} z + \alpha}{\bar{\alpha} e^{i\theta} z + 1}.$$

$\square$

Note also that

$$m(z) = e^{i\theta}\left(\frac{z + \alpha e^{-i\theta}}{\bar{\alpha} e^{i\theta} z + 1}\right) = e^{i\theta} m_{\alpha e^{-i\theta}}(z),$$

and that

$$m_\alpha'(z) = \frac{1 - \alpha\bar{\alpha}}{(\bar{\alpha} z + 1)^2}.$$

In particular, the quantities

$$m_\alpha'(0) = 1 - |\alpha|^2, \; m_\alpha'(\alpha) = \frac{1 - |\alpha|^2}{(1 + |\alpha|^2)^2} \text{ and } m_\alpha'(-\alpha) = \frac{1}{1 - |\alpha|^2}$$

are all real and positive.

Exercises

25.5.1 Use Schwarz' lemma to give another proof of Liouville's theorem.

25.5.2 Show that if m is a conformal automorphism of the upper half-plane H_+ for which $m(i) = i$ then there exists $0 \le \theta \le 2\pi$ such that

$$m(z) = M_\theta(z) = \frac{\cos\theta\, z + \sin\theta}{-\sin\theta\, z + \cos\theta}.$$

25.5.3 Show that the group of conformal automorphisms of H_+ is generated by the automorphisms of the previous exercise, together with translations T_a (with $a \in \mathbf{R}$) and dilations D_r.

25.6 Some more conformal transformations

We can consider other conformal transformations than Möbius transformations. First, the function exp is univalent on the strip $S = \{z = x + iy :$

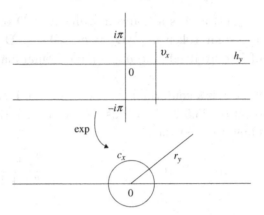

Figure 25.6a.

$-\pi < y < \pi\}$, and defines a conformal transformation of S onto the cut plane $C_0 = \mathbf{C} \setminus (-\infty, 0]$; its inverse is the principal logarithm. The lines $h_y = \{z = x + iy : x \in \mathbf{R}\}$ and $v_x = \{z = x + iy : -\pi < y < \pi\}$ are orthogonal; they are transformed to the ray $r_y = \{z = re^{iy} : r > 0\}$ and the punctured circle $c_x = \{z = e^x e^{i\theta} : -\pi < \theta < \pi\}$.

Secondly, a related conformal transformation is obtained by considering the map $z \to z^\alpha$, where α is real and positive. If $0 < \beta < \pi$, let P_β denote the sector $\{z = re^{i\theta} : r > 0, -\beta < \theta < \beta\}$. If $0 < \alpha\beta < \pi$ then the map $z \to z^\alpha$ is a conformal transformation of the sector P_β onto $P_{\alpha\beta}$. Rays are mapped to rays and circular arcs to circular arcs.

A third interesting example is provided by the function $f(z) = \frac{1}{2}(z+1/z)$, for $z \in \mathbf{C} \setminus \{0\}$. This is not univalent, since $f(z) = f(1/z)$. On the other hand, if $z = re^{i\theta}$, with $r > 0$, then

$$f(z) = a_r \cos\theta + ib_r \sin\theta, \text{ where } a_r = \frac{1}{2}\left(r + \frac{1}{r}\right) \text{ and } b_r = \frac{1}{2}\left(r - \frac{1}{r}\right).$$

Thus f is a one-one mapping of the circle $\{z : |z| = r\}$ onto the ellipse

$$E_r = \left\{w = u + iv : \frac{u^2}{a_r^2} + \frac{v^2}{b_r^2} = 1\right\},$$

from which it follows that f is univalent on the punctured disc $\mathbf{D} \setminus \{0\}$, and is also univalent on the domain $\{z \in \mathbf{C} : |z| > 1\}$, and that f maps each conformally onto the domain $\mathbf{C} \setminus [-1, 1]$. Note also that $f(re^{i\theta}) \to \cos\theta \in [-1, 1]$ as $r \nearrow 1$ and as $r \searrow 1$.

If $z = re^{i\theta} \in \mathbf{C} \setminus \{0\}$, then $f(z) \in H_+$ if and only if $b_r \sin\theta > 0$, and so f defines a conformal transformation of $\mathbf{D}_+ = \mathbf{D} \cap H_+$ onto H_-.

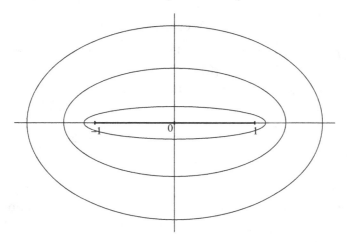

Figure 25.6b.

Many conformal transformations can be obtained by composing these transformations with other Möbius transformations. For example, let S be the semi-infinite strip $\{z = x+iy : 0 < x < 1, y > 0\}$. We shall show that the mapping $z \to \cos \pi z$ is a conformal transformation of S onto H_-. First, let $m_1(z) = i\pi z$. m_1 is a conformal transformation of S onto the semi-infinite strip $\{z = x + iy : x < 0, 0 < y < \pi\}$. The function $\exp$ is a conformal transformation of $m_1(S)$ onto $\mathbf{D}_+$, and the function f, defined above, is a conformal transformation of $\mathbf{D}_+$ onto H_-. Thus $f \circ \exp \circ m_1$ is a conformal transformation of S onto H_-. It is a straightforward matter to verify that $\cos \pi z = (f \circ \exp \circ m_1)(z)$, for $z \in S$.

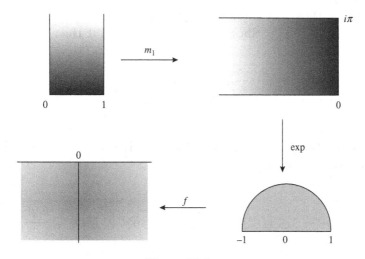

Figure 25.6c.

Exercises

25.6.1 Let $f(z) = \frac{1}{2}(z + 1/z)$, for $z \in \mathbf{D} \setminus \{0\}$. Calculate the inverse mapping $f^{-1} : \mathbf{C} \setminus [-1, 1] \to \mathbf{D} \setminus \{0\}$.

25.6.2 Find a conformal mapping of the domain

$$U = \{z : |z - 2| < 2 < |2z - 2|\}$$

onto $\mathbf{D}$.

25.6.3 Show that the function $\tanh z$ defines a conformal mapping of the strip $\{z = x + iy : 0 < y < \pi/2\}$ onto the upper half-plane H_+. What is the inverse mapping?

25.6.4 Show that the function $\cos z$ defines a conformal mapping of the strip $\{z = x + iy : 0 < x < \pi\}$ onto $\mathbf{C} \setminus ((-\infty, -1] \cup [1, \infty))$. What is the inverse mapping?

25.6.5 Show that the function $g(z) = 4z/(1 + z)^2$ is univalent on $\mathbf{D}$. What is $g(\mathbf{D})$? [Hint: Express g as the composition of Möbius transformations and other mappings.]

25.6.6 Find conformal mappings of the following domains onto $\mathbf{D}$. (The mappings may be expressed as compositions of holomorphic functions.)

(i) $U_1 = \mathbf{D} \cap H_+$
(ii) $U_2 = \mathbf{D} \setminus (-1, 0]$
(iii) $U_4 = \mathbf{D} \cap \{z : |z + 1| > \sqrt{2}\}$

25.6.7 Suppose that f is a holomorphic function on $\mathbf{D}$, taking values in $\mathbf{D}$, and that $f(0) = c$. Show that if $0 < |z| = r < 1$ then

$$\frac{1 - |f(z)|}{1 - r} \geq \frac{1 - |c|}{1 + r|c|}.$$

25.6.8 Suppose that f is a holomorphic function on $\mathbf{D}$ whose real part is positive. Show that

$$\frac{1 - |z|}{1 + |z|} \leq |f(z)| \leq \frac{1 + |z|}{1 - |z|} \text{ and that } |\Im f(z)| \leq \frac{2|z|}{1 - |z|^2}.$$

25.6.9 Verify that $\cos \pi z = (f \circ \exp \circ m_1)(z)$, for $z \in S$, where m_1, f and S are defined above.

25.7 The space $H(U)$ of holomorphic functions on a domain U

We now establish further properties of the space $H(U)$ of holomorphic functions on a domain U. Recall (Theorem 22.6.10) that $H(U)$ is a closed linear subspace of the space $(C(U), d)$ where d is a complete metric defining the topology of local uniform convergence.

Theorem 25.7.1 *The mapping $f \to f' : (H(U), d) \to (H(U), d)$ is continuous.*

Proof   It is enough to show that if $f_n \to f$ in $(H(U), d)$ then $f_n' \to f'$ in $(H(U), d)$, and so it is enough to show that if $M_r(z_0)$ is a closed neighbourhood of an element z_0 of U then $f_n' \to f'$ uniformly on $M_r(z_0)$. There exists $s > r$ such that $M_s(z_0) \subseteq U$. If $w \in M_r(z_0)$ then

$$f_n'(w) - f'(w) = \frac{1}{2\pi i} \int_{\kappa_s(z_0)} \frac{f_n(z) - f(z)}{(z - w)^2} \, dz.$$

But if $z \in [\kappa_s(z_0)]$ and $w \in M_r(z_0)$ then $|z - w| > s - r$, so that

$$\left| \frac{f_n(z) - f(z)}{(z - w)^2} \right| \leq \left| \frac{f_n(z) - f(z)}{(s - r)^2} \right| ;$$

hence

$$|f_n'(w) - f'(w)| \leq \frac{s}{(s - r)^2} \sup_{z \in [\kappa_s(z_0)]} |f_n(z) - f(z)|,$$

and $f_n' \to f'$ uniformly on $M_r(z_0)$ as $n \to \infty$. $\square$

Recall that a subset A of $C(U)$ is *locally uniformly bounded* if

$$\sup\{|f(z)| : f \in A, \ z \in K\} < \infty,$$

for each compact subset K of U.

Theorem 25.7.2 (Montel's theorem)   *A subset A of $H(U)$ is compact if and only if it is closed and locally uniformly bounded.*

Proof   In one direction it is easy. If A is compact, then it is certainly closed. If K is a compact subset of U, the restriction map $\pi_K : C(U) \to C(K)$ is continuous, and so $\pi(A)$ is a compact subset, and therefore a bounded subset, of $C(K)$.

Suppose conversely that A is closed and locally uniformly bounded. We use the local Arzelà–Ascoli theorem (Volume II, Theorem 15.8.4). It is sufficient to show that A is equicontinuous. Suppose that $z_0 \in U$ and that $0 < \epsilon < 1$. There exists $s > 0$ such that $M_s(z_0) \subseteq U$. Since A is locally uniformly bounded,

$$L = \sup\{|f(z)| : f \in A, z \in M_s(z_0)\} < \infty.$$

Let $r = s\epsilon/(2L+2)$. Then $0 < r < s/2$, so that $s - r > s/2$. If $f \in A$ and $w \in N_r(z_0)$, then, using Cauchy's integral formula,

$$|f(w) - f(z_0)| = \frac{1}{2\pi} \left| \int_{\kappa_s(z_0)} \frac{f(z)}{z-w} \, dz - \int_{\kappa_s(z_0)} \frac{f(z)}{z-z_0} \, dz \right|$$

$$= \frac{1}{2\pi} \left| \int_{\kappa_s(z_0)} \frac{(w-z_0)f(z)}{(z-w)(z-z_0)} \, dz \right|$$

$$\leq \frac{Ls}{(s-r)s}|w - z_0| \leq \frac{2Lr}{s} < \epsilon.$$

Thus A is equicontinuous at z_0. $\qquad\square$

The set $\mathrm{Uni}(U)$ of univalent functions on U also has remarkable properties.

Theorem 25.7.3 *Let $\mathrm{Uni}(U)$ be the set of univalent functions on a domain U, and let $\mathrm{Con}(U)$ be the set of constant functions on U. Then $\mathrm{Uni}(U) \cup \mathrm{Con}(U)$ is closed in $H(U)$.*

Proof We must show that if $(f_n)_{n=1}^{\infty}$ is a sequence of univalent functions which converges in $H(U)$ to f, and if f is not a constant, then f is univalent. Suppose not, so that there exist distinct $z_1, z_2 \in U$ such that $f(z_1) = f(z_2) = v_0$. Then z_1 and z_2 are zeros of the non-constant holomorphic function $f - v_0$. Since the zeros of $f - v_0$ are isolated, there exist disjoint closed discs $M_{r_1}(z_1)$ and $M_{r_2}(z_2)$ in U such that $f(z) - v_0 \neq 0$ for $z \in M_{r_1}^*(z_1) \cup M_{r_2}^*(z_2)$. We use Rouché's theorem to show that for sufficiently large n the function $f_n - v_0$ has a zero in each of $N_{r_1}(z_1)$ and $N_{r_2}(z_2)$, contradicting the fact that f_n is univalent. Let $m = \inf\{|f(z)| : z \in \mathbf{T}_{r_1}(z_1) \cup \mathbf{T}_{r_2}(z_2)\}$; then $m > 0$. Since $f_n \to f$ in $H(U)$, there exists n_0 such that $|f_n(z) - f(z)| < m$ for $z \in \mathbf{T}_{r_1}(z_1) \cup \mathbf{T}_{r_2}(z_2)$, for $n \geq n_0$. Thus

$$|(f_n(z) - v_0) - (f(z) - v_0)| < m \leq |f(z) - v_0|$$

for $z \in \mathbf{T}_{r_1}(z_1) \cup \mathbf{T}_{r_2}(z_2)$. By Rouché's theorem, if $n \geq n_0$ then $f_n - v_0$ has a zero in each of $N_{r_1}(z_1)$ and $N_{r_2}(z_2)$. $\qquad\square$

Exercises

25.7.1 Suppose that f is a non-constant holomorphic function on a domain U which has k zeros, counted according to multiplicity. Show that there is a neighbourhood N of f in $H(U)$ such that if $g \in N$ then g has at least k zeros, counted according to multiplicity.

25.7.2 Give an example of a univalent function f on the half-space $H_r = \{z \in \mathbf{C} : |\arg z| < \pi/2\}$ which is the limit in $H(H_r)$ of a sequence of non-univalent holomorphic functions.

25.8 The Riemann mapping theorem

We end by proving a truly remarkable theorem.

Theorem 25.8.1 (The Riemann mapping theorem) *Suppose that U is a simply connected domain which is a proper subset of $\mathbf{C}$, and that $z_0 \in U$. Then there exists a unique conformal transformation f of U onto $\mathbf{D}$ with the properties that $f(z_0) = 0$ and $f'(0)$ is real and positive.*

Proof   First, let us prove uniqueness. If f_1 and f_2 are conformal transformations of U onto $\mathbf{D}$ which satisfy the requirements of the theorem, then $\phi = f_2 \circ f_1^{-1}$ is a conformal automorphism of $\mathbf{D}$, $\phi(0) = 0$, and $\phi'(0)$ is real and positive, so that ϕ is the identity mapping, by Theorem 25.5.3. Thus $f_1 = f_2$.

It is existence that is the real problem. To prove this, we use the fact that if g is a holomorphic function on U which has no zeros in U, then g has a holomorphic square root h on U: $(h(z))^2 = g(z)$ for all $z \in U$ (Corollary 22.7.2). If we set $s(z) = z^2$ for $z \in \mathbf{C}$ then $s \circ h = g$. Note that if g is univalent, then h is univalent, and s is univalent on $h(U)$.

Let us describe structure of the proof. We consider the set G of univalent functions g on U taking values in $\mathbf{D}$, for which $g(z_0) = 0$ and $g'(z_0)$ is real and positive. First we show that G is non-empty. Next we show that $\{g'(z_0) : g \in G\}$ is bounded. We then use a compactness argument to show that there exists $f \in G$ such that

$$f'(z_0) = \sup\{g'(z_0) : g \in G\}.$$

Finally we show that $f(U) = \mathbf{D}$, so that f satisfies the conclusions of the theorem.

First we show that G is non-empty. There exists $z_1 \in \mathbf{C} \setminus U$. The univalent function $a(z) = z - z_1$ does not have a zero in U, and so it has a univalent square root: there exists a univalent function h on U such that $(h(z))^2 = z - z_1$, for $z \in U$. Let $z_2 = h(z_0)$. By the open mapping theorem, $h(U)$ is open in $\mathbf{C}$, and so there exists $r > 0$ such that $N_r(z_2) \subseteq h(U)$. We show that $-N_r(z_2) = N_r(-z_2)$ is disjoint from $h(U)$. If $w = h(z) \in h(U)$, then

$$s(-w) = (-w)^2 = w^2 = s(w) = z - z_1;$$

since s is univalent on $h(U)$ and $w \in h(U)$, $-w \notin h(U)$. Thus $h(U)$ is contained in the outside of $\mathbf{T}_r(-z_2)$. The Möbius transformation $m(z) = r/(z + z_2)$ maps the outside of $\mathbf{T}_r(-z_2)$ conformally onto $\mathbf{D} \setminus \{0\}$, mapping z_2 to $r/2z_2$, and the Möbius transformation $m_{-r/2z_2}$ is a conformal automorphism of $\mathbf{D}$, mapping $r/2z_2$ to 0. Thus $j = m_{-r/2z_2} \circ m \circ h$ is a conformal transformation of U onto a subset of $\mathbf{D}$, and $j(z_0) = 0$. Since j is univalent, $j'(z_0) \neq 0$. Let $g_0 = e^{-i\theta} j$, where $\theta = \arg(j'(z_0))$. Then $g_0 \in G$.

Let us set $l = g_0'(z_0)$, and let us set $G_l = \{g \in G : g'(z_0) \geq l\}$. Thus G_l is a non-empty subset of G; we shall show that it is a compact subset of $H(U)$. By Montel's theorem, the set

$$F = \{g \in H(U) : g(U) \subseteq \overline{\mathbf{D}}\}$$
$$= \{g \in H(U) : \sup_{z \in K} |g(z)| \leq 1 \text{ for } K \text{ compact}, K \subseteq U\}$$

is a compact subset of $H(U)$. By Theorem 25.7.1, the mapping $g \to g'(z_0)$ is continuous on $H(U)$, and so the set

$$F_l = \{g \in F : g'(z_0) \text{ is real, and } g'(z_0) \geq l\}$$

is closed, and is therefore compact. If $g \in F_l$, then g is not constant and so, by the open mapping theorem,

$$F_l = \{g \in H(U) : g(U) \subseteq \mathbf{D}, g'(z_0) \text{ is real, and } g'(z_0) \geq l\}.$$

Finally,
$$G_l = F_l \cap \{g \in H(U) : g \text{ is univalent or constant}\}$$

is a closed subset of F_l, and is therefore compact. Since the mapping $g \to g'(z_0)$ is continuous on $H(U)$, there therefore exists $f \in G_l$ for which $f'(z_0) = \sup\{g'(z_0) : g \in G_l\}$. We shall show that $f(U) = \mathbf{D}$, so that f satisfies the requirements of the theorem.

Suppose not, so that there exists $b \in \mathbf{D} \setminus f(U)$. The Möbius transformation m_{-b} moves b to 0 and $f(z_0)$ to $-b$. Thus $0 \notin m_{-b}f(U)$. Since $m_{-b}f(U)$ is simply connected, there exists a holomorphic square root function h on it; $(h(z))^2 = z$ for $z \in m_{-b}f(U)$. Let $c = h(-b)$, so that $c \in \mathbf{D}$, $c^2 = -b$ and

$h(m_{-b}f(z_0)) = c$. The Möbius transformation m_{-c} then moves c to 0. Let $\phi = \arg c$ and let $g = e^{i\phi}m_{-c} \circ h \circ m_{-b} \circ f$. Then g is a univalent mapping of U into $\mathbf{D}$ and $g(z_0) = 0$.

Since $h(z)^2 = z$, $h'(z) = 1/2h(z)$, so that $h'(-b) = 1/2c$. Further,

$$m'_{-b}(0) = 1 - b\bar{b} \text{ and } m'_{-c}(c) = 1/(1 - c\bar{c}).$$

Applying the chain rule,

$$g'(0) = e^{i\phi}m'_{-c}(c).h'(-b).m'_{-b}(0).f'(z_0)$$

$$= e^{i\phi}\left(\frac{1}{(1 - c\bar{c})}\right)\left(\frac{1}{2c}\right)(1 - b\bar{b})\,f'(z_0)$$

$$= \left(\frac{1}{2|c|}\right)\left(\frac{1 - |b|^2}{1 - |b|}\right)f'(z_0) = \frac{1 + |b|}{2|c|}f'(z_0).$$

But $2|c| < 1 + |c|^2 = 1 + |b|$. Thus $g \in G$, and $g'(0) > f'(0)$, giving a contradiction. $\square$

Corollary 25.8.2   *The group of conformal automorphisms of U is isomorphic to the group of conformal automorphisms of $\mathbf{D}$.*

Proof   The mapping $m \to f^{-1} \circ m \circ f$ is an isomorphism of group of conformal automorphisms of $\mathbf{D}$ onto the group of conformal automorphisms of U. $\square$

Corollary 25.8.3   *If U_1 and U_2 are simply connected domains which are proper subsets of $\mathbf{C}$ then there is a conformal transformation ϕ of U_1 onto U_2.*

Proof   Take $\phi = f_2^{-1} \circ f_1$, where f_1 is a conformal transformation of U_1 onto $\mathbf{D}$ and f_2 is a conformal transformation of U_2 onto $\mathbf{D}$. $\square$

Inspection of the proof of the Riemann mapping theorem shows that it depends on the fact that if g is a holomorphic function on the simply connected domain U, and if g has no zeros in U then g has a square root; no other consequences of simple connectivity are used. Thus we have the following result.

Proposition 25.8.4   *A domain U is simply connected if and only if whenever g is a holomorphic function on U which has no zeros in U then g has a square root.*

Proof   The condition is necessary, by Corollary 22.7.2. If it is satisfied, then either $U = \mathbf{C}$ or U is homeomorphic to the simply connected domain $\mathbf{D}$. $\square$

26

Applications

We now apply the theory that we have developed to obtain further results. These are interesting in themselves (although they are only the first of many such important results), but they are principally intended to illustrate how the theory is used in practice.

26.1 Jensen's formula

Our aim is to show that the growth of an entire function f is related to the location of the zeros of f. For this, we need Jensen's formula.

Theorem 26.1.1 *Suppose that $r > 1$ and that f is a meromorphic function on $\mathbf{D}_r = \{z : |z| < r\}$ which has no zeros or poles in the set $\mathbf{D}_r \setminus \mathbf{D} = \{z : 1 \leq |z| < r\}$ or at 0. Then*

$$\log|f(0)| = -\sum_{s \in S_f} k_f(s) \log|s| + \sum_{\zeta \in Z_f} l_f(\zeta) \log|\zeta| + \frac{1}{2\pi} \int_{-\pi}^{\pi} \log|f(e^{it})|\, dt,$$

where $k_f(s)$ is the order of the pole at s and $l_f(\zeta)$ is the order of the zero at ζ.

Proof   Suppose first that $S_f \cup Z_f$ is empty. There then exists a holomorphic branch of $\log z$ on $f(\mathbf{D}_r)$, so that $\log f$ is a holomorphic function on $\mathbf{D}_r$. Thus $\log f(0) = \frac{1}{2\pi} \int_{-\pi}^{\pi} \log f(e^{it})\, dt$, by Cauchy's integral formula, and the result follows by taking the real part of this equation.

Secondly, suppose that $S_f \cup Z_f$ is not empty. We use Möbius functions to remove the zeros and poles, so that we can again appeal to Cauchy's integral formula. Recall that if $w \in \mathbf{D}$ then the Möbius function

$$m_{-w}(z) = \frac{z - w}{1 - \overline{w}z}$$

is an automorphism of $\mathbf{D}$, with a simple zero at w and a simple pole at $1/\overline{w}$, and that if $z = e^{it} \in \mathbf{T}$ then

$$|m_w(e^{it})| = \left|\frac{e^{it} - w}{1 - e^{it}\overline{w}}\right| = \left|\frac{e^{it} - w}{e^{-it} - \overline{w}}\right| = 1.$$

Let $\rho = \sup\{|z| : z \in S_f \cup Z_f\}$ and let $\tau = \min(r, 1/\rho)$. Let

$$g(z) = f(z) \cdot \left(\prod_{s \in S_f} m_{-s}(z)^{k_f(s)}\right) \cdot \left(\prod_{\zeta \in Z_f} m_{-\zeta}(z)^{-l_f(\zeta)}\right).$$

Then g has removable singularities at the points of $S_f \cup Z_f$. We remove them; the resulting function, again called g, is then a holomorphic function on $\mathbf{D}_\tau$ with no zeros. Further, $|f(e^{it})| = |g(e^{it})|$ for $e^{it} \in \mathbf{T}$. Thus $\log|g(0)| = \frac{1}{2\pi}\int_{-\pi}^{\pi}\log|g(e^{it})|\,dt = \frac{1}{2\pi}\int_{-\pi}^{\pi}\log|f(e^{it})|\,dt$, by the first case. Since

$$\log|g(0)| = \log|f(0)| + \sum_{s \in S_f} k_f(s)\log|m_s(0)| - \sum_{\zeta \in Z_f} l_f(\zeta)\log|m_\zeta(0)|$$

$$= \log|f(0)| + \sum_{s \in S_f} k_f(s)\log|s| - \sum_{\zeta \in Z_f} l_f(\zeta)\log|\zeta|,$$

the result follows.   □

Corollary 26.1.2 *Suppose that $r > u > 0$ and that f is a meromorphic function on $\mathbf{D}_r = \{z : |z| < r\}$ which has no zeros or poles in the set $\mathbf{D}_r \setminus \mathbf{D}_u = \{z : u \le |z| < r\}$ or at 0. Then*

$$\log|f(0)| = \sum_{s \in S_f} k_f(s)\log\left|\frac{u}{s}\right| - \sum_{\zeta \in Z_f} l_f(\zeta)\log\left|\frac{u}{\zeta}\right| + \frac{1}{2\pi}\int_{-\pi}^{\pi}\log|f(ue^{it})|\,dt,$$

where $k_f(s)$ is the order of the pole at s and $l_f(\zeta)$ is the order of the zero at ζ.

Proof Apply the theorem to the function $f(z/u)$.   □

Suppose that f is an entire function. We set $n_f(t)$ to be the number of zeros, counted according to multiplicity, in $\mathbf{D}_t$. Thus $n(t)$ is a piecewise constant increasing function on $[0, \infty)$.

Theorem 26.1.3 *Suppose that f is an entire function and that $f(0) = 1$. If f has no zeros on $\mathbf{T}_u$ then*

$$\int_0^u \frac{n(t)}{t}\,dt = \frac{1}{2\pi}\int_{-\pi}^{\pi}\log|f(ue^{it})|\,dt.$$

Proof   Since

$$\int_0^u \frac{n(t)}{t}\,dt = \sum_{\zeta \in \mathbf{Z}_f \cap \mathbf{D}_u} \int_{|\zeta|}^u \frac{l_f(\zeta)}{t}\,dt = \sum_{\zeta \in \mathbf{Z}_f \cap \mathbf{D}_u} l_f(\zeta) \log \left| \frac{u}{\zeta} \right|,$$

this follows from Corollary 26.1.2.   □

26.2 The function $\pi \cot \pi z$

Note that the function $\pi \cot \pi z$ is a periodic meromorphic function of period 1, with singular set $\mathbf{Z}$, and with residue 1 at each pole.

Proposition 26.2.1   *Let $R_\alpha = \mathbf{C} \setminus \cup_{n=-\infty}^\infty N_\alpha(n)$, for $0 < \alpha < \frac{1}{2}$. Then the function $\pi \cot \pi z$ is bounded on R_α.*

Proof   By periodicity, it is enough to show that the function $\pi \cot \pi z$ is bounded on the set $S_\alpha = R_\alpha \cap \{x + iy : 0 \le x \le 1\}$. It is continuous on the compact set $K_\alpha = S_\alpha \cap \{x + iy : |y| \le 1\}$, and is therefore bounded on it. It is therefore sufficient to show that the function $\pi \cot \pi z$ is bounded on the set $L = \{x + iy : 0 \le x \le 1, |y| > 1\}$. If $z = x + iy \in L$, then

$$\pi \cot \pi z = i\pi \frac{e^{i\pi x}e^{-\pi y} + e^{-i\pi x}e^{\pi y}}{e^{i\pi x}e^{-\pi y} - e^{-i\pi x}e^{\pi y}},$$

so that, since $3e^{-2\pi} < 1$,

$$|\pi \cot \pi z| \le \pi \frac{e^{\pi|y|} + e^{-\pi|y|}}{e^{\pi|y|} - e^{-\pi|y|}} \le \pi \frac{1 + e^{-2\pi}}{1 - e^{-2\pi}} \le 2\pi.$$

□

Theorem 26.2.2   *If $z \in \mathbf{C} \setminus \mathbf{Z}$ then*

$$\pi \cot \pi z = \frac{1}{z} + 2 \sum_{j=1}^\infty \frac{z}{z^2 - j^2} = \lim_{k \to \infty} \left(\sum_{j=-k}^k \frac{1}{z - j} \right),$$

the sum and limit converging locally uniformly on $\mathbf{C} \setminus \mathbf{Z}$.

Proof   Note that neither of the series $\sum_{j=1}^\infty 1/(z - j)$ and $\sum_{j=1}^\infty 1/(z + j)$ converges.

   The sum

$$\frac{1}{z} + 2 \sum_{j=1}^\infty \frac{z}{z^2 - j^2}$$

converges locally uniformly on $\mathbf{C} \setminus \mathbf{Z}$ to a meromorphic function g, periodic with period 1, and with simple poles on $\mathbf{Z}$, with residue 1 at each point. Thus the function $\pi \cot \pi z - g(z)$ has removable singularities at the integers: removing the singularities, we obtain an entire function f. We show that g is bounded on R_α; by periodicity, it is enough to show that it is bounded on S_α. Since it is continuous, it is bounded on K_α, and it is therefore enough to show that it is bounded on L. If $z \in L$ then the real part of z^2 is negative, so that $|z^2 - n^2| \geq \max(|z|^2, n^2)$. Let k be the integral part of $|z|$. Then

$$\left| \sum_{j=1}^{k} \frac{z}{z^2 - j^2} \right| \leq \frac{k|z|}{|z|^2} \leq 1$$

and

$$\sum_{j=k+1}^{\infty} \left| \frac{z}{z^2 - j^2} \right| \leq |z| \sum_{j=k+1}^{\infty} \frac{1}{j(j-1)} = \frac{|z|}{k} \leq 2,$$

which gives the result.

Consequently, the function f is a bounded entire function, and is therefore constant, by Liouville's theorem. Finally, $\pi \cot \pi/2 = 0$ and

$$g(1/2) = \lim_{k \to \infty} \left(\sum_{j=-k}^{k} \frac{1}{\frac{1}{2} - j} \right) = \lim_{k \to \infty} \frac{1}{k + \frac{1}{2}} = 0,$$

so that $f = 0$.   □

We can use this theorem, together with the residue theorem, to calculate certain infinite sums.

Corollary 26.2.3 *Suppose that f is a meromorphic function with a finite singular set disjoint from $\mathbf{Z}$, for which*

$$N_R = \sup\{|zf(z)| : |z| = R\} \to 0 \text{ as } R \to \infty.$$

Let $g(z) = \pi f(z) \cot \pi z$. Then

$$\sum_{s \in S_f} \operatorname{res}_g(s) = -f(0) - \sum_{j=1}^{\infty} (f(j) + f(-j)).$$

Proof The function g has simple poles on $\mathbf{Z}$, and the residue at j is $f(j)$. Suppose that $k \in \mathbf{N}$ and that $k > \sup\{|z| : z \in S_f\}$. By the residue theorem,

$$\int_{\kappa_{k+1/2}} \pi f(z) \cot \pi z \, dz = 2\pi i \left(\sum_{s \in S_f} \operatorname{res}_g(s) + \sum_{j=-k}^{k} f(j) \right).$$

By Proposition 26.2.1, $M = \sup_{k \in \mathbf{N}} \left(\sup_{|z|=k+\frac{1}{2}} |\pi \cot \pi z| \right) < \infty$, and so

$$\left| \int_{\kappa_{k+1/2}(0)} \pi f(z) \cot \pi z \, dz \right| \le 2\pi M N_{k+1/2} \to 0$$

as $k \to \infty$, from which the result follows.   □

Example 26.2.4   If $0 < a < 1$ then

$$\sum_{j=-\infty}^{\infty} \frac{1}{(j-a)^2} = \left(\frac{\pi}{\sin \pi a} \right)^2.$$

Let $f(z) = 1/(z-a)^2$. Then $\pi \cot \pi z/(z-a)^2$ has a pole of order 2 at a, with residue $-(\pi/\sin \pi a)^2$, so that $\sum_{j=-\infty}^{\infty} 1/(j-a)^2 = (\pi/\sin \pi a)^2$.

In particular, putting $a = 1/2$, it follows that

$$\sum_{n=0}^{\infty} \frac{1}{(2n+1)^2} = \frac{1}{4} \sum_{n=0}^{\infty} \frac{1}{(n+\frac{1}{2})^2} = \frac{1}{8} \sum_{n=-\infty}^{\infty} \frac{1}{(n+\frac{1}{2})^2} = \pi^2/8.$$

Since

$$\sum_{n=1}^{\infty} \frac{1}{n^2} = \sum_{n=0}^{\infty} \frac{1}{(2n+1)^2} + \sum_{n=1}^{\infty} \frac{1}{(2n)^2} = \sum_{n=0}^{\infty} \frac{1}{(2n+1)^2} + \frac{1}{4} \sum_{n=1}^{\infty} \frac{1}{n^2},$$

it follows that $\sum_{n=1}^{\infty} 1/n^2 = \pi^2/6$. We can also obtain this result directly. The function $(\pi \cot \pi z)/z^2$ has a pole of order 3 at 0, and straightforward calculations show that the residue is $-\pi^2/3$. Thus we again find that $\sum_{n=1}^{\infty} 1/n^2 = \pi^2/6$.

26.3 The functions $\pi \operatorname{cosec} \pi z$

The function $g(z) = \sin \pi z$ is an entire function. It is periodic, with period 2. Its zero set is $\mathbf{Z}$ and $g'(z) = \pi \cos \pi z$, so that $g'(n) = (-1)^n \pi$, for $n \in \mathbf{Z}$. Thus the function $\pi \operatorname{cosec} \pi z = \pi/\sin \pi z$ is a meromorphic function on $\mathbf{C}$, with singular set $\mathbf{Z}$; the residue at n is $(-1)^n$.

Proposition 26.3.1   Let $z = x + iy$. If $k \in \mathbf{Z}$ and $x = k + \frac{1}{2}$, then $|\pi \operatorname{cosec} \pi z| \le 2\pi e^{-\pi|y|}$ and if $|y| \ge 1$ then $|\pi \operatorname{cosec} \pi z| \le 4\pi e^{-\pi|y|}$.

Proof   If $|x| = k + \frac{1}{2}$ then

$$|\pi \operatorname{cosec} \pi z| = \pi/\cosh y \le 2\pi e^{-\pi|y|}.$$

If $|y| \geq 1$ then

$$|\pi \cosec \pi z| = \left| \frac{2i\pi}{e^{i\pi x - \pi y} - e^{-i\pi x + \pi y}} \right| \leq \frac{2\pi}{e^{\pi |y|} - e^{-\pi |y|}} \leq 4\pi e^{-\pi |y|}.$$

$\square$

Theorem 26.3.2 *If $z \in \mathbf{C} \setminus \mathbf{Z}$ then*

$$\pi \cosec \pi z = \sum_{j=-\infty}^{\infty} \frac{(-1)^j}{z - j},$$

and the double series converges locally uniformly on $\mathbf{C} \setminus \mathbf{Z}$.

Proof First observe that

$$\sum_{j=1}^{2k} \frac{(-1)^j}{z - j} = \sum_{j=1}^{k} \left(\frac{1}{z - 2j} - \frac{1}{z - (2j - 1)} \right) = \sum_{j=1}^{k} \frac{1}{(z - 2j)(z - (2j - 1))},$$

and

$$\sum_{j=1}^{2k} \frac{(-1)^j}{z + j} = \sum_{j=1}^{k} \left(\frac{1}{z + 2j} - \frac{1}{z + (2j - 1)} \right)$$

$$= -\sum_{j=1}^{k} \frac{1}{(z + 2j)(z + (2j - 1))},$$

so that each of the series $\sum_{j=1}^{\infty}(-1)^j/(z - j)$ and $\sum_{j=1}^{\infty}(-1)^j/(z + j)$ converges locally uniformly on $\mathbf{C} \setminus \mathbf{Z}$, and so the double series converges locally uniformly on $\mathbf{C} \setminus \mathbf{Z}$.

Now $\cosec u = \cot u/2 - \cot u$. It therefore follows from Theorem 26.2.2 that

$$\pi \cosec \pi z = \left(\frac{2}{z} + 2\sum_{j=1}^{\infty} \frac{z/2}{(z/2)^2 - j^2} \right) - \left(\frac{1}{z} + 2\sum_{j=1}^{\infty} \frac{z}{z^2 - j^2} \right)$$

$$= \frac{1}{z} + 2 \left(\sum_{j=1}^{\infty} \frac{2z}{z^2 - (2j)^2} - \sum_{j=1}^{\infty} \frac{z}{z^2 - j^2} \right)$$

$$= \frac{1}{z} + 2\sum_{j=1}^{\infty} \frac{(-1)^j z}{z^2 - j^2} = \sum_{j=-\infty}^{\infty} \frac{(-1)^j}{z - j}.$$

$\square$

We can require weaker conditions on the decay of f when we consider infinite sums, using the function $\pi \operatorname{cosec} \pi z$ instead of $\pi \cot \pi z$.

Proposition 26.3.3   *Suppose that f is a meromorphic function with a finite singular set disjoint from $\mathbf{Z}$, for which*

$$M_R = \sup\{|f(z)| : |z| \ge R\} \to 0 \text{ as } R \to \infty.$$

Let $h(z) = \pi f(z) \operatorname{cosec} \pi z$. Then

$$\sum_{s \in S_f} \operatorname{res}_h(s) = -f(0) - \sum_{j=1}^{\infty} (-1)^j (f(j) + f(-j)).$$

Proof   The function h has simple poles on $\mathbf{Z}$, the residue at j being $(-1)^j f(j)$. Here it is convenient to consider square contours $\gamma_{k+1/2}$, with vertices at $(\pm 1 \pm i)(k + \frac{1}{2})$. Let $\nu_{k+1/2} = \sup\{|f(z)| : z \in \gamma_{k+1/2}\}$; then $\nu_{k+1/2} \to 0$ as $k \to \infty$. By the residue theorem,

$$\int_{\gamma_{k+1/2}} \pi f(z) \operatorname{cosec} \pi z \, dz = 2\pi i \left(\sum_{s \in S_f} \operatorname{res}_g(s) + \sum_{j=-k}^{k} (-1)^j f(j) \right).$$

Using Proposition 26.3.1, it follows that

$$\left| \int_{\gamma_{k+1/2}} h(z) \, dz \right| \le \nu_{k+1/2} \left(4 \int_0^{k+\frac{1}{2}} 4\pi e^{-\pi t} \, dt + 4\pi e^{-\pi(k+1/2)} (4k + 2) \right)$$

$$\le 32 \nu_{k+1/2} \to 0$$

as $k \to \infty$, and so the result follows.   □

Example 26.3.4   If $a \in \mathbf{R}$ and $a \ne 0$, then

$$\frac{\pi}{\sinh \pi a} = \frac{1}{a} + 2a \sum_{j=1}^{\infty} \frac{(-1)^j}{j^2 + a^2}.$$

Take $f(z) = 1/(z - ia)$. The residue of $h(z) = \pi f(z) \operatorname{cosec} \pi z$ at ia is $\pi / i \sinh ia$, so that

$$\frac{\pi}{i \sinh \pi a} = -\frac{1}{-ia} - \sum_{j=1}^{\infty} (-1)^j \left(\frac{1}{j - ia} + \frac{1}{-j - ia} \right).$$

Multiply by i, and simplify the summands.

Recall that the beta function B on $(0, \infty) \times (0, \infty)$ is defined as $B(x, y) = \int_0^1 t^{x-1}(1-t)^{y-1} \, dt$.

Corollary 26.3.5 *If $0 < x < 1$ then $B(x, 1-x) = \pi \operatorname{cosec} \pi x$.*

Proof   For each is equal to $\sum_{j=-\infty}^{\infty} \frac{(-1)^j}{x-j}$. (See Volume I, Section 10.3.) □

Exercises

26.3.1 Show that if $0 < a < 1$ then

$$\pi \operatorname{cosec} \pi a = \frac{1}{a} - \sum_{j=1}^{\infty} \frac{2(-1)^j a}{j^2 - a^2}.$$

Show that when $a = \frac{1}{2}$ then this formula reduces to the familiar formula

$$\frac{\pi}{4} = 1 - \frac{1}{3} + \frac{1}{5} - \frac{1}{7} + \cdots.$$

26.3.2 Show that if $0 < a < 1$ then

$$\sum_{n=-\infty}^{\infty} \frac{(-1)^n}{(n-a)^2} = \frac{\pi^2 \cos \pi a}{\sin^2 \pi a}.$$

26.3.3 Calculate the sum

$$1 - \frac{1}{3^2} - \frac{1}{5^2} + \frac{1}{7^2} + \frac{1}{9^2} - \cdots.$$

26.4 Infinite products

Suppose that F is a meromorphic function on $\mathbf{C}$, with nonzero poles $\{s_1, s_2, \ldots\}$ and zeros $\{\zeta_1, \zeta_2, \ldots\}$ listed in order of increasing modulus. Suppose that k_j is the order of the pole s_j and that l_j is the order of the zero ζ_j. Then, as in Section 23.5, the function $f(z) = F'(z)/F(z)$ is a meromorphic function on $\mathbf{C}$, with simple poles on $S_F \cup Z_F$, the residue at s_j being $-k_j$ and the residue at ζ_j being l_j. Again, let (r_n) be an increasing unbounded sequence of positive numbers for which $\mathbf{T}_{r_n} \cap (S_F \cup Z_F)$ is empty, and let $M_n = \sup\{|f(z)| : z \in \mathbf{T}_{r_n}\}$. Then there are finitely many poles and zeros of f inside $\mathbf{T}_{r_n}$: let them be $\{s_1, s_2, \ldots, s_{j_n}\} \cup \{\zeta_1, \zeta_2, \ldots, \zeta_{i_n}\}$.

Theorem 26.4.1   *Suppose that F is a meromorphic function on $\mathbf{C}$ with the properties described above, that $M_n \to 0$ as $n \to \infty$, and that $0 \notin S_F \cup Z_F$. Suppose that $w \in \mathbf{C} \setminus (S_F \cup Z_F)$. Then*

$$F(w) = F(0). \lim_{n\to\infty} \left(\prod_{i=1}^{i_n} \left(1 - \frac{w}{\zeta_i}\right)^{l_i} . \prod_{j=1}^{j_n} \left(1 - \frac{w}{s_j}\right)^{-k_j} \right).$$

The limit exists locally uniformly on $\mathbf{C} \setminus (S_F \cup Z_F)$.

Proof   Applying Theorem 23.5.6, we see that

$$\frac{F'(w)}{F(w)} = f(w) = - \lim_{n\to\infty} \left(\sum_{j=1}^{j_n} \frac{k_j}{w - s_j} - \sum_{i=1}^{i_n} \frac{l_i}{w - \zeta_i} \right),$$

the limit existing locally uniformly. Suppose that $z_0 \in \mathbf{C} \setminus (S_F \cup Z_F)$, that $K = M_\delta(z_0) \subseteq \mathbf{C} \setminus (S_F \cup Z_F)$ and that $w \in K$. Integrating along a rectifiable path in $\mathbf{C} \setminus (S_F \cup Z_F)$ from 0 to z_0, and in K from z_0 to w we see that

$$\log F(w) = \log F(0) - \lim_{n\to\infty} \left(\sum_{j=1}^{j_n} k_j \log_K \left(1 - \frac{w}{s_j}\right) - \sum_{i=1}^{i_n} l_i \log_K \left(1 - \frac{w}{\zeta_i}\right) \right),$$

where $\log_K$ is appropriately defined for $w \in K$, and that the convergence is uniform on K. Applying the exponential function, the result follows.   □

If the sequence $(M_n)_{n=1}^\infty$ is bounded, but not a null sequence, we must appeal to Theorem 23.5.7.

Theorem 26.4.2   *Suppose that F is a meromorphic function on $\mathbf{C}$ with the properties described above, that $(M_n)_{n=1}^\infty$ is a bounded sequence, and that $0 \notin S_F \cup Z_F$. Suppose that $w \in \mathbf{C} \setminus (S_F \cup Z_F)$. Then*

$$F(w) = F(0). \lim_{n\to\infty} \left(\prod_{i=1}^{i_n} \left(\left(1 - \frac{w}{\zeta_i}\right)^{l_i} e^{l_i w / \zeta_i} \right) . \prod_{j=1}^{j_n} \left(\left(1 - \frac{w}{s_j}\right)^{-k_j} e^{-k_j w / s_j} \right) \right).$$

The limit exists locally uniformly on $\mathbf{C} \setminus (S_F \cup Z_F)$.

Proof Using Theorem 23.5.7, and arguing as above,

$$\log F(w) - \log F(0)$$

$$= -\lim_{n \to \infty} \left(\sum_{j=1}^{j_n} k_j \left(\log_K \left(1 - \frac{w}{s_j} \right) + \frac{w}{s_j} \right) - \sum_{i=1}^{i_n} l_i \left(\log_K \left(1 - \frac{w}{\zeta_i} \right) + \frac{w}{\zeta_i} \right) \right),$$

and exponentiation again gives the result. □

Corollary 26.4.3 *If, in addition, F is an even function, with zeros $\{\zeta_1', \zeta_2', \ldots\}$ and poles $\{s_1', s_2', \ldots\}$ in the half space $H_r = \{z = x + iy : x > 0\}$ (listed in order of increasing modulus) then*

$$F(w) = F(0). \lim_{n \to \infty} \left(\prod_{i=1}^{i_n} \left(1 - \frac{w^2}{\zeta_i'^2} \right)^{l_i} \cdot \prod_{j=1}^{j_n} \left(1 - \frac{w^2}{s_j'^2} \right)^{-k_j} \right).$$

Proof Pair the zeros ζ_i' and $-\zeta_i'$, and the poles s_j' and $-s_j'$. □

Example 26.4.4 (Euler's product formula)

$$\sin \pi z = \pi z \prod_{j=1}^{\infty} \left(1 - \frac{z^2}{n^2} \right)$$

$$= \pi z \left(\prod_{j=1}^{\infty} \left(\left(1 - \frac{z}{n} \right) e^{z/n} \right) \right) \cdot \left(\prod_{j=1}^{\infty} \left(\left(1 + \frac{z}{n} \right) e^{-z/n} \right) \right),$$

and each of the products converges locally absolutely uniformly on $\mathbf{C} \setminus \mathbf{Z}$.

Proof Let $F(z) = (\sin \pi z)/\pi z$. Then $F'(z)/F(z) = \pi \cot \pi z - 1/z$, and so, taking $r_n = n + \frac{1}{2}$, the sequence $(M_n)_{n=1}^{\infty}$ is bounded. We can apply Theorem 26.4.2, and Corollary 26.4.3. Corollary 26.4.3 gives the first equation. Since $0 < 1 - (1 - w)e^w < w^2$ for $0 < w < 1$, it follows from Proposition 20.5.2 that each of the products

$$\prod_{j=1}^{\infty} \left(\left(1 - \frac{z}{n} \right) e^{z/n} \right) \text{ and } \prod_{j=1}^{\infty} \left(\left(1 + \frac{z}{n} \right) e^{-z/n} \right)$$

converges locally absolutely uniformly on $\mathbf{C} \setminus \mathbf{Z}$. Thus

$$\lim_{n\to\infty} \left(\prod_{i=1}^{n} \left(1 - \frac{z}{n}\right) e^{z/n} \right) \cdot \left(\prod_{i=1}^{n} \left(1 + \frac{z}{n}\right) e^{-z/n} \right)$$

$$= \left(\lim_{n\to\infty} \prod_{i=1}^{n} \left(1 - \frac{z}{n}\right) e^{z/n} \right) \cdot \left(\lim_{n\to\infty} \prod_{i=1}^{n} \left(1 + \frac{z}{n}\right) e^{-z/n} \right),$$

so that the second equation follows from Theorem 26.4.1. □

26.5 *Euler's product formula*

(This section can be omitted on a first reading.)

In Example 26.4.4, we established Euler's product formula for $\sin \pi z$. The proof depended in an essential way on the residue theorem. Euler established his formula long before Cauchy established the residue theorem, and it is of interest to prove Euler's theorem in a more elementary way. In Euler's time, rigorous analysis had not been developed, but we shall proceed accurately, making use of Weierstrass' uniform M test for products (Volume II, Corollary 14.2.10).

Let us set $\omega_n = e^{2\pi i/n}$, for $n \in \mathbf{N}$. Then $\omega_n^n = 1$ and the roots of the polynomial $X^n - 1$ are $1, \omega_n, \omega_n^2, \ldots, \omega_n^{n-1}$, so that

$$X^n - 1 = (X - 1) \prod_{j=1}^{n-1} (X - \omega_n^j).$$

Note that ω_n^{n-j} is the complex conjugate of ω_n^j, so that

$$(X - \omega_n^j)(X - \overline{\omega_n^j}) = X^2 - 2\cos(2\pi j/n)X + 1.$$

Thus if $n = 2k + 1$ is odd then the homogeneous polynomial $X^n - Y^n$ can be factorized as a product of real polynomials

$$X^n - Y^n = (X - Y) \prod_{j=1}^{k} (X^2 - 2\cos(2\pi j/n)XY + Y^2),$$

while if $n = 2k$ is even then we have the factorization

$$X^n - Y^n = (X - Y)(X + Y) \prod_{j=1}^{k-1} (X^2 - 2\cos(2\pi j/n)XY + Y^2).$$

These factorizations are very useful, and we use them to establish *Euler's product formula*.

Theorem 26.5.1 *If $z \in \mathbf{C}$ then*

$$\sin \pi z = \pi z \prod_{j=1}^{\infty} \left(1 - \frac{z^2}{j^2} \right),$$

and the product converges locally uniformly.

Proof   Suppose that $z \in \mathbf{C}$ and that $n = 2k + 1$ is an odd natural number greater than $|z|$. If log is the principal value of Log in the right half-plane, then

$$n \log \left(1 + \frac{z}{n} \right) - z = \frac{z^2}{n} \left(-\frac{1}{2} + \frac{z}{3n} - \frac{z^2}{4n^2} + \cdots \right),$$

so that

$$\left| n \log \left(1 + \frac{z}{n} \right) - z \right| \le \frac{|z|^2}{2n} \left(1 + \frac{|z|}{n} + \left(\frac{|z|}{n} \right)^2 + \cdots \right) = \frac{|z|^2}{2(n - |z|)}$$

and $(1 + z/n)^n \to e^z$ as $n \to \infty$. Thus if we set

$$\sin_n(z) = \frac{1}{2i} \left(\left(1 + \frac{iz}{n} \right)^n - \left(1 - \frac{iz}{n} \right)^n \right)$$

it follows that $\sin_n(z) \to \sin z$ as $n \to \infty$.

Since $(1 + iw)^2 + (1 - iw)^2 = 2(1 - w^2)$ and $(1 + iw)(1 - iw) = 1 + w^2$, applying the formula above we find that

$$(1 + iw)^n - (1 - iw)^n$$

$$= 2iw \prod_{j=1}^{k} \left(2(1 - w^2) - 2\cos(2\pi j/n)(1 + w^2) \right)$$

$$= 2iw \prod_{j=1}^{k} \left((2 - 2\cos(2\pi j/n)) - (2 + 2\cos(2\pi j/n))w^2 \right)$$

$$= 2iw A_n \prod_{j=1}^{k} \left(1 - \frac{1 + \cos(2\pi j/n)}{1 - \cos(2\pi j/n)} w^2 \right)$$

$$= 2iw A_n \prod_{j=1}^{k} \left(1 - w^2 \cot^2 \frac{\pi j}{n} \right),$$

where A_n is a constant. Comparing the coefficients of w on the two sides of the equation, we see that $A_n = n$.

Setting $w = \pi z/n = \pi z/(2k+1)$ we see that

$$\sin_{2k+1}(\pi z) = \pi z \prod_{j=1}^{k} \left(1 - \frac{\pi^2 z^2}{(2k+1)^2} \cot^2 \frac{\pi j}{2k+1} \right).$$

We now appeal to Weierstrass' uniform M test for products (Volume II, Corollary 14.2.10). Let $\overline{\mathbf{N}} = N \cup \{\infty\}$ be the one-point compactification of $\mathbf{N}$. If $k \in \overline{\mathbf{N}}$ let

$$f_j(k) = 0 \text{ for } k < j,$$

$$= \frac{\pi^2 z^2}{(2k+1)^2} \cot^2 \frac{\pi j}{2k+1} \text{ for } j \le k < +\infty,$$

$$= \frac{z^2}{j^2} \text{ for } k = +\infty.$$

Since $\theta \cot \theta \to 1$ as $\theta \to 0$, it follows that each f_j is continuous on $\overline{\mathbf{N}}$. Further, $\theta \cot \theta$ is a decreasing function on $(0, \pi/2)$ (verify this!), so that $\|f_j\|_\infty \le |z^2|/j^2$ and $\sum_{j=1}^{\infty} \|f_j\|_\infty < \infty$. Thus the conditions of Weierstrass' uniform M-test for products are satisfied, and so the product $\prod_{j=1}^{J}(1-f_j(k))$ converges uniformly to a continuous function g_z on $\overline{\mathbf{N}}$ as $J \to \infty$. But

$$\pi z g_z(k) = \pi z \prod_{j=1}^{\infty} (1 - f_j(k)) = \sin_{2k+1}(\pi z)$$

for $k \in \mathbf{N}$, and

$$\pi z g_z(\infty) = \pi z \prod_{j=1}^{\infty} \left(1 - \frac{z^2}{j^2} \right),$$

so that, since $g_z(k) \to g_z(\infty)$ as $k \to \infty$,

$$\sin \pi z = \lim_{k \to \infty} \sin_{2k+1}(\pi z) = \pi z \prod_{j=1}^{\infty} \left(1 - \frac{z^2}{j^2} \right).$$

Finally, the product converges locally uniformly, since if $|z| \le R$ then $|z^2/j^2| \le R^2/j^2$. $\qquad \square$

We can use this to give a proof of Theorem 26.2.2 which does not depend upon the residue theorem.

Corollary 26.5.2 *If $z \in \mathbf{C} \setminus \mathbf{Z}$ then*

$$\pi \cot \pi z = \frac{1}{z} + 2 \sum_{j=1}^{\infty} \frac{z}{z^2 - j^2} = \frac{1}{z} + \sum_{j=1}^{\infty} \left(\frac{1}{z-j} - \frac{1}{z+j} \right),$$

and the convergence is uniform on the compact subsets of $\mathbf{C} \setminus \mathbf{Z}$.

Proof First consider the case where $x \in (0,1)$, so that $0 < \sin \pi x \le 1$. Since the function log is continuous on $(0,1]$,

$$\log \sin \pi x = \log \pi x + \sum_{j=1}^{\infty} \log \left(1 - \frac{x^2}{j^2} \right).$$

Now

$$\frac{d}{dx} \log \left(1 - \frac{x^2}{j^2} \right) = \frac{2x}{x^2 - j^2},$$

and $\sum_{j=1}^{\infty} 2x/(x^2 - j^2)$ converges uniformly on compact subsets of $(0,1)$. We now appeal to Corollary 12.1.7 of Volume II. This implies that

$$\pi \cot \pi x = \frac{d}{dx} \log \sin \pi x = \frac{1}{x} + 2 \sum_{j=1}^{\infty} \frac{x}{x^2 - j^2} = \frac{1}{x} + 2 \sum_{j=1}^{\infty} \left(\frac{1}{x-j} - \frac{1}{x+j} \right),$$

and that the convergence is uniform on the compact subsets of $(0,1)$.

The series on the right also converges locally uniformly on $\mathbf{C} \setminus \mathbf{Z}$ to a holomorphic function f on $\mathbf{C} \setminus \mathbf{Z}$. Since $f(x) = \pi \cot \pi x$ for $x \in (0,1)$, it follows that $f(z) = \pi \cot \pi z$ for $z \in \mathbf{C} \setminus \mathbf{Z}$. □

Exercises

26.5.1 Suppose that $n = 2k$. Show that

$$X^n + Y^n = \prod_{j=1}^{k} \left(X^2 - 2\cos((2j-1)\pi/2k)XY + Y^2 \right).$$

Argue as in Theorem 26.5.1 to show that

$$\cos \pi z = \prod_{j=1}^{\infty} \left(1 - \frac{4z^2}{(2j-1)^2} \right).$$

26.5.2 Obtain the same result, by using the formula $\sin 2x = 2 \sin x \cos x$, and carefully using Euler's product formula for $\sin x$.

26.5.3 We define $\sec z = 1/\cos z$ for $z \neq (2n-1)\pi/2$. Show that if $z \neq w$ and $w \neq 0$ then $(1-z/w)^{-1} = 1 + z/(w-z)$. Use this to establish the following identities:

$$\pi \operatorname{cosec} \pi z = \frac{1}{z} \prod_{j=1}^{\infty} \left(1 + \frac{z^2}{j^2 - z^2}\right) \quad \text{for } z \in \mathbf{C} \setminus \mathbf{Z};$$

$$\pi \sec \pi z = \prod_{j=1}^{\infty} \left(1 + \frac{4z^2}{(2j-1)^2 - 4z^2}\right) \quad \text{for } z - \tfrac{1}{2} \in \mathbf{C} \setminus \mathbf{Z};$$

$$\tan \pi z = \pi z \prod_{j=1}^{\infty} \left(1 + \frac{(4j-1)z^2}{j^2((2j-1)^2 - 4z^2)}\right) \quad \text{for } z - \tfrac{1}{2} \in \mathbf{C} \setminus \mathbf{Z};$$

$$\cot \pi z = \frac{1}{z} \prod_{j=1}^{\infty} \left(1 - \frac{(4j-1)z^2}{(2j-1)^2(j^2 - z^2)}\right) \quad \text{for } z \in \mathbf{C} \setminus \mathbf{Z};$$

$$\frac{\sin \pi z}{\sin \pi w} = \frac{z}{w} \prod_{j=1}^{\infty} \left(1 + \frac{w^2 - z^2}{j^2 - w^2}\right) \quad \text{for } w \in \mathbf{C} \setminus \mathbf{Z}.$$

Show that the products for $\pi \operatorname{cosec} \pi z$ and $\cot \pi z$ converge uniformly on the compact subsets of $\mathbf{C} \setminus \mathbf{Z}$ and that the products for $\pi \sec \pi z$ and $\tan \pi z$ converge uniformly on the compact subsets of $\mathbf{C} \setminus (\mathbf{Z} + 1/2)$.

26.5.4 Show that $(d/dx) \log \tan x = 2/\sin 2x$, for $x \notin \pi \mathbf{Z}$. Establish the following identities:

$$\tan \pi z = \sum_{j=1}^{\infty} \frac{8z^2}{(2j-1)^2 - 4z^2}$$

$$= \sum_{j=1}^{\infty} \left(\frac{1}{j - \tfrac{1}{2} - x} - \frac{1}{j - \tfrac{1}{2} + x}\right) \quad \text{for } z - 1/2 \in \mathbf{C} \setminus \mathbf{Z};$$

$$\operatorname{cosec} \pi z = \frac{1}{z} + 2 \sum_{j=1}^{\infty} (-1)^j \left(\frac{z}{z^2 - j^2}\right)$$

$$= \frac{1}{z} + \sum_{j=1}^{\infty} (-1)^j \left(\frac{1}{z+j} + \frac{1}{z-j}\right) \quad \text{for } z \in \mathbf{C} \setminus \mathbf{Z}.$$

26.6 Weierstrass products

Suppose that $\zeta_1, \ldots, \zeta_N$ are distinct non-zero complex numbers, and that $l_1, \ldots, l_N$ are natural numbers. Then the polynomial function

$$p(z) = \prod_{j=1}^{N}(1 - \frac{z}{\zeta_j})^{l_j}$$

has zeros at $\zeta_1, \ldots, \zeta_N$, with multiplicities $l_1, \ldots, l_N$. If, further, $l_0 \in \mathbf{N}$ then $z^{l_0}p(z)$ also has a zero at 0, with multiplicity l_0. (The fact that we have to consider 0 separately is a rather trivial nuisance, but is one that will recur.)

Suppose that U is a domain and that Z is an infinite discrete subspace of $U \setminus \{0\}$. We can write $Z = \{\zeta_1, \zeta_2, \ldots\}$ where the terms are distinct, and are arranged in order of increasing modulus. Suppose that $(l_j)_{j=1}^{\infty}$ is a sequence in $\mathbf{N}$. Can we find a holomorphic function f on U with zero set Z_f equal to Z, and with the multiplicity of the zero at each ζ_j equal to l_j? A first attempt might be to try $f(z) = \prod_{j=1}^{\infty}\left(1 - \frac{z}{\zeta_j}\right)^{l_j}$; but as Euler's product formula shows, the product need not converge. On the other hand, the inclusion of an exponential term in each factor of Euler's product formula produced a product which converges locally uniformly. Weierstrass showed that if suitable exponential terms are included in each factor, then a locally uniformly convergent product results.

First, let us describe the exponential terms that we shall need. We introduce several entire functions. Suppose that $n \in \mathbf{Z}^+$ and that $w \in \mathbf{C}$. Let

$$\lambda_0(w) = 0,$$
$$\lambda_n(w) = w + w^2/2 + \cdots + w^n/n, \text{ for } n > 0,$$
$$d_n(w) = e^{\lambda_n(w)},$$
$$E_n(w) = (1 - w)d_n(w),$$
$$g_n(w) = 1 - E_n(w).$$

Note that if $|w| < 1$ then $\lambda_n(w) \to -\log(1-w)$ as $n \to \infty$, so that $d_n(w) \to 1/(1-w)$, $E_n(w) \to 1$ and $g_n(w) \to 0$ as $n \to \infty$.

The entire functions E_n are called *elementary factors*. Suppose that U is a domain and that m is a Möbius function on U which does not have a singularity in U. The holomorphic function $E_n \circ m$ on U is called a *Weierstrass factor*, and a product of Weierstrass factors which converges locally uniformly on a domain is called a *Weierstrass product*, as is the holomorphic function which it defines. We shall answer the question above by constructing functions which are Weierstrass products.

We need to know how quickly $g_n(w)$ converges to 0 as $n \to \infty$.

Proposition 26.6.1   *If $n \in \mathbf{N}$ and $|w| \leq 1$ then $|g_n(w)| \leq |w|^{n+1}$.*

Proof   Each of the functions described above is an entire function. Since $d_n(0) = 1$, d_n has a Taylor series expansion $d_n(w) = 1 + \sum_{j=1}^{\infty} a_j w^j / j!$. Since all the coefficients of the Taylor expansion of e^w and all the coefficients in the definition of $\lambda_n(w)$ are positive, it follows that $a_j > 0$ for $j \in \mathbf{N}$. Since $E_n(0) = 1$, E_n has a Taylor series expansion $E_n(w) = 1 + \sum_{j=1}^{\infty} b_j w^j / j!$. Let us consider the derivative of E_n:

$$
\begin{aligned}
E_n'(w) &= -d_n(w) + (1-w)d_n'(w) \\
&= -d_n(w) + (1-w)\lambda_n'(w)d_n(w) \\
&= -w^n d_n(w).
\end{aligned}
$$

We draw two conclusions from this. First, $b_j = 0$ for $1 \leq j \leq n$. Secondly, $b_j < 0$ for $j > n+1$. Thus

$$
0 = E_n(1) = 1 + \sum_{j=n+1}^{\infty} \frac{b_j}{j!} = 1 - \sum_{j=n+1}^{\infty} \frac{|b_j|}{j!},
$$

so that $\sum_{j=n+1}^{\infty} |b_j|/j! = 1$. Consequently, if $|w| \leq 1$ then

$$
|g_n(w)| = |1 - E_n(w)| = |w|^{n+1} \left| \sum_{j=n+1}^{\infty} \frac{b_j w^{j-(n+1)}}{j!} \right|
$$

$$
\leq |w|^{n+1} \sum_{j=n+1}^{\infty} \frac{|b_j|}{j!} = |w|^{n+1}.
$$

$\square$

We begin with the simplest case, when $U = \mathbf{C}$.

Theorem 26.6.2   *Suppose that Z is an infinite closed discrete subspace of $\mathbf{C}$, that $0 \notin Z$ and that $(l_j)_{j=1}^{\infty}$ is a sequence of natural numbers. Let $Z = \{\zeta_1, \zeta_2, \ldots\}$, where the terms are listed in order of increasing modulus. Write $Z = \{\eta_1, \eta_2, \ldots\}$, where each ζ_j is repeated l_j times, and the terms are listed in order of increasing modulus. If $(p_n)_{n=1}^{\infty}$ is a sequence in $\mathbf{N}$ for which*

$$
\sum_{n=1}^{\infty} \left(\frac{r}{|\eta_n|} \right)^{p_n+1} < \infty \text{ for all } r > 0,
$$

then the product

$$\prod_{n=1}^{\infty} E_{p_n}\left(\frac{w}{\eta_n}\right) = \lim_{N\to\infty} \prod_{n=1}^{N} E_{p_n}\left(\frac{w}{\eta_n}\right)$$

converges locally uniformly to an entire function f on $\mathbf{C}$, for which $Z_f = Z$ and the zero at ζ_j has multiplicity l_j, for $j \in \mathbf{N}$.

Proof   We begin with two remarks. First, $|\eta_j| \to \infty$ as $j \to \infty$, and so the condition holds if we take $p_n = n$ for $n \in \mathbf{N}$. But it is desirable to take p_n small; for example, if $Z = \mathbf{Z} \setminus \{0\}$, the we can take $p_n = 1$ for all n, as in Euler's product formula. Secondly, the infinite product does not converge in the strict sense of infinite products, since there are terms which are zero at points of Z.

We show that the product converges locally uniformly. Suppose that K is a compact subset of $\mathbf{C}$. Let $r = \sup\{|z| : z \in K\}$. There exists n_0 such that $|\eta_n| > r$ for $n \geq n_0$. If $w \in K$ and $n \geq n_0$ then

$$\left|g_{p_n}\left(\frac{w}{\eta_n}\right)\right| \leq \left(\frac{r}{|\eta_n|}\right)^{p_n+1} \quad \text{for } w \in K,$$

so that the sum $\sum_{n=n_0}^{\infty} g_{p_n}(w/\eta_n)$ converges uniformly on K. It therefore follows from Proposition 20.5.2 that the infinite product

$$\prod_{n=n_0}^{\infty}\left(1 - g_{p_n}\left(\frac{w}{\eta_n}\right)\right) = \prod_{n=n_0}^{\infty} E_{p_n}\left(\frac{w}{\eta_n}\right)$$

converges uniformly on K to a continuous function, not taking the value 0. Consequently $\prod_{n=1}^{\infty} E_{p_n}(w/\eta_n)$ tends locally uniformly to an entire f on $\mathbf{C}$, $Z_f = Z$, and each zero ζ_j has multiplicity l_j, for $j \in \mathbf{N}$.   □

We can easily deal with the case where $0 \in Z$: let $h(z) = z^{l_0} f(z)$. Then h also has a zero, with multiplicity l_0, at 0.

Example 26.6.3   The product

$$W_1(z) = \prod_{n=1}^{\infty} E_1\left(\frac{z}{n}\right) = \prod_{n=1}^{\infty}\left(\left(1 - \frac{z}{n}\right)e^{-z/n}\right).$$

The product converges, since $\sum_{n=1}^{\infty}(1/n^2) < \infty$, and so W_1 is an entire function with zero set $\mathbf{N}$. Each of these zeros is a simple zero. Euler's product formula can then be written as $\sin \pi z = \pi z W_1(z) W_1(-z)$. We shall consider this function further in the next section.

Next we consider the case where U is a proper subset of $\mathbf{C}$ and Z is bounded. The idea of the proof is the same, but the details are rather more complicated.

Theorem 26.6.4 *Suppose that U is a domain which is a proper subset of $\mathbf{C}$, that Z is an infinite bounded closed discrete subspace of U and that $(l_j)_{j=1}^{\infty}$ is a sequence of natural numbers. Let $Z = \{\zeta_1, \zeta_2, \ldots\}$, where the terms are distinct. Then $d(\zeta_j, \partial U) \to 0$ as $j \to \infty$. Write $Z = \{\eta_1, \eta_2, \ldots\}$, where each ζ_j is repeated l_j times, and the terms are listed so that $(d(\eta_n, \partial U))_{n=1}^{\infty}$ is a decreasing sequence. The sequence $(d(\eta_n, \partial U))_{n=1}^{\infty}$ is a null sequence. For each n, there exists $\delta_n \in \partial U$ such that $|\eta_n - \delta_n| = d(\eta_n, \partial U)$. If $(p_n)_{n=1}^{\infty}$ is a sequence in $\mathbf{N}$ for which*

$$\sum_{n=1}^{\infty} \left(\frac{d(\eta_n, \partial U)}{r} \right)^{p_n + 1} < \infty \text{ for all } r > 0,$$

then the product

$$\prod_{n=1}^{\infty} E_{p_n} \left(\frac{\eta_n - \delta_n}{w - \delta_n} \right)$$

converges locally uniformly to a holomorphic function $f(w)$ on U, for which $Z_f = Z$ and the zero at ζ_j has multiplicity l_j, for $j \in \mathbf{N}$.

Proof If $r > 0$ then the set $\{\zeta \in Z : d(\zeta, \partial U) \geq r\}$ is a bounded closed subset of U, and is therefore finite. Thus $d(\eta_n, \partial U) \to 0$ as $n \to \infty$. If $\eta_n \in Z$ then $\{\delta \in \partial U : |\eta_n - \delta| \leq 2d(\eta_n, \partial U)\}$ is a compact set, so that there exists $\delta_n \in \partial U$ for which $|\eta_n - \delta_n| = d(\eta_n, \partial U)$. Suppose that K is a compact subset of U. Let $r = \inf\{d(w, \partial U) : w \in K\}$. Then $r > 0$, and so $|(\eta_n - \delta_n)/(w - \delta_n)| \leq d(\eta_n, \partial U)/r$, for $w \in K$. Thus

$$\sum_{n=1}^{\infty} \left| g_{p_n} \left(\frac{\eta_n - \delta_n}{w - \delta_n} \right) \right| \leq \sum_{n=1}^{\infty} \left(\frac{d(\eta_n, \partial U)}{r} \right)^{p_n + 1} < \infty,$$

so that the product

$$\prod_{n=1}^{\infty} E_{p_n} \left(\frac{\eta_n - \delta_n}{w - \delta_n} \right)$$

converges locally uniformly to a holomorphic function $f(w)$ on U, for which $Z_f = Z$ and the zero at ζ_j has multiplicity l_j, for $j \in \mathbf{N}$. $\square$

Example 26.6.5 *Blaschke products.*

Theorem 26.6.4 applies when U is a bounded domain, and in particular, it applies when $U = \mathbf{D}$. For example, suppose that $(\zeta_n)_{n=1}^{\infty}$ is a sequence of distinct non-zero elements in $\mathbf{D}$ and $(l_n)_{n=1}^{\infty}$ is a sequence in $\mathbf{N}$ for which $\sum_{n=1}^{\infty} l_n(1 - |\zeta_n|) < \infty$. Then, writing $\zeta_n = r_n e^{i\theta_n}$, the function $f(w) = \prod_{n=1}^{\infty}((w - \zeta_n)/(w - e^{i\theta_n}))^{l_n}$ is a holomorphic function f on $\mathbf{D}$ for which $Z_f = Z$ and the zero at ζ_j has multiplicity l_j, for $j \in \mathbf{N}$. This function has some unfortunate features, since $|(w - \zeta_n)/(w - e^{i\theta_n})| \to \infty$ as $w \to e^{i\theta_n}$.

Let us replace each δ_n by $\gamma_n = 1/\bar{\zeta}_n$. Then $\sum_{n=1}^{\infty} l_n|\zeta_n - \gamma_n| < \infty$, and so the product

$$\prod_{n=1}^{\infty} \left(1 - \frac{\zeta_n - \gamma_n}{w - \gamma_n}\right)^{l_n} = \prod_{n=1}^{\infty} \left(\frac{\zeta_n - w}{\gamma_n - w}\right)^{l_n}$$

$$= \prod_{n=1}^{\infty} (\bar{\zeta}_n)^{l_n} \left(\frac{\zeta_n - w}{1 - \bar{\zeta}_n w}\right)^{l_n}$$

also converges locally uniformly to a holomorphic function $f(w)$ on $\mathbf{D}$ for which $Z_f = Z$ and the zero at ζ_j has multiplicity l_j, for $j \in \mathbf{N}$. But $\prod_{n=1}^{\infty} (1/|\zeta_n|)^{l_n}$ also converges, and so the product

$$\prod_{n=1}^{\infty} \left(\frac{|\zeta_n|}{\zeta_n}\right)^{l_n} \cdot \left(\frac{\zeta_n - w}{1 - \bar{\zeta}_n w}\right)^{l_n}$$

converges to a function $B(w)$, with the same properties. This function is called a *Blaschke product*. Each function $(\zeta_n - w)/(1 - \bar{\zeta}_n w)$ is a Möbius function which is an automorphism of $\mathbf{D}$ and which is a homeomorphism of $\overline{\mathbf{D}}$. Consequently $|B(w)| < 1$ for $w \in \mathbf{D}$. As we shall see in Part Six, $B(w)$ also behaves well as w approaches the boundary $\mathbf{T}$ of $\mathbf{D}$.

We now return to our original problem, and consider the general case.

Theorem 26.6.6 *Suppose that U is a domain which is a proper subset of $\mathbf{C}$, that Z is an infinite closed discrete subspace of U and that $(l_j)_{j=1}^{\infty}$ is a sequence of natural numbers. Let $Z = \{\zeta_1, \zeta_2, \ldots\}$, where the terms are distinct, and let $Z = \{\eta_1, \eta_2, \ldots\}$, where each ζ_j is repeated l_j times. Then there exists a sequence $(m_n)_{n=1}^{\infty}$ of Möbius functions such that $\prod_{n=1}^{\infty} E_n(m_n(w))$ converges locally uniformly to a holomorphic function f on U, for which $Z_f = Z$ and the zero at ζ_j has multiplicity l_j, for $j \in \mathbf{N}$.*

Proof Note that if Z is bounded, then the result follows from Theorem 26.6.4. For then, with the notation of Theorem 26.6.4, $|\eta_n - \delta_n| \to 0$, so that

$$\sum_{n=1}^{\infty} \left|g_n\left(\frac{\eta_n - \delta_n}{w - \delta_n}\right)\right| \leq \sum_{n=1}^{\infty} \left|\left(\frac{\eta_n - \delta_n}{w - \delta_n}\right)^{n+1}\right| < \infty;$$

consequently, the product converges locally uniformly.

Suppose first that Z is bounded, and that U is unbounded. We retain the notation of Theorem 26.6.4. We show that $f(w) \to 1$ as $w \to \infty$. Let $R = \mathrm{Sup}\{|\zeta| : \zeta \in Z\}$. Then $|z| \leq R$ for all $z \in \overline{Z}$, and so there exists $z \in \partial U$ with $|z| \leq R$. Consequently $|\eta_n - \delta_n| \leq 2R$ for all $n \in \mathbf{N}$. Suppose that $0 < \epsilon < \frac{1}{4}$. There exists $S > 0$ such that if $|w| > S$ then $2R/(|w| - R) < \epsilon$. For such w, $|\eta_n - \delta_n|/|w - \delta_n| < \epsilon$ for all n, so that

$$\sum_{n=1}^{\infty} g_n \left(\frac{|\eta_n - \delta_n|}{|w - \delta_n|} \right) \leq \sum_{n=1}^{\infty} \epsilon^{n+1} < \epsilon.$$

Consequently, $|f(w) - 1| < 2\epsilon$, by Proposition 20.5.2.

Now return to the general case. Let w_0 be an element of U. There exists $r > 0$ such that $M_r(w_0) \subseteq U$. Let $T(w) = r/(w - w_0)$, for $w \in U \setminus \{w_0\}$. Then T maps $U \setminus \{w_0\}$ conformally onto $V = T(U \setminus \{w_0\})$ and $T(Z) \subseteq \mathbf{D}$. Thus there exists a Weierstrass product which converges locally uniformly on V to a holomorphic function f, for which $Z_h = T(Z)$, and the zeros have the appropriate multiplicity. Let $f(z) = h(T(z))$ for $z \in U \setminus \{w_0\}$. Then f has a removable singularity at w_0; setting $f(w_0) = 1$, we obtain a holomorphic function on U with the required properties.   □

Theorem 26.6.7 (The Weierstrass factorization theorem)   *Suppose that f is a non-constant holomorphic function on a simply connected domain U. There exist a holomorphic function h and a Weierstrass product w on U such that $f = e^h w$.*

Proof   Let Z_f be the zero set of f. There exists a Weierstrass product w on U with zero set Z_f, and with zeros with the same multiplicity as the zeros of f. Thus the function f/w has removable singularities at the points of Z_f. Let g be the function obtained by removing the singularities. Then g is a holomorphic function on U with no zeros. By Theorem 22.7.1, there exists a continuous branch of $\log g$ on U. Let $h = \log g$. Then $f = e^h w$.   □

In the case where f is an entire function for which $f(0) \neq 0$, with zero set $\{\zeta_1, \zeta_2, \dots\}$, we can take w to be $\prod_{n=1}^{\infty} (E_{p_n}(z/\zeta_n))^{l_n}$, where l_n is the multiplicity of the zero ζ_n, and the sequence $(p_n)_{n=1}^{\infty}$ is chosen in such a way that the Weierstrass product converges locally uniformly. If $f(0) = 0$, we must include a factor z^{l_0}, where l_0 is the multiplicity of the zero at 0.

We can use these results to construct meromorphic functions with given zeros and poles. If S is a closed discrete subspace of a domain U and k is a mapping from S to $\mathbf{N}$, we can construct a holomorphic function h on U with zero set S, where the zero at $s \in S$ has multiplicity $k(s)$: then $1/h$ is a meromorphic function on U with no zeros, and with singular set S, the pole at $s \in S$ having order $k(s)$. If Z is a closed discrete subspace of U disjoint from S and l is a mapping from Z to $\mathbf{N}$, we can construct a holomorphic function g on U with zero set Z, where the zero at $\zeta \in Z$ has multiplicity $l(\zeta)$: then $f = g/h$ is a meromorphic function on U with given zeros and poles.

In fact, meromorphic functions can be constructed with more strongly prescribed properties at the poles.

Theorem 26.6.8 (The Mittag–Leffler theorem) *Suppose that S is a closed discrete subspace of a domain U and that p is a mapping from S into the space of complex polynomials of positive degree. Then there exists a meromorphic function f on U such that the principal part of f at s is $p_s(1/(z - s))$.*

Proof We shall only prove this in the case where $U = \mathbf{C}$ or $\mathbf{D}$: the proof in the general case requires results about the approximation of holomorphic functions by rational functions. The proof involves sums, rather than products.

First we consider the case where $0 \notin S$. Let $(r_n)_{n=1}^{\infty}$ be a strictly increasing sequence such that $\inf_{s \in S} |s| > r_1$, and such that $r_n \to \infty$ (if $U = \mathbf{C}$) or $r_n \to 1$ (if $U = \mathbf{D}$) as $n \to \infty$. Let $\mathbf{D}_n = \{z : |z| \leq r_n\}$, let $A_n = \mathbf{D}_{n+1} \setminus \mathbf{D}_n$ and let $S_n = S \cap A_n$, for $n \in \mathbf{N}$. The function $f_n(s) = \sum_{s \in S_n} p_s(1/(z - s))$ has poles in A_n with the correct principal parts, and is holomorphic in a neighbourhood of $\mathbf{D}_n$. Thus the Taylor series expansion of f_n about 0 has radius of convergence greater than r_n. It follows, by taking sufficiently many terms, that there is a polynomial g_n such that $\sup_{z \in \mathbf{D}_n} |f_n(z) - g_n(z)| < 1/2^n$. But then the function $h_n = f_n - g_n$ has the correct principal parts in A_n, and the series $\sum_{n=1}^{\infty} h_n$ converges locally uniformly on $U \setminus S$ to a meromorphic function with the required properties.

If $0 \in S$, we simply add $p_0(1/z)$ to the function obtained for the set $S \setminus \{0\}$. $\qquad\square$

Corollary 26.6.9 *Suppose that f is a meromorphic function on a domain U, and that S_f is the disjoint union of A and B. Then there exist meromorphic functions g and h such that $f = g + h$, $S_g = A$ and $S_h = B$.*

Proof By the theorem, there exists g with $S_g = A$ such that $f - g$ has removable singularities at the points of A. Remove them, and set $h = f - g$. $\qquad\square$

Corollary 26.6.10 *Suppose that f and g are holomorphic functions on a domain U, and that $Z_f \cap Z_g = \emptyset$. Then there exist holomorphic functions h and k on U such that $hf + kg = 1$.*

Proof The function $1/fg$ is meromorphic on U, with singular set $Z_f \cup Z_g$. By the preceding corollary, we can write $1/fg = a + b$, with $S_a = Z_f$ and $S_b = Z_g$. Let $k = af$. If $\zeta \in Z_f$, then $1/g$ and af are both holomorphic in a neighbourhood of ζ, and so $k = 1/g - af$ is holomorphic in a neighbourhood of ζ. Since it is holomorphic elsewhere, k is a holomorphic function on U. Similarly, $h = bg$ is a holomorphic function on U. Finally, $1 = fgb + fga = hf + kg$. $\qquad\square$

Exercises

26.6.1 Suppose that U is a domain other than $\mathbf{C}$. Show that there is a closed discrete subspace Z of U such that $\overline{Z} = Z \cup \partial U$. Construct a holomorphic function f on U with the property that if V is a domain which contains U as a proper subset, then f cannot be extended to a holomorphic function on V.

26.6.2 Construct a holomorphic function B on $\mathbf{D}$ with the property that for each $z \in \mathbf{T}$ there exist sequences $(z_n)_{n=1}^{\infty}$ and $(w_n)_{n=1}^{\infty}$ in $\mathbf{D}$ such that $z_n \to z$, $w_n \to z$, $B(z_n) \to 0$ and $B(w_n) \to 1$ as $n \to \infty$.

26.7 The gamma function revisited

In Volume I, Section 10.5, we established properties of the gamma function, considered as a function of a real variable. Here we consider it as a function of a complex variable.

If $z = x + iy$ and $t > 0$, then $|t^{z-1}e^{-t}| = t^{x-1}e^{-t}$, so that the integral

$$\lim_{\epsilon \to 0, R \to \infty} \left(\int_{\epsilon}^{R} t^{z-1}e^{-t}\, dt \right)$$

converges locally uniformly on the right half-plane $H_r = \{z = x + iy : x > 0\}$ to a holomorphic function Γ on H_r. We can however extend Γ further. We split the defining integral into two. Let

$$\Gamma_0(z) = \lim_{\epsilon \to 0} \left(\int_{\epsilon}^{1} t^{z-1}e^{-t}\, dt \right)$$

$$\text{and } \Gamma_1(z) = \lim_{R \to \infty} \left(\int_{1}^{R} t^{z-1}e^{-t}\, dt \right).$$

The integral for Γ_1 converges locally uniformly on $\mathbf{C}$, and so Γ_1 is an entire function.

Suppose that $z = x + iy$, with $x > 1$. Since

$$e^{-t} = \sum_{n=0}^{\infty}(-1)^n t^n / n!,$$

and since the series converges uniformly on $[0,1]$,

$$\Gamma_0(z) = \lim_{\epsilon \to 0}\left(\sum_{n=0}^{\infty}(-1)^n \left(\int_{\epsilon}^{1} t^{z-1}\frac{t^n}{n!}\, dt\right)\right)$$

$$= \sum_{n=0}^{\infty}(-1)^n \lim_{\epsilon \to 0}\left(\int_{\epsilon}^{1} \frac{t^{z+n-1}}{n!}\, dt\right) = \sum_{n=0}^{\infty}\frac{(-1)^n}{n!(z+n)}.$$

Now this series converges locally uniformly on $\mathbf{C} \setminus \{0, -1, -2, \dots\}$, and so it defines a holomorphic function on $\mathbf{C} \setminus \{0, -1, -2, \dots\}$; we again denote this function by Γ_0. Further, omitting the term $(-1)^n / n!(z+n)$, we see that Γ_0 has a simple pole at $-n$, with residue $(-1)^n / n!$. Thus, if we set $\Gamma(z) = \Gamma_0(z) + \Gamma_1(z)$ for $z \in \mathbf{C} \setminus \{0, -1, -2, \dots\}$, we obtain a meromorphic function on $\mathbf{C}$.

Proposition 26.7.1 *If $z \in \mathbf{C} \setminus \{0, -1, -2, \dots\}$ then $\Gamma(z+1) = z\Gamma(z)$.*

Proof Let $f(z) = \Gamma(z+1) - z\Gamma(z)$ for $z \in \mathbf{C} \setminus \{0, -1, -2, \dots\}$. Then f is a holomorphic function. By Proposition 10.5.1 of Volume I, $f(x) = 0$ for $x \in (0, \infty)$, and so the zeros of f are not isolated. Thus $f = 0$. □

Proposition 26.7.2 *If $z \in \mathbf{C} \setminus \mathbf{Z}$ then $\Gamma(z)\Gamma(1-z) = \pi\operatorname{cosec}\pi z$.*

Proof Proposition 10.5.4 of Volume I stated that if x and y are real and positive then $\Gamma(x)\Gamma(y) = B(x, y)\Gamma(x+y)$; in particular, if $x \in (0,1)$ then

$$\Gamma(x)\Gamma(1-x) = B(x, 1-x) = \pi\operatorname{cosec}\pi x,$$

by Corollary 26.3.5. Thus $\Gamma(z)\Gamma(1-z) - \pi\operatorname{cosec}\pi z$ is a holomorphic function on $\mathbf{C} \setminus \mathbf{Z}$ which vanishes on $(0,1)$, and is therefore zero. □

Corollary 26.7.3 *The function $\Gamma(z)$ has no zeros in $\mathbf{C} \setminus \{0, -1, -2, \dots\}$.*

Proof If $z \in \mathbf{C} \setminus \mathbf{Z}$ and $\Gamma(z) = 0$ then $\pi\operatorname{cosec}\pi z = 0$; but $\pi\operatorname{cosec}\pi z$ has no zeros. If $z \in \mathbf{N}$, then $\Gamma(z) = (z-1)! \neq 0$. □

Corollary 26.7.4 $\Gamma(\frac{1}{2}) = \sqrt{\pi}$.

Proof   For $\pi \operatorname{cosec} \pi/2 = \pi$.   □

We can see this another way. Setting $t = s^2/2$,

$$\Gamma(\tfrac{1}{2}) = \int_0^\infty t^{-1/2} e^{-t}\, dt = \sqrt{2} \int_0^\infty e^{-s^2/2}\, ds = \sqrt{\pi}.$$

We can also extend the beta function. Let $B_w(z) = \Gamma(z)\Gamma(w)/\Gamma(z+w)$ for $w \notin \{0, -1, -2, \ldots\}$ and $z \notin \{0, -1, -2, \ldots\} \cup \{-w, -1-w, -2-w, \ldots\}$. Then B_w is a meromorphic function of z. If $n \in \mathbf{Z}^+$ then B_w has a simple pole at $-n$, with residue $(-1)^n \Gamma(w)/n!\Gamma(w-n)$. On the other hand, the function $\Gamma_w(z) = \Gamma(z+w)$ has a simple pole at $-n-w$, and so B_w has a removable singularity at $-n-w$. Thus B_w can be extended to be a meromorphic function on $\mathbf{C} \setminus \{0, -1, -2, \ldots\}$ and B_w then has zero set $\{-n-w : n \in \mathbf{Z}^+\}$. We therefore define $B(z, w)$ to be $B_w(z)$ for $z, w \in \mathbf{C} \setminus \{0, -1, -2, \ldots\}$; if z and w are real and positive, this agrees with the previous definition of the beta function.

The function $1/\Gamma$ is an entire function, with simple zeros at $0, -1, -2, \ldots,$ and we can apply the Weierstrass factorization theorem to it. What is the result?

Theorem 26.7.5   *(i) Let*

$$L(z) = \frac{e^{-\gamma z}}{z W_1(-z)} = \frac{e^{-\gamma z}}{z \prod_{n=1}^\infty ((1 + z/n)e^{-z/n})},$$

where γ is Euler's constant. Then $L = \Gamma$.

(ii) Let $L_n(z) = (n-1)! n^z / z(z+1) \ldots (z+n-1)$. Then $L_n(z) \to \Gamma(z)$ locally uniformly on the domain $U = \mathbf{C} \setminus \{0, -1, -2, \ldots\}$.

Proof   First we show that $L_n(z) \to L(z)$ locally uniformly on U. Now

$$L_n(z) = \frac{n^z}{z(1+z)\ldots(1+\frac{z}{n-1})} = \frac{e^{z(\log n - (1 + \frac{1}{2} + \cdots \frac{1}{n-1}))}}{z E_1(-z) E_1(\frac{-z}{2}) \ldots E_1(-\frac{z}{n-1})};$$

since

$$z\left(\log n - \left(1 + \tfrac{1}{2} + \cdots + \frac{1}{n-1}\right)\right) \to \gamma z$$

$$\text{and } E_1(-z) E_1\left(\frac{-z}{2}\right) \ldots E_1\left(-\frac{z}{n-1}\right) \to W_1(-z)$$

locally uniformly on U as $n \to \infty$, the result follows.

Now $L_n(1) = 1$ and $L_n(z+1) = nzL_n(z)/(z+n)$, so that $L(1) = 1$ and $L(z+1) = zL(z)$, for $z \in U$. Let T_1 be the strip $\{z = x + iy : 1 \leq x \leq 2\}$. If $z = x + iy \in T_1$, then $|1 + z/n| \geq |1 + x/n|$ and $|n^z| = n^x$, so that $|L_n(z)| \leq L_n(x)$ and $|L(z)| \leq L(x)$. Since L is bounded on $[1,2]$, it follows that L is bounded on T_1. Similarly

$$|\Gamma(z)| = \left| \int_0^\infty t^{z-1} e^{-t} \, dt \right| \leq \int_0^\infty t^{x-1} e^{-t} \, dt = \Gamma(x),$$

so that Γ is also bounded on T_1.

Now let $F(z) = L(z) - \Gamma(z)$. The function F is a meromorphic function on U which satisfies $F(z+1) = zF(z)$ for $z \in U$ and is bounded on T_1. Since $L(1) = \Gamma(1) = 1$, $F(1) = 0$. Using the equation $F(z+1) = zF(z)$, it follows that $F(z) \to F'(1)$ as $z \to 0$, so that F has a removable singularity at 0. Using the equation $F(z+1) = zF(z)$ repeatedly, it then follows that $F(z) \to (-1)^n F(0)/n!$ as $z \to -n$, so that all the singularities are removable; removing them, F becomes an entire function. We must show that $F = 0$. Let T_0 be the strip $\{z = x + iy : 0 \leq x \leq 1\}$. Since F is continuous, F is bounded on the set $\{z = x + iy \in T_0 : |y| \leq 1\}$. If $z = x + iy \in T_0$ and $|y| > 1$, then $|F(z)| = |F(z+1)/z| \leq |F(z+1)|$, and so F is bounded on T_0.

Let us now set $G(z) = F(z)F(1-z)$. Then G is an entire function which is bounded on T_0, and $G(z) = G(1-z)$. Further,

$$G(z+1) = F(z+1)F(-z) = zF(z)F(-z) = -F(z)F(1-z) = -G(z)$$

so that $G(z+2) = G(z)$ and $G(-z) = -G(1-z) = -G(z)$. Thus G is periodic, with period 2, and is bounded on $T_0 \cup T_1$. It is therefore a bounded entire function, and so is constant, by Liouville's theorem. Since $F(1) = 0$, $G = 0$, and so $F(z)F(1-z) = 0$ for all z. This implies that $F = 0$; for if not, then $Z_G = Z_F \cup (1 - Z_F)$ would be countable. □

We shall see in Part Six (Exercise 29.1.3) that this theorem can be proved more directly, once the Lebesgue integral has been introduced.

Exercises

26.7.1 Show that if $z \in \mathbf{C} \setminus (-\infty, 0]$ then

$$\log \Gamma(z) = \log z - \gamma z - \sum_{n=1}^\infty \left(\log \left(1 + \frac{z}{n} \right) - \frac{z}{n} \right).$$

26.7.2 Let $\Psi(z) = \Gamma'(z)/\Gamma(z)$, for $z \in U = \mathbf{C} \setminus \{0, -1, -2, \ldots\}$.

(i) Show that

$$\Psi(z) = -\gamma - \frac{1}{z} + \sum_{n=1}^{\infty} \frac{z}{n(z+n)},$$

and that the sum converges locally uniformly in U.

(ii) What is the singular set of Ψ? What is the order of each pole? What is the residue there?

(iii) Evaluate $\Psi(1)$.

(iv) Show that $\Psi(z+1) = \Psi(z) + 1/z$. What is $\lim_{n\to\infty}(\Psi(n) - \log n)$?

(v) Show that if $z \in \mathbf{C} \setminus \mathbf{Z}$ then $\Psi(z) - \Psi(1-z) = -\pi \cot \pi z$.

26.8 Bernoulli numbers, and the evaluation of $\zeta(2k)$

Euler not only showed that $\zeta(2) = \sum_{j=1}^{\infty} j^{-2} = \pi^2/6$, but also evaluated $\zeta(2k) = \sum_{j=1}^{\infty} j^{-2k}$, for $k \in \mathbf{N}$, in terms of the *Bernoulli numbers*. We begin by considering the function $B(z) = z/(e^z - 1)$. The entire function $(e^z - 1)/z$ has a removable singularity at 0 and zeros at $2\pi i \mathbf{Z} \setminus \{0\}$. Consequently $B(z)$ is a meromorphic function on $\mathbf{C}$, with simple poles at $2\pi i \mathbf{Z} \setminus \{0\}$. We denote its power series expansion about 0 as

$$B(z) = \frac{z}{e^z - 1} = 1 + \sum_{j=1}^{\infty} \frac{B_j}{j!} z^j, \text{ for } |z| < 1;$$

the series has radius of convergence 2π. The coefficients $(B_j)_{j=1}^{\infty}$ are the *Bernoulli numbers*. Note that $B_1 = -1/2$. Now consider

$$B(z) + \frac{z}{2} = \frac{z(e^z + 1)}{2(e^z - 1)} = \frac{z}{2} \cdot \frac{e^{z/2} + e^{-z/2}}{e^{z/2} - e^{-z/2}} = \frac{z}{2} \coth \frac{z}{2}.$$

This is an even function, and so $B_{2k+1} = 0$ for $k \in \mathbf{N}$. If $z \neq 0$ then

$$\left(B(z) + \frac{z}{2}\right)\left(\frac{e^z - 1}{z}\right) = \frac{e^z + 1}{2}$$

so that

$$\left(1 + \sum_{j=2}^{\infty} \frac{B_j}{j!} z^j\right) \cdot \left(1 + \sum_{j=1}^{\infty} \frac{z^j}{(j+1)!}\right) = \frac{1}{2}\left(2 + \sum_{j=1}^{\infty} \frac{z^j}{j!}\right).$$

Multiplying the two series, and equating the coefficient of z^{2k}, we obtain the equation

$$\frac{B_{2k}}{(2k)!} = -\sum_{j=1}^{k-1} \frac{B_{2j}}{(2j)!(2k-2j+1)!} + \frac{1}{2(2k)!} - \frac{1}{(2k+1)!}.$$

Consequently,

$$(2k+1)B_{2k} = -\sum_{j=1}^{k-1} \binom{2k+1}{2j} B_{2j} + k - \frac{1}{2}.$$

Thus

$$B_2 = \frac{1}{6}, \; B_4 = -\frac{1}{30}, \; B_6 = \frac{1}{42}, \; B_8 = -\frac{1}{30}, \; B_{10} = \frac{5}{66}, \; B_{12} = -\frac{691}{2730}.$$

Note that it follows from the recurrence relation that the Bernoulli numbers are rational numbers. The form of B_{12} suggests that there is no obvious pattern for them. A formula for B_k is given in Exercise 26.8.2.

Putting $z = 2iw$ in the formula above, we see that if $|w| < \pi$ then

$$w \cot w = \sum_{k=0}^{\infty} (-4)^k B_{2k} \frac{w^{2k}}{(2k)!}.$$

Theorem 26.8.1 *If* $k \in \mathbf{N}$ *then*

$$\zeta(2k) = 1 + \frac{1}{2^{2k}} + \frac{1}{3^{2k}} + \cdots = (-1)^{k-1} \frac{2^{2k-1}\pi^{2k} B_{2k}}{(2k)!} = \frac{2^{2k-1}\pi^{2k}|B_{2k}|}{(2k)!}.$$

Proof If $|w| < \pi$ then

$$\frac{w^2}{j^2\pi^2 - w^2} = \frac{w^2}{j^2\pi^2}\left(1 - \frac{w^2}{j^2\pi^2}\right)^{-1} = \sum_{k=1}^{\infty} \left(\frac{w^2}{j^2\pi^2}\right)^k.$$

It therefore follows from Theorem 26.2.2 that

$$w \cot w = 1 - 2 \sum_{j=1}^{\infty} \frac{w^2}{j^2 \pi^2 - w^2}$$

$$= 1 - 2 \sum_{j=1}^{\infty} \left(\sum_{k=1}^{\infty} \left(\frac{w^2}{j^2 \pi^2} \right)^k \right)$$

$$= 1 - 2 \sum_{k=1}^{\infty} \frac{w^{2k}}{\pi^{2k}} \left(\sum_{j=1}^{\infty} \frac{1}{j^{2k}} \right)$$

$$= 1 - 2 \sum_{k=1}^{\infty} \frac{w^{2k}}{\pi^{2k}} \zeta(2k),$$

the change of order of summation being justified, since

$$\sum_{j=1}^{\infty} \left(\sum_{k=1}^{\infty} \left| \frac{w^2}{j^2 \pi^2} \right|^k \right) < \infty.$$

The result now follows by equating the coefficients of w^{2k} in the two power series for $w \cot w$.   □

Note that this implies that the Bernoulli numbers B_{2k} alternate in sign.

What about the values of $\zeta(2k+1)$? They remain a mystery. It was not until 1979 that the French mathematician Roger Apéry showed, to great acclaim, that $\zeta(3)$ is irrational.

Exercises

26.8.1 Use Stirling's formula to show that

$$\frac{(e\pi)^{2k}|B_{2k}|}{(2k)^{2k+\frac{1}{2}}} \to 4\sqrt{\pi} \text{ as } k \to \infty.$$

26.8.2 Show that

$$B_k = \sum_{j=1}^{k} \left(\sum_{l=1}^{j} (-1)^l \binom{j}{l} \frac{l^k}{j+1} \right),$$

for $k \in \mathbf{N}$. (I don't know how difficult this is!)

26.9 The Riemann zeta function revisited

Since $|n^{-z}| = n^{-x}$ for $z = x + iy$, the series

$$\zeta(z) = \sum_{n=1}^{\infty} \frac{1}{n^z}$$

converges locally uniformly to a holomorphic function ζ on the open half-space $H_1 = \{x + iy : x > 1\}$. Can we extend ζ to a meromorphic function on $\mathbf{C}$? If so, what are its properties?

In order to answer this, we need to establish relations between ζ and the gamma function Γ.

Proposition 26.9.1   *If $z \in H_1$ then*

$$\Gamma(z)\zeta(z) = \int_0^{\infty} \frac{t^{z-1}}{e^t - 1} \, dt.$$

Proof   Making the change of variables $t = nu$,

$$\Gamma(z) = \int_0^{\infty} t^{z-1} e^{-t} \, dt = n^z \int_0^{\infty} u^{z-1} e^{-nu} \, du,$$

so that if $z \in H_1$ then

$$\Gamma(z)\zeta(z) = \sum_{n=1}^{\infty} n^{-z} \Gamma(z) = \sum_{n=1}^{\infty} \left(\int_0^{\infty} t^{z-1} e^{-nt} \, dt \right)$$

$$= \int_0^{\infty} \left(\sum_{n=1}^{\infty} t^{z-1} e^{-nt} \right) dt$$

$$= \int_0^{\infty} \frac{t^{z-1}}{e^t - 1} \, dt.$$

(Justify the interchange of addition and integration!)   □

The function $1/(e^w - 1)$ is a meromorphic function on $\mathbf{C}$, with simple poles on the set $\{2\pi i j : j \in \mathbf{Z}\}$. Recall that if $z = re^{i\theta}$ with $0 < \theta < 2\pi$ then $w_{(\pi)}^{z-1} = r^{z-1} e^{i(z-1)\theta}$. The function $f_z(w) = w_{(\pi)}^{z-1}/(e^w - 1)$ is meromorphic on the cut plane $\mathbf{C}_\pi = \mathbf{C} \setminus [0, \infty)$. If $j \in \mathbf{N}$ then the residue at $2\pi i j$ is $(2\pi j)^{z-1} e^{i(z-1)\pi/2}$, and the residue at $-2\pi i j$ is $(2\pi j)^{z-1} e^{i(z-1)\pi} e^{i(z-1)\pi/2}$. If $0 < r < 2\pi < R$, let us set

$$I_{r,R}(z) = \int_R^r e^{2\pi i z} \frac{t^{z-1}}{e^t - 1} \, dt + \int_{\kappa_r(0)\leftarrow} f_z(w) \, dw + \int_r^R \frac{t^{z-1}}{e^t - 1} \, dt.$$

Then $I_{r,R}$ is an entire function on $\mathbf{C}$, which converges locally uniformly as $R \to \infty$ to the entire function

$$I_r(z) = \int_\infty^r e^{2\pi i z} \frac{t^{z-1}}{e^t - 1}\, dt + \int_{\kappa_r(0)\leftarrow} f_z(w)\, dw + \int_r^\infty \frac{t^{z-1}}{e^t - 1}\, dt.$$

Further, it follows from Cauchy's theorem that I_r does not depend on r. We therefore denote I_r by I.

Theorem 26.9.2 Let $\widetilde{\zeta}(z) = ie^{-i\pi z}I(z)\Gamma(1-z)/2\pi$ for $z \in \mathbf{C} \setminus \mathbf{N}$. Then $\widetilde{\zeta}$ has removable singularities at $2, 3, \dots$ and a simple pole at 1, with residue 1. If $z \in H_1 \setminus \mathbf{N}$, then $\widetilde{\zeta}(z) = \zeta(z)$, so that $\widetilde{\zeta}$ extends ζ to a meromorphic function on $\mathbf{C}$.

Proof If $z = x + iy \in H_1$ and $0 < r|w| \le \frac{1}{2}$ then

$$|e^w - 1| \ge |w| - \sum_{j=2}^\infty \frac{|w|^j}{j!} \ge \frac{|w|}{2},$$

so that $|f_z(w)| \le 2r^{x-2}$, and

$$\left| \int_{\kappa_r(0)\leftarrow} f_z(w)\, dw \right| \le 4\pi r^{x-1} \to 0 \text{ as } r \to 0.$$

Thus if $z \in H_1$ then

$$I(z) = (1 - e^{2\pi i z}) \int_0^\infty \frac{t^{z-1}}{e^t - 1}\, dt = (1 - e^{2\pi i z})\Gamma(z)\zeta(z).$$

In particular, $I(n) = 0$ for $n \in \{2, 3, 4, \dots\}$ Recall that $\Gamma(z)\Gamma(1-z) = \pi\operatorname{cosec}\pi z$. Thus if $z \in H_1 \setminus \mathbf{N}$ then

$$\widetilde{\zeta}(z) = \frac{\sin(\pi z)I(z)\Gamma(1-z)}{\pi(1 - e^{2\pi i z})} = \widetilde{\zeta}(z).$$

Since Γ has poles at $0, -1, -2, \dots$, the function ζ appears to have singularities at $1, 2, 3, \dots$. Since $\widetilde{\zeta}(z) = \zeta(z)$ for $z \in H_1 \setminus \mathbf{N}$, the singularities at $2, 3, \dots$ are all removable, and there is a single simple pole at 1, with residue 1. $\square$

We now write ζ for $\widetilde{\zeta}$, and obtain a functional equation for ζ.

Theorem 26.9.3 If $z \ne 1$ then

$$\zeta(z) = 2^z \pi^{z-1} \sin(\tfrac{1}{2}\pi z)\Gamma(1-z)\zeta(1-z).$$

Proof   As usual it is only necessary to establish this identity for all real z in an interval in $\mathbf{R}$. We consider $z = s \in (-1,0)$. Let S_k be the sum of the residues of $f_s(w)$ in the annulus $A_k = \{z : \pi < |z| < (2k+1)\pi\}$. It then follows from the residue theorem that

$$I_{\pi,(2k+1)\pi}(s) + \int_{\kappa_{(2k+1)\pi}(0)} f_s(w)\, dw = 2\pi i S_k$$

$$= 2\pi i (1 - e^{i\pi s}) \sum_{j=1}^{k} (2j\pi i)^{s-1}$$

$$= (2\pi i)^s (1 - e^{i\pi s}) \sum_{j=1}^{k} j^{s-1}.$$

Suppose that $w = u + iv$ and that $e^w = z = x + iy$. If $|w| = (2k+1)\pi$ then either $(2k + \frac{1}{2})\pi < |v| < (2k + \frac{3}{2})\pi$, in which case $x \leq 0$ and $|e^w - 1| \geq 1$, or $|u| \geq \pi/2$, in which case either $x \geq e^{\pi/2} > 4$ or $x < e^{-\pi/2} < \frac{1}{4}$. Thus $|e^w - 1| > \frac{1}{2}$, so that

$$\left| \int_{\kappa_{(2k+1)\pi}(0)} f_s(w)\, dw \right| \leq 4\pi [(2k+1)]^{s-1} \to 0 \text{ as } k \to \infty.$$

Consequently

$$I(s) = (2\pi i)^s (1 - e^{i\pi s}) \sum_{j=1}^{\infty} j^{s-1} = (2\pi i)^s (1 - e^{i\pi s})\zeta(1 - s).$$

Combining this with the equation $\zeta(s) = ie^{-i\pi s} I(s)\Gamma(1 - s)/2\pi$, the result follows.   □

What can we say about $\zeta(z)$ for $z \in \mathbf{C} \setminus H_1$?

Proposition 26.9.4   *If $k \in \mathbf{N}$ then $\zeta(-k) = (-1)^k B_{k+1}/(k+1)$.*

Proof

$$I(-k) = -\int_{\kappa_r(0)} \frac{w^{-k-1}}{e^w - 1}\, dw$$

$$= -\int_{\kappa_r(0)} w^{-k-2} \left(1 + \sum_{j=1}^{\infty} \frac{B_j}{j!} w^j \right) dw$$

$$= -\int_{\kappa_r(0)} w^{-k-2}\, dw - \sum_{j=1}^{\infty} \frac{B_j}{j!} \int_{\kappa_r(0)} w^{j-k-2}\, dw.$$

(Again, justify the interchange of addition and integration!)

All the terms vanish, except for the term where $j = k+1$, so that $I(-k) = -2\pi i B_{k+1}/(k+1)!$. Thus, applying the formula of Theorem 26.9.2,

$$\zeta(-k) = ie^{\pi ik}I(-k)\Gamma(k+1) = (-1)^k B_{k+1}/(k+1).$$

□

Note that this implies that $\zeta(-2k) = 0$ for $k \in \mathbf{N}$. These zeros are the *trivial zeros* of ζ. If $\zeta \in H_1$ then

$$\zeta(z) = \prod_p \frac{1}{1 - p^z},$$

where the product is taken over all primes, and so there are no zeros in H_1. In fact, it can be shown that all the other zeros lie in the *critical strip* $\{x + iy : 0 < x < 1\}$. In 1857, Riemann conjectured that all the zeros lie on the *critical line* $\{x + iy : x = 1/2\}$. This is still *the* great unsolved problem of mathematics, and here is a good place to stop.[1]

Exercises

26.9.1 Let $p_1, p_2, \ldots$ be the sequence of primes, and suppose that $z \in H_1$.
    (i) Show that $1/(1 - p_n^{-z}) = \sum_{j=0}^{\infty} p_n^{-jz}$.
    (ii) Suppose that $\prod_{m=1}^{n} 1/(1 - p_m^{-z}) = 1 + \sum_{j=1}^{\infty}(a_j^{(n)})^{-jz}$. When is $a_j^{(n)} = 0$?
    (iii) Show that the product $\prod_{m=1}^{\infty} 1/(1 - p_m^{-z})$ converges locally uniformly on H_1 to $\zeta(z)$.

26.9.1 Show that if $z \in H_1$ then $\zeta(z)^2 = \sum_{n=1}^{\infty} \tau(n)/n^z$, where $\tau(n)$ is the number of divisors of n.

26.9.2 Show that if $z \in H_1$ then $\zeta(z)\zeta(z + 1) = \sum_{n=1}^{\infty} \sigma(n)/n^{z+1}$, where $\sigma(n)$ is the sum of the divisors of n.

26.9.3 Show that if $z \in H_1$ then $\zeta(z) = \zeta(z + 1)\sum_{n=1}^{\infty} \phi(n)/n^{z+1}$, where $\phi(n)$ is the number of positive integers less than n which are coprime to n.

---

[1] There are many accounts of the analytic properties of the Riemann zeta function: see, for example, G. Tenenbaum, *Introduction to Analytic and Probabilistic Number Theory*, Cambridge University Press, 1995.

Part Six

Measure and Integration

Part Six

27

Lebesgue measure on **R**

27.1 Introduction

In Volume I, we developed properties of the Riemann integral. This is very satisfactory when we wish to integrate continuous or monotonic functions, and is a useful precursor for the complex path integrals that we considered in Part Five, but it has serious shortcomings. It can only be applied to a rather small class of functions, and it is not good for taking limits. Let us give two examples to illustrate this; they also indicate how the shortcomings will be overcome.

First, let us recall the definition of a fat Cantor set. Suppose that $\epsilon = (\epsilon_j)_{j=1}^{\infty}$ is a sequence of positive numbers, for which $\sum_{j=1}^{\infty} \epsilon_j = \sigma < 1$. Let $\sigma_n = \sum_{j=1}^{n} \epsilon_j$. We set $C_0^{(\epsilon)} = [0, 1]$, and define a decreasing sequence $(C_n^{(\epsilon)})_{n=0}^{\infty}$ of closed subsets of $[0, 1]$ recursively. The set $C_n^{(\epsilon)}$ is the union of 2^n closed intervals, each of length $(1 - \sigma_n)/2^n$; the set $C_{n+1}^{(\epsilon)}$ is obtained by removing an open interval of length $\epsilon_{n+1}/2^n$ from the middle of each of these intervals. Then the fat Cantor set $C^{(\epsilon)}$ is the intersection $\cap_{n=0}^{\infty} C_n^{(\epsilon)}$; it is a perfect subset of $[0, 1]$ with empty interior. Let $U^{(\epsilon)} = [0, 1] \setminus C^{(\epsilon)}$ and let $U_n^{(\epsilon)} = [0, 1] \setminus C_n^{(\epsilon)}$; $(U_n^{(\epsilon)})_{n=1}^{\infty}$ is an increasing sequence of open subsets of $[0, 1]$ with union $U^{(\epsilon)}$. The indicator function of $C^{(\epsilon)}$ is not Riemann integrable (Volume I, Example 8.3.11). That is, $C^{(\epsilon)}$ is a perfect compact set which is not Jordan measurable (Volume II, Section 18.3), and consequently $U^{(\epsilon)}$ is a bounded open set which is not Jordan measurable. On the other hand, each of the sets $U_n^{(\epsilon)}$ is a finite union of open intervals, and is therefore Jordan measurable: $v(U_n^{(\epsilon)}) = \sum_{j=1}^{n} \epsilon_j$. This suggests that the size of $U^{(\epsilon)}$ should be $\sum_{j=1}^{\infty} \epsilon_j$, the sum of the lengths of the disjoint intervals whose union is U.

The second example is rather simpler. The set $\mathbf{Q} \cap [0, 1]$ is a countable dense subset of $[0, 1]$, and its indicator function f is not Riemann integrable.

Let $(r_j)_{j=1}^{\infty}$ be an enumeration of $\mathbf{Q} \cap [0,1]$, and let f_n be the indicator function of the set $\{r_1, \ldots, r_n\}$. Then f_n is Riemann integrable, and its Riemann integral is 0. The sequence $(f_n)_{n=1}^{\infty}$ increases pointwise to f. This suggests that the integral of f should be 0, and that the size of $\mathbf{Q} \cap [0,1]$ should be 0.

How do we resolve these difficulties? The trouble with the Riemann integral, and with Jordan content, is that very simple functions (step functions) and very simple sets (finite unions of cells) are used to provide approximations. Lebesgue's fundamental insight was to see that it is easy to define the size of a bounded open subset O of **R** as the sum of the lengths of the disjoint intervals whose union is O. The size of a compact set is then defined by taking complements. Open sets are then used to measure the size of a bounded subset A of **R** from the outside, and compact sets to measure the size from the inside. If the two values coincide (and this is not always the case) then A is *Lebesgue measurable*, and the common value is the *Lebesgue measure* $\lambda(A)$ of A. In this chapter, we develop these ideas in some detail. This reveals one of the unfortunate features of measure theory; much of it develops by taking many small steps, rather than one big one.

27.2 The size of open sets, and of closed sets

We begin by considering a non-empty open subset U of **R**. Recall (Volume I, Theorem 5.3.3) that U is the union of a finite or infinite sequence of disjoint open intervals I_j. We define the *size* $l(U)$ of U to be the sum $\sum_j l(I_j)$, where $l(I_j)$ is the length of I_j (here summation is over a finite set $\{1, \ldots, n\}$ or over **N**). Since the summands are all positive, the sum does not depend upon the order in which the terms are listed. Then $0 < l(U) \leq \infty$. We define $l(\emptyset) = 0$. The size of U can be infinite, even if U does not contain an infinite or semi-infinite interval; for example if $U = \cup_{n=1}^{\infty}(n, n + \frac{1}{n})$, $l(U) = \infty$. This can cause some inconvenience; the next result shows how this can be avoided.

Proposition 27.2.1 *If U is a bounded open subset of **R** and $U \subseteq (a,b)$, then $l(U) \leq b - a$.*

Proof Suppose first that $U = \cup_{j=1}^{n} I_j$ is a finite union of disjoint open intervals, and that $I_j = (a_j, b_j)$. We order the intervals from left to right, so that

$$a \leq a_1 < b_1 \leq a_2 < b_2 \leq \ldots \leq a_n < b_n \leq b.$$

Then

$$l(U) = \sum_{j=1}^{n}(b_j - a_j) = -a_1 + \sum_{j=1}^{n-1}(b_j - a_{j+1}) + b_n \le b_n - a_1 \le b - a.$$

If $U = \cup_{j=1}^{\infty}I_j$ then

$$l(U) = \sup_{n \in \mathbf{N}} \sum_{j=1}^{n} l(I_j) = \sup_{n \in \mathbf{N}} l(\cup_{j=1}^{n}I_j) \le b - a.$$

□

Thus if an open set is bounded, it has finite size. The converse is not true; for example, if $U = \cup_{n=1}^{\infty}(n + \frac{1}{n+1}, n + \frac{1}{n})$ then $l(U) = 1$.

We now establish some fundamental properties of the size of open sets. The results of this theorem lie at the heart of the theory of Lebesgue measure and the Lebesgue integral: you should take note of this as the theory develops. Most of the proofs only involve simple manipulation; the exception is the proof of (ii), which involves topological properties of $\mathbf{R}$.

Theorem 27.2.2 *Suppose that U, $(U_n)_{n=1}^{\infty}$ and V are open subsets of $\mathbf{R}$, and that $U = \cup_{n=1}^{\infty}U_n$.*

(i) If $U \subseteq V$ then $l(U) \le l(V)$.
(ii) If $(U_n)_{n=1}^{\infty}$ is an increasing sequence, then $l(U) = \lim_{n \to \infty} l(U_n)$.
(iii) $l(U) + l(V) = l(U \cup V) + l(U \cap V)$.
(iv) $l(U) \le \sum_{n=1}^{\infty} l(U_n)$.
(v) If $U_i \cap U_j = \emptyset$ for $i \ne j$ then $l(U) = \sum_{n=1}^{\infty} l(U_n)$.

Proof (i) If $l(V) = \infty$, there is nothing to prove. Otherwise, suppose that $U = \cup_j I_j$ and $V = \cup_k J_k$ are representations as unions of disjoint intervals. Since each I_j is connected, it is contained in some J_k. Then, using Proposition 27.2.1,

$$l(U) = \sum_{j} l(I_j) = \sum_{k}\left(\sum\{l(I_j) : I_j \subseteq J_k\}\right) \le \sum_{k} l(J_k) = l(V).$$

(ii) Since $U_n \subseteq U$ for all $n \in \mathbf{N}$, $l(U_n) \le l(U)$, for each $n \in \mathbf{N}$, by (i), and so $\sup_n l(U_n) \le l(U)$. It is the converse inequality that is important. We consider the case where $l(U) < \infty$ and $U = \cup_j I_j$, where $(I_j) = ((a_j, b_j))$ is a finite or infinite sequence of disjoint open intervals. Suppose that $\epsilon > 0$. There exists $J \in \mathbf{N}$ such that $\sum_{j=1}^{J} l(I_j) \ge l(U) - \epsilon/2$. Choose $\eta > 0$ so that $\eta < \epsilon/4J$ and $\eta < (b_j - a_j)/2$ for $1 \le j \le J$. Let

$$L_j = (a_j + \eta, b_j - \eta) \text{ and let } K_j = [a_j + \eta, b_j - \eta],$$

and let $L = \cup_{j=1}^{J} L_j$, $K = \cup_{j=1}^{J} K_j$. Then K is a compact subset of **R**. Since (U_n) is an increasing sequence of open sets which covers K, there exists $N \in \mathbf{N}$ such that $L \subseteq K \subseteq U_N$. Thus if $n \geq N$ then

$$l(U_n) \geq l(U_N) \geq l(L) = \sum_{j=1}^{J} (b_j - a_j - 2\eta) \geq \sum_{j=1}^{J} (b_j - a_j) - \epsilon/2 \geq l(U) - \epsilon.$$

Thus $l(U_n) \to l(U)$ as $n \to \infty$.

When $l(U) = \infty$ it is necessary to make some straightforward modifications to the proof; the details are left to the reader.

(iii) The result holds when U and V are open intervals, and a straightforward inductive argument shows that the result holds when U and V are finite unions of open intervals. In general, $U = \cup_{n=1}^{\infty} U_n$ and $V = \cup_{n=1}^{\infty} V_n$, where $(U_n)_{n=1}^{\infty}$ and $(V_n)_{n=1}^{\infty}$ are increasing sequences of open sets, each of which is a finite union of open intervals. Since $U \cup V = \cup_{n=1}^{\infty} (U_n \cup V_n)$ and $U \cap V = \cup_{n=1}^{\infty} (U_n \cap V_n)$,

$$l(U) + l(V) = \lim_{n \to \infty} (l(U_n) + l(V_n))$$
$$= \lim_{n \to \infty} (l(U_n \cup V_n) + l(U_n \cap V_n)) = l(U \cup V) + l(U \cap V).$$

(iv) Let $W_n = \cup_{j=1}^{n} U_j$. Then $l(W_{n+1}) \leq l(W_n) + l(U_{n+1})$, by (iii), and so $l(W_n) \leq \sum_{j=1}^{n} l(U_j) \leq \sum_{j=1}^{\infty} l(U_j)$. Since $(W_n)_{n=1}^{\infty}$ is an increasing sequence of open sets whose union is U,

$$l(U) = \lim_{n \to \infty} l(W_n) \leq \sum_{j=1}^{\infty} l(U_j),$$

by (ii).

(v) In this case, $l(W_{n+1}) = l(W_n) + l(U_{n+1})$, by (iii), so that $l(W_n) = \sum_{j=1}^{n} l(U_j)$ and $l(U) = \sum_{j=1}^{\infty} l(U_j)$. $\qquad\square$

Corollary 27.2.3 *Suppose that K is a compact subset of* **R**, *and that U_1 and U_2 are bounded open subsets of* **R**, *each containing K. Then*

$$l(U_1) + l(U_2 \setminus K) = l(U_2) + l(U_1 \setminus K).$$

Proof   Since

$$U_1 \cup (U_2 \setminus K) = U_1 \cup U_2 \text{ and } U_1 \cap (U_2 \setminus K) = (U_1 \cap U_2) \setminus K,$$

$$l(U_1) + l(U_2 \setminus K) = l(U_1 \cup U_2) + l((U_1 \cap U_2) \setminus K).$$

Exchanging U_1 and U_2,

$$l(U_2) + l(U_1 \setminus K) = l(U_1 \cup U_2) + l((U_1 \cap U_2) \setminus K),$$

which gives the result. $\qquad\square$

If K is a compact subset of $\mathbf{R}$ we define the *size* $s(K)$ of K to be $l(U) - l(U \setminus K)$, where U is a bounded open set containing K; Corollary 27.2.3 shows that $s(K)$ does not depend upon the choice of U. Note that $s(K) < l(U)$.

Here are some easy examples: you should verify the details.

1. $s([a, b]) = b - a$.
2. If F is a finite set, then $s(F) = 0$
3. If C is Cantor's ternary set, then $s(C) = 0$.
4. If $C^{(\epsilon)}$ is the fat Cantor set described in the previous section, then $s(C^{(\epsilon)}) = 1 - \sigma$.

The following theorem follows from Theorem 27.2.2 by taking complements.

Theorem 27.2.4 *Suppose that K, $(K_n)_{n=1}^{\infty}$ and L are compact subsets of $\mathbf{R}$.*

(i) *If $K \subseteq L$ then $s(K) \le s(L)$.*
(ii) *If $(K_n)_{n=1}^{\infty}$ is a decreasing sequence, and $K = \cap_{n=1}^{\infty} K_n$ then $s(K) = \lim_{n \to \infty} s(K_n)$.*
(iii) *$s(K) + s(L) = s(K \cup L) + s(K \cap L)$.*

The size of open and closed sets behaves well under translation, scaling and reversal:

$$l(a + U) = l(U) = l(-U), \text{ and } l(cU) = cl(U) \text{ for } c > 0,$$

and corresponding results hold for the size of compact sets. On the other hand, there are no good results concerning addition. For example, the Cantor ternary set has size 0, while $C + C = [0, 2]$, so that $s(C + C) = 2$. Similarly, if $U = \cup_{n=1}^{\infty}(n + \frac{1}{n+1}, n + \frac{1}{n})$ then $\lambda(U) = 1$, while $\lambda(U + V) = \infty$, for any non-empty open set V.

Exercises

27.2.1 Suppose that U and V are bounded open subsets of $\mathbf{R}$, each of which is the finite union of disjoint open intervals. Use Riemann integration to show that $l(U) + l(V) = l(U \cup V) + l(U \cap V)$.

27.2.2 Suppose that U and V are non-empty open subsets of **R**. Show that
$l(U) + l(V) \leq l(U + V)$.

27.2.3 Let $\mathcal{U}$ be the set of non-empty open subsets of $(0, 1)$. Show that there
does not exist $K \in \mathbf{R}^+$ such that $l(U + V) \leq K(l(U) + l(V))$ for all
U, V in $\mathcal{U}$.

27.2.4 Let $U = \cup_{n=1}^\infty (n + \frac{1}{n+1}, n + \frac{1}{n})$. Show that $l(U) = 1$, and that
$l(U + V) = \infty$, for any non-empty open set V.

27.2.5 Suppose that U is an open subset of **R**, that K is a compact subset of U and that V is an open subset of U. Show that $s(K) \leq
s(K \setminus V) + l(V)$.

27.3 Inner and outer measure

We now use open sets to measure the size of a bounded subset of **R** from
the outside, and use compact sets to measure the size from the inside. We
restrict attention to bounded subsets of **R**, to avoid problems with infinite
values; we shall come to these later.

Suppose that A is a bounded subset of **R**. We set

$$\lambda^*(A) = \inf\{l(U) : U \text{ open and bounded}, A \subseteq U\},$$

$$\lambda_*(A) = \sup\{s(K) : K \text{ compact}, K \subseteq A\}.$$

The quantity $\lambda^*(A)$ is the *outer measure* of A, and $\lambda_*(A)$ is the *inner measure*
of A.

Proposition 27.3.1   *If A is a bounded subset of* **R** *then* $\lambda_*(A) \leq \lambda^*(A)$.

Proof   If $K \subseteq A \subseteq U$, where K is compact and U is bounded and open then
$s(K) = l(U) - l(U \setminus K) < l(U)$. Letting K vary, we see that $\lambda_*(A) \leq l(U)$.
Letting U vary, $\lambda_*(A) \leq \lambda^*(A)$.   □

If A is a bounded subset of **R** for which $\lambda_*(A) = \lambda^*(A)$, we say that A is
Lebesgue measurable, or, simply, *measurable*, and set $\lambda(A) = \lambda_*(A) = \lambda^*(A)$.
The quantity $\lambda(A)$ is the *Lebesgue measure* of A.

Theorem 27.3.2   *(i) A bounded open subset U of* **R** *is Lebesgue measurable, and* $\lambda(U) = l(U)$.
(ii) A compact subset K of **R** *is Lebesgue measurable, and* $\lambda(K) = s(K)$.

Proof   (i) If V is a bounded open subset of **R** which contains U, then
$l(U) \leq l(V)$; hence $\lambda^*(U) = l(U)$. If $U = \emptyset$ then $l(U) = 0$, so that $\lambda^*(U) =
\lambda^*(U) = 0$. Next, suppose that $U = \cup_{j=1}^n I_j$ is a finite disjoint union of non-
empty open intervals $I_j = (a_j, b_j)$, and that $\epsilon > 0$. Choose $\eta > 0$ so that

$\eta < \epsilon/2n$ and $\eta < \min\{(b_j - a_j)/2 : 1 \leq j \leq n\}$. Let $K = \cup_{j=1}^{n}[a_j + \eta, b_j - \eta]$. Then K is a compact subset of U, and

$$s(K) = \sum_{j=1}^{n}(b_j - a_j - 2\eta) > \sum_{j=1}^{n}(b_j - a_j) - \epsilon = l(U) - \epsilon.$$

Since ϵ is arbitrary, $\lambda_*(U) = l(U) = \lambda^*(U)$.

Finally suppose that $U = \cup_{j=1}^{\infty}I_j$ is an infinite disjoint union of non-empty open intervals $I_j = (a_j, b_j)$, and that $\epsilon > 0$. There exists $n_0 \in \mathbf{N}$ such that $\sum_{j=1}^{n_0}l(I_j) > \sum_{j=1}^{\infty}l(I_j) - \epsilon/2 = l(U) - \epsilon/2$. Let $U_{n_0} = \cup_{j=1}^{n_0}I_j$. By the previous case, there exists a compact subset K of U_{n_0} such that $s(K) > l(U_{n_0}) - \epsilon/2 = \sum_{j=1}^{n_0}l(I_j) - \epsilon/2$. Hence $s(K) > l(U) - \epsilon$. Since ϵ is arbitrary, $\lambda_*(U) = l(U) = \lambda^*(U)$.

(ii) As in (i), $\lambda_*(K) = s(K)$. There is a bounded open set U such that $K \subseteq U$. Suppose that $\epsilon > 0$. There exists a compact subset L of $U \setminus K$ such that $s(L) > l(U \setminus K) - \epsilon$. Let $V = U \setminus L$. Then

$$l(V) = l(U) - s(L) < l(U) - l(U \setminus K) + \epsilon = s(K) + \epsilon.$$

Thus $\lambda^*(K) = s(K) = \lambda_*(K)$. $\qquad\qquad\qquad\qquad\qquad\qquad\qquad\square$

We now establish results which correspond to Theorems 27.2.2 and 27.2.4.

Theorem 27.3.3 *Suppose that A, $(A_n)_{n=1}^{\infty}$ and B are bounded subsets of $\mathbf{R}$, and that $A = \cup_{n=1}^{\infty}A_n$.*

(i) *If $A \subseteq B$ then $\lambda_*(A) \leq \lambda_*(B)$ and $\lambda^*(A) \leq \lambda^*(B)$.*
(ii) *$\lambda^*(A) \leq \sum_{n=1}^{\infty}\lambda^*(A_n)$.*
(iii) *If $A_i \cap A_j = \emptyset$ for $i \neq j$ then $\lambda_*(A) \geq \sum_{n=1}^{\infty}\lambda_*(A_n)$.*
(iv) *$\lambda^*(A \cup B) + \lambda^*(A \cap B) \leq \lambda^*(A) + \lambda^*(B)$.*
(v) *$\lambda_*(A \cup B) + \lambda_*(A \cap B) \geq \lambda_*(A) + \lambda_*(B)$.*

Proof   (i) follows immediately from the definitions.

(ii) Suppose that $\epsilon > 0$. For each $n \in \mathbf{N}$ there exists a bounded open set U_n such that $A_n \subseteq U_n$ and $l(U_n) < \lambda^*(A_n) + \epsilon/2^n$. Let $U = \cup_{n=1}^{\infty}U_n$. Then $A \subseteq U$, and, using Theorem 27.2.2,

$$\lambda^*(A) \leq l(U) \leq \sum_{n=1}^{\infty}l(U_n) < \sum_{n=1}^{\infty}\lambda^*(A_n) + \epsilon.$$

Since ϵ is arbitrary, the result follows.

(iii) Suppose that $\epsilon > 0$. For each $n \in \mathbf{N}$ there exists a compact set K_n such that $K_n \subseteq A_n$ and $s(K_n) > \lambda_*(A_n) - \epsilon/2^n$. Let $L_n = \cup_{i=1}^{n}K_i$.

Then $L_n \subseteq A$ and $s(L_n) = \sum_{i=1}^{n} s(K_i) \geq \sum_{i=1}^{n} \lambda_*(A_i) - \epsilon$. Thus $\lambda_*(A) \geq \sum_{i=1}^{n} \lambda_*(A_i) - \epsilon$. Since this holds for all $n \in \mathbf{N}$, $\lambda_*(A) \geq \sum_{i=1}^{\infty} \lambda_*(A_i) - \epsilon$. Since this holds for all $\epsilon > 0$, the result follows.

(iv) Suppose that $\epsilon > 0$. There exist bounded open sets U and V such that $A \subseteq U$, $B \subseteq V$, $l(U) < \lambda^*(A) + \epsilon/2$ and $l(V) < \lambda^*(B) + \epsilon/2$. Then

$$\lambda^*(A) + \lambda^*(B) \geq l(U) + l(V) - \epsilon$$
$$= l(U \cup V) + l(U \cap V) - \epsilon \geq \lambda^*(A \cup B) + \lambda^*(A \cap B) - \epsilon.$$

Since this holds for all $\epsilon > 0$, the result follows.

(v) The proof of this is similar to the proof of (iv), and is left as an exercise for the reader.   □

As with length, if A is a bounded subset of **R**, then

$$\lambda^*(a + A) = \lambda^*(A) = \lambda^*(-A), \text{ and } \lambda^*(cA) = c\lambda^*(A) \text{ for } c > 0,$$

and similar results hold for inner measure.

Exercises

27.3.1 Suppose that A is a subset of a bounded open subset U of **R**. Show that $\lambda_*(A) = l(U) - \lambda^*(U \setminus A)$.

27.3.2 Suppose that A and B are bounded disjoint subsets of **R**. Show that

$$\lambda_*(A \cup B) \leq \lambda_*(A) + \lambda^*(B) \leq \lambda^*(A \cup B).$$

27.4 Lebesgue measurable sets

Here are the fundamental properties of Lebesgue measurable sets, and Lebesgue measure.

Theorem 27.4.1 *Suppose that A, $(A_n)_{n=1}^{\infty}$ and B are Lebesgue measurable subsets of a bounded interval I. Let $C = \cup_{n=1}^{\infty} A_n$ and let $D = \cap_{n=1}^{\infty} A_n$.*

(i) The sets $A \cup B$ and $A \cap B$ are Lebesgue measurable and

$$\lambda(A \cup B) + \lambda(A \cap B) = \lambda(A) + \lambda(B).$$

(ii) $A \setminus B$ is Lebesgue measurable, and $\lambda(A) = \lambda(A \setminus B) + \lambda(A \cap B)$.
(iii) If $B \subseteq A$ then $\lambda(B) = \lambda(A) - \lambda(A \setminus B) \leq \lambda(A)$.
(iv) (Countable additivity) If $A_i \cap A_j = \emptyset$ for $i \neq j$, then C is measurable, and $\lambda(C) = \sum_{n=1}^{\infty} \lambda(A_n)$.

(v) (Upwards continuity) *If* $(A_n)_{n=1}^{\infty}$ *is an increasing sequence then* C *is measurable, and* $\lambda(C) = \sup_{n \in \mathbf{N}} \lambda(A_n)$.

(vi) *The set* C *is measurable, and* $\sup_{n \in \mathbf{N}} \lambda(A_n) \leq \lambda(C) \leq \sum_{n=1}^{\infty} \lambda(A_n)$.

(vii) (Downwards continuity) *If* $(A_n)_{n=1}^{\infty}$ *is a decreasing sequence, then* D *is measurable, and then* $\lambda(D) = \inf_{n \in \mathbf{N}} \lambda(A_n)$.

(viii) *The set* D *is measurable, and* $\lambda(D) \leq \inf_{n \in \mathbf{N}} \lambda(A_n)$.

Proof

(i) It follows from Theorem 27.3.3 (iv) and (v) that

$$\lambda^*(A \cup B) + \lambda^*(A \cap B) = \lambda_*(A \cup B) + \lambda_*(A \cap B).$$

Thus

$$0 \leq \lambda^*(A \cup B) - \lambda_*(A \cup B) = \lambda_*(A \cap B) - \lambda^*(A \cap B) \leq 0.$$

Consequently $\lambda^*(A \cup B) = \lambda_*(A \cup B)$ and $\lambda^*(A \cap B) = \lambda_*(A \cap B)$.

(ii) Suppose that $\epsilon > 0$. There exist bounded open sets U and V and compact sets K and L such that $K \subseteq A \subseteq U$ and $L \subseteq B \subseteq V$, and such that $\lambda(K) > \lambda(U) - \epsilon$ and $\lambda(L) > \lambda(V) - \epsilon$. Then, using Exercise 27.2.5,

$$\lambda^*(A \setminus B) \leq \lambda(U \setminus L) = \lambda(U) - \lambda(L) \leq \lambda(A) - \lambda(B) + 2\epsilon,$$

$$\text{and } \lambda^*(A \setminus B) \geq \lambda(K \setminus V) \geq \lambda(K) - \lambda(V) \geq \lambda(A) - \lambda(B) - 2\epsilon.$$

Since $\epsilon > 0$ is arbitrary, $\lambda^*(A \setminus B) = \lambda_*(A \setminus B) = \lambda(A) - \lambda(B)$.

(iii) is an immediate consequence.

(iv) By Theorem 27.3.3 (ii) and (iii),

$$\lambda^*(C) \leq \sum_{n=1}^{\infty} \lambda^*(A_n) = \sum_{n=1}^{\infty} \lambda(A_n) = \sum_{n=1}^{\infty} \lambda_*(A_n) \leq \lambda_*(C) \leq \lambda^*(C),$$

so that all the terms are equal. Thus C is measurable, and $\lambda(C) = \sum_{n=1}^{\infty} \lambda(A_n)$.

(v) Let $E_1 = A_1$ and let $E_{n+1} = A_{n+1} \setminus A_n$ for $n \in \mathbf{N}$. Each E_n is Lebesgue measurable, by (ii), and C is the disjoint union of the sequence $(E_n)_{n=1}^{\infty}$. Hence C is Lebesgue measurable, and $\lambda(C) = \sum_{n=1}^{\infty} \lambda(E_n)$. Since $\lambda(A_n) = \sum_{j=1}^{n} \lambda(E_j)$, by (i), the result follows.

(vi) Let $G_n = \cup_{j=1}^{n} A_j$, for $n \in \mathbf{N}$. Then G_n is Lebesgue measurable, and $\lambda(G_n) = \sum_{j=1}^{n} \lambda(E_j) \leq \sum_{j=1}^{n} \lambda(A_j)$, by (i) and (ii). Since $(G_n)_{n=1}^{\infty}$ is an increasing sequence and $C = \cup_{n=1}^{\infty} G_n$, C is Lebesgue measurable and $\lambda(C) = \sup_{n \in \mathbf{N}} \lambda(G_n) \leq \sum_{n=1}^{\infty} \lambda(A_n)$, by (v). Since $A_n \subseteq C$ for all $n \in \mathbf{N}$, $\sup_{n \in \mathbf{N}} \lambda(A_n) \leq \lambda(C)$.

(vii) This is a matter of taking relative complements. Let $F_n = A_1 \setminus A_n$, for $n \in \mathbf{N}$. Then $(F_n)_{n=1}^{\infty}$ is an increasing sequence of Lebesgue measurable sets, with union the bounded set $A_1 \setminus D$. Thus $A_1 \setminus D$ is Lebesgue measurable, and

$$\lambda(A_1 \setminus D) = \sup_{n \in \mathbf{N}} \lambda(A_1 \setminus A_n) = \lambda(A_1) - \inf_{n \in \mathbf{N}} \lambda(A_n),$$

by (iii) and (v). Using (iii) again,

$$\lambda(D) = \lambda(A_1) - \lambda(A_1 \setminus D) = \inf_{n \in \mathbf{N}} \lambda(A_n).$$

(viii) Let $H_n = \cap_{j=1}^{n} A_j$, for $n \in \mathbf{N}$. Then H_n is Lebesgue measurable, and $\lambda(H_n) \leq \lambda(A_n)$, by (i) and (ii). Since $(H_n)_{n=1}^{\infty}$ is a decreasing sequence and $D = \cap_{n=1}^{\infty} H_n$, D is Lebesgue measurable, and $\lambda(D) = \inf_{n \in \mathbf{N}} \lambda(H_n) \leq \inf_{n \in \mathbf{N}} \lambda(A_n)$, by (vii).

$\square$

Exercises

27.4.1 Suppose that A and B are bounded subsets of **R**, and that A is Lebesgue measurable. Show that $\lambda^*(B) = \lambda^*(A \cap B) + \lambda^*(B \setminus A)$.

27.4.2 Suppose that A is a bounded Lebesgue measurable subset of **R**, that $\lambda(A) > 0$, and that $0 < r < 1$. Show that there is a non-empty open interval I such that $\lambda(A \cap I) > rl(I)$. [Hint: Consider an open subset U of **R** such that $A \subseteq U$ and $l(U) < (1/r)\lambda(A)$.]

27.4.3 Suppose that A is a bounded Lebesgue measurable subset of **R**, and that $\lambda(A) > 0$. Let $A - A = \{a_1 - a_2 : a_1, a_2 \in A\}$. Choose r in $(1/2, 1)$; by the previous question, there is a non-empty open interval I such that $\lambda(A \cap I) > rl(I)$. Suppose that $|x| < (2r - 1)l(I)$. What is the length of $I \cup (I + x)$? What is $\lambda((A \cap I) + x)$? Can $A \cap I$ and $(A \cap I) + x$ be disjoint? Deduce that 0 is in the interior of $A - A$.

27.5 Lebesgue measure on R

So far we have only defined Lebesgue measure on the bounded subsets of **R**. We now use the fact that **R** is a countable union of disjoint bounded intervals, or that **R** is the union of an increasing sequence of bounded intervals, to extend the definition to unbounded sets. Let us count the unit intervals in **R**: we set

$$
I_k = \begin{cases} (l, l+1] & \text{if } k = 2l+1, \\ (-l, -l+1] & \text{if } k = 2l. \end{cases}
$$

We also set $J_k = (-k, k] = \cup_{j=1}^{2k} I_j$: $(J_k)_{k=1}^{\infty}$ is an increasing sequence of bounded intervals whose union is **R**. We say that a subset A of **R** is Lebesgue measurable if $A \cap I_k$ is Lebesgue measurable, (or, equivalently, if $A \cap J_k$ is Lebesgue measurable) for all $k \in \mathbf{N}$, and denote the set of Lebesgue measurable subsets of **R** by $\mathcal{L}(\mathbf{R})$. If $A \in \mathcal{L}(\mathbf{R})$, we define the Lebesgue measure $\lambda(A)$ to be

$$
\lambda(A) = \sum_{k=1}^{\infty} \lambda(A \cap I_k) = \sup_{k \in \mathbf{N}} \lambda(A \cap J_k).
$$

Thus $0 \leq \lambda(A) \leq \infty$. This definition clearly extends the definitions of Lebesgue measurability, and of the Lebesgue measure, of a bounded set.

Most, but not all, of Theorem 27.4.1 extends to this case.

Theorem 27.5.1 *Suppose that A, $(A_n)_{n=1}^{\infty}$ and B are Lebesgue measurable subsets of* **R**. *Let $C = \cup_{n=1}^{\infty} A_n$ and let $D = \cap_{n=1}^{\infty} A_n$.*

(i) *The sets $A \cup B$ and $A \cap B$ are Lebesgue measurable and*

$$
\lambda(A \cup B) + \lambda(A \cap B) = \lambda(A) + \lambda(B).
$$

(ii) *$A \setminus B$ is Lebesgue measurable, and $\lambda(A) = \lambda(A \setminus B) + \lambda(A \cap B)$.*
(iii) *If $B \subseteq A$ then $\lambda(B) \leq \lambda(A)$.*
(iv) *(Countable additivity) If $A_i \cap A_j = \emptyset$ for $i \neq j$, then C is measurable, and $\lambda(C) = \sum_{n=1}^{\infty} \lambda(A_n)$.*
(v) *(Upwards continuity) If $(A_n)_{n=1}^{\infty}$ is an increasing sequence then C is measurable, and $\lambda(C) = \sup_{n \in \mathbf{N}} \lambda(A_n)$.*
(vi) *The set C is measurable, and $\sup_{n \in \mathbf{N}} \lambda(A_n) \leq \lambda(C) \leq \sum_{n=1}^{\infty} \lambda(A_n)$.*
(vii) *(Downwards continuity) If $(A_n)_{n=1}^{\infty}$ is a decreasing sequence, and if $\lambda(A_1) < \infty$, then D is measurable, and then $\lambda(D) = \inf_{n \in \mathbf{N}} \lambda(A_n)$.*
(viii) *The set D is measurable, and $\lambda(D) \leq \inf_{n \in \mathbf{N}} \lambda(A_n)$.*

Proof We prove (iv) and (vii), and leave the other parts as easy exercises for the reader.

(iv) Since $C \cap I_k = \bigcup_{n=1}^{\infty}(A_n \cap I_k)$ for each $k \in$ **N**, C is Lebesgue measurable. If $A_i \cap A_j = \emptyset$ for $i \neq j$ then

$$\sum_{n=1}^{\infty} \lambda(A_n) = \sum_{n=1}^{\infty} \left(\sum_{k=1}^{\infty} \lambda(A_n \cap I_k) \right) = \sum_{k=1}^{\infty} \left(\sum_{n=1}^{\infty} \lambda(A_n \cap I_k) \right)$$

$$= \sum_{k=1}^{\infty} \lambda((\bigcup_{n=1}^{\infty} A_n) \cap I_k) = \lambda(\bigcup_{n=1}^{\infty} A_n),$$

the change of order of summation being justified, as all the summands are non-negative.

(vii) Since $D \cap I_k = \bigcap_{n=1}^{\infty}(A_n \cap I_k)$ for each $k \in$ **N**, D is Lebesgue measurable. Suppose that $(A_n)_{n=1}^{\infty}$ is a decreasing sequence, and that $\lambda(A_1) < \infty$. Certainly, $\lambda(D) \leq \inf_{n \in \mathbf{N}} \lambda(A_n)$, since $D \subseteq A_n$ for $n \in$ **N**. Suppose that $\epsilon > 0$. There exists k such that $\lambda(A_1 \setminus J_k) < \epsilon$. If $n \in$ **N** then $A_n \setminus J_k \subseteq A_1 \setminus J_k$, so that $\lambda(A_n \setminus J_k) < \epsilon$ and $\lambda(A_n \cap J_k) > \lambda(A_n) - \epsilon$. Thus

$$\lambda(D) \geq \lambda(D \cap J_k) = \inf_{n \in \mathbf{N}} \lambda(A_n \cap J_k) \geq \inf_{n \in \mathbf{N}} \lambda(A_n) - \epsilon.$$

Since this holds for all $\epsilon > 0$, $\lambda(D) \geq \inf_{n \in \mathbf{N}} \lambda(A_n)$.   □

Note carefully that downwards continuity requires that $\lambda(A_1) < \infty$. If $A_n = (n, \infty)$ then $(A_n)_{n=1}^{\infty}$ is a decreasing sequence, and $\bigcap_{n=1}^{\infty} A_n = \emptyset$; $\lambda(A_n) = \infty$ and $\lambda(A_n)$ does not tend to $\lambda(\bigcap_{n=1}^{\infty} A_n) = 0$. This phenomenon will recur. If $\lambda(A) = \infty$ then A must be unbounded, but the converse does not hold. For example,

$$\lambda \left(\bigcup_{n=1}^{\infty} \left(n, n + \frac{1}{n(n+1)} \right) \right) = \sum_{n=1}^{\infty} \frac{1}{n(n+1)} = 1.$$

Thus infinite values can cause problems: these problems will continue to recur.

27.6 A non-measurable set

Is every bounded subset of **R** measurable? It depends! If we assume that the axiom of choice holds, we can give an example of a subset C of the interval $I = [0, 1)$ which is not Lebesgue measurable. On the other hand, Solovay has shown that, starting from the axiom system ZF, but denying the axiom of choice, it is possible to construct a model of the real numbers for which every subset of the real line is Lebesgue measurable. In general, mathematicians are reluctant to give up the axiom of choice, and prefer to accept that not all subsets of **R** are Lebesgue measurable.

Let α be an irrational number in I. We define a bijection $\gamma : I \to I$ by setting $\gamma(x) = x + \alpha$ (mod 1). Thus γ translates the interval $[0, 1 - \alpha)$ by an amount α onto the interval $[\alpha, 1)$ and translates the interval $[1 - \alpha, 1)$ by a negative amount $\alpha - 1$ onto the interval $[0, \alpha)$. The bijection γ generates a group $\Gamma = \{\gamma^n : n \in \mathbf{Z}\}$ of bijections of I onto itself. Since α is irrational, if $x \in I$ then $\gamma^m(x) \neq \gamma^n(x)$ for $m \neq n$, and the orbit $O_x = \{\gamma^n(x) : n \in \mathbf{Z}\}$ of x is a countably infinite set. It is an easy exercise to show that O_x is a dense subset of $[0, 1)$.

Using the axiom of choice, we pick one element out of each orbit. That is to say, there exists a subset C of I such that $C \cap O_x$ is a singleton set, for each $x \in [0, 1)$. We claim that C is not Lebesgue measurable.

Suppose, if possible, that C is Lebesgue measurable, with measure $\lambda(C)$. Then $\lambda(C) = \lambda(C \cap [0, 1 - \alpha)) + \lambda(C \cap [1 - \alpha, 1))$. Since

$$\gamma(C) = \gamma(C \cap [0, 1 - \alpha)) \cup \gamma(C \cap [1 - \alpha, 1)),$$

$\gamma(C)$ is Lebesgue measurable, and $\lambda(\gamma(C)) = \lambda(C)$. Similarly, $\gamma^{-1}(C)$ is Lebesgue measurable, and $\lambda(\gamma^{-1}(C)) = \lambda(C)$. Iterating, $\gamma^n(C)$ is Lebesgue measurable, for all $n \in \mathbf{Z}$, and $\lambda(\gamma^n(C)) = \lambda(C)$.

Now the sets $\gamma^n(C)$ are disjoint, and $\cup_{n \in \mathbf{N}} \gamma^n(C) = [0, 1)$, by the construction. Thus

$$1 = \lambda([0, 1)) = \sum_{-\infty}^{\infty} \lambda(\gamma^n(C)).$$

If $\lambda(C) = 0$, then $\sum_{-\infty}^{\infty} \lambda(\gamma^n(C)) = 0$, and if $\lambda(C) > 0$, then $\sum_{-\infty}^{\infty} \lambda(\gamma^n(C)) = \infty$; in either case we obtain a contradiction.

In fact, we can say more. Let $A = \{\gamma^{2n}(0) : n \in \mathbf{Z}\}$ and let $B = \{\gamma^{2n+1}(0) : n \in \mathbf{Z}\}$. Then A and B are dense in $[0, 1)$. Let $P = C + A$ (mod 1) and let $Q = C + B$ (mod 1). Then $[0, 1)$ is the disjoint union of P and Q, and $Q = \gamma(P)$. Suppose that K is a compact subset of P. Since $(K - K) \cap B = \emptyset$ (why?), and since B is dense in $[0, 1)$, it follows from Exercise 27.4.3 that $s(K) = 0$. Thus $\lambda_*(P) = 0$ and $\lambda_*(Q) = \lambda_*(\gamma(P)) = 0$. Consequently if E is any Lebesgue measurable subset of I of positive measure, $\lambda_*(P \cap E) = 0$ and $\lambda^*(P \cap E) = \lambda(E) - \lambda_*(B \cap E) = \lambda(E)$. Thus $A \cap E$ is not Lebesgue measurable. Now let $D = \cup_{n \in \mathbf{Z}}(P + n)$. Then D is a subset of **R** with the property that if E is *any* Lebesgue measurable subset E of **R** with positive measure, then $D \cap E$ is not Lebesgue measurable.

Exercises

27.6.1 We recall the construction of the Cantor–Lebesgue function, described in Volume I (Exercise 6.3.9). At the jth stage in the construction of Cantor's ternary set C, 2^{j-1} intervals, each of length $1/3^j$, are removed. List these intervals from left to right as $I_{1,j}, \ldots, I_{2^{j-1},j}$: that is, $\sup(I_{i,j}) < \inf(I_{i+1,j})$ for $1 \le i < 2^{j-1}$. Define a function f on $[0,1] \setminus C$ by setting $f(x) = (2i-1)/2^j$ for $x \in I_{i,j}$. Set $f(1) = 1$, and if $x \in C$ and $x \ne 1$, set $f(x) = \inf\{f(y) : y > x, y \in [0,1] \setminus C\}$. Then f is a continuous increasing function on $[0,1]$. This is the *Cantor–Lebesgue function*.

Now let $g(x) = f(x) + x$. Show that g is a uniformly continuous homeomorphism of $[0,1]$ onto $[0,2]$. Show that $\lambda(g(C)) = 1$. There exists a subset D of $g(C)$ which is not Lebesgue measurable. Let $F = g^{-1}(D)$. Show that F is Lebesgue measurable. Thus the uniformly continuous image of a Lebesgue measurable set need not be Lebesgue measurable.

28

Measurable spaces and measurable functions

28.1 Some collections of sets

Since we have seen that, if we assume that the axiom of choice holds, then not every subset of $\mathbf{R}$ is Lebesgue measurable, it is sensible to consider the properties that the collection of Lebesgue measurable sets possesses. We use the ideas that result from this to provide a setting for more general measures than Lebesgue measures. We make several definitions. Throughout this section, X is a non-empty set.

A set R of subsets of X is called a *ring* if

(i) the empty set belongs to R,
(ii) if $A, B \in R$ then $A \cup B \in R$, and
(iii) if $A, B \in R$ then $A \setminus B \in R$.

Proposition 28.1.1 *If R is a ring and $A, B \in R$ then $A \cap B \in R$ and $A \triangle B \in R$.*

Proof For $A \cap B = B \setminus (B \setminus A)$ and $A \triangle B = (A \setminus B) \cup (B \setminus A)$. □

Example 28.1.2 Three examples of rings.

The collection of finite subsets of X is a ring.

The collection of subsets of $\mathbf{R}$ of the form $A = \cup_{j=1}^{n} I_j$, where each $I_j = (a_j, b_j]$ is a bounded half-open half-closed interval in $\mathbf{R}$, is a ring.

The collection of bounded subsets of $\mathbf{R}^d$ which are Jordan measurable is a ring.

A set F of subsets of X is called a *field*, or *algebra*, if it is a ring, and if, in addition, $X \in F$.

Example 28.1.3   Two examples of a field.

Suppose that $(a, b]$ is a bounded half-open half-closed interval in $\mathbf{R}$. The collection of sets of the form $A = \cup_{j=1}^{n} I_j$, where each $I_j = (a_j, b_j]$ is a half-open half-closed interval contained in $(a, b]$, is a field.

The collection of subsets of a compact cell C in $\mathbf{R}^d$ which are Jordan measurable is a field.

A set S of subsets of X is called a *σ-ring* if it is a ring, and if, in addition,
(iv) if $(A_j)_{j=1}^{\infty}$ is a sequence of sets in S then $\cap_{j=1}^{\infty} A_j \in S$.

Example 28.1.4   Four examples of σ-rings.

  (i) The collection of countable subsets of X is a σ-ring.
 (ii) The collection of bounded Lebesgue measurable subsets of $\mathbf{R}$ is a σ-ring.
(iii) The collection of Lebesgue measurable subsets of $\mathbf{R}$ of finite measure is a σ-ring.
(iv) The collection of Lebesgue measurable subsets of $\mathbf{R}$ of measure 0 is a σ-ring.

The first result is an immediate consequence of the definition. The others are consequences of Theorems 27.4.1 and 27.5.1. Other characterizations of σ-rings are given in Exercise 28.1.2.

A set S of subsets of X is called a *σ-field*, or *σ-algebra*, if it is a σ-ring, and if, in addition, $X \in S$.

Example 28.1.5   The collection $\mathcal{L}$ of Lebesgue measurable subsets of $\mathbf{R}$ is a σ-field.

This is a consequence of Theorem 27.5.1.

Proposition 28.1.6   *Suppose that Σ is a non-empty collection of subsets of a set X. The following are equivalent.*

  (i) Σ is a σ-field.
 (ii) (a) if $A \in \Sigma$ then $X \setminus A \in \Sigma$, and
      (b) if $(A_n)_{n=1}^{\infty}$ is a sequence of disjoint elements of Σ then $\cup_{n=1}^{\infty} A_n \in \Sigma$.
(iii) (a) if $A \in \Sigma$ then $X \setminus A \in \Sigma$, and
      (b) if $(A_n)_{n=1}^{\infty}$ is an increasing sequence in Σ then $\cup_{n=1}^{\infty} A_n \in \Sigma$.

Proof   An easy exercise for the reader.   □

A *measurable space* is a pair (X, Σ), where X is a set and Σ is a σ-field of subsets of X. σ-fields and measurable spaces are the natural setting for measure theory. Here are some basic facts concerning them.

Proposition 28.1.7 *Suppose that $\mathcal{F}$ is a set of subsets of a set S. Then there is a smallest σ-field $\sigma(\mathcal{F})$ of subsets of S which contains $\mathcal{F}$.*

Proof   Let $\boldsymbol{s}$ be the collection of those σ-fields of subsets of S which contain $\mathcal{F}$. It is non-empty, since $P(S) \in \boldsymbol{s}$. Let

$$\sigma(\mathcal{F}) = \{A : A \in \Sigma, \text{ for all } \Sigma \in \boldsymbol{s}\}.$$

Then it is easy to verify that $\sigma(\mathcal{F})$ is a σ-field, and that it belongs to $\boldsymbol{s}$. It is then clearly the smallest element of $\boldsymbol{s}$.   □

The σ-field $\sigma(\mathcal{F})$ is called the *σ-field generated by $\mathcal{F}$*. An important feature of this proposition is that its proof is indirect, and gives no indication of the structure of sets in $\sigma(\mathcal{F})$. This fact gives a particular flavour to much of measure theory.

Proposition 28.1.8 *Suppose that (X, Σ) is a measurable space, that Y is a set and that $f : X \to Y$ is a mapping. Then $\{A \subseteq Y : f^{-1}(A) \in \Sigma\}$ is a σ-field.*

Proof   Suppose that $f^{-1}(A) \in \Sigma$, and that $(A_n)_{n=1}^{\infty}$ is a sequence of subsets of Y such that $f^{-1}(A_n) \in \Sigma$ for $n \in \mathbf{N}$. Then $f^{-1}(Y) = X \in \Sigma$, $f^{-1}(Y \backslash A) = X \setminus f^{-1}(A) \in \Sigma$ and $f^{-1}(\cup_{n \in \mathbf{N}} A_n) = \cup_{n \in \mathbf{N}} f^{-1}(A_n) \in \Sigma$.   □

A mapping $f : (X_1, \Sigma_1) \to (X_2, \Sigma_2)$ from a measurable space (X_1, Σ_1) to a measurable space (X_2, Σ_2) is said to be *measurable* if $f^{-1}(A) \in \Sigma_1$ for each $A \in \Sigma_2$.

Corollary 28.1.9 *Suppose that $\mathcal{F}$ is a set of subsets of Y, and that $f^{-1}(A) \in \Sigma$ for $A \in \mathcal{F}$. Then $f^{-1}(A) \in \Sigma$ for $A \in \sigma(\mathcal{F})$; the mapping $f : (X, \Sigma) \to (Y, \sigma(\mathcal{F}))$ is measurable.*

Proof   For $\{A \subseteq Y : f^{-1}(A) \in \Sigma\}$ is a σ-field containing $\mathcal{F}$, and so $\sigma(\mathcal{F}) \subseteq \{A \subseteq Y : f^{-1}(A) \in \Sigma\}$. Thus $f^{-1}(A) \in \Sigma$ for $A \in \sigma(\mathcal{F})$.   □

Exercises

28.1.1 Suppose that (X, τ) is a topological space and that $x \in X$. Show that the collection of sets $R = \{A : x \notin \overline{A}\}$ is a ring of subsets of X.

28.1.2 Show that the intersection of a collection of σ-fields is a σ-field. Show that the union of two σ-fields need not be a σ-field.

28.1.3 Suppose that R is a ring of subsets of a set X. Show that the following are equivalent:

(i) R is a σ-ring;

    (ii) if $(A_j)_{j=1}^{\infty}$ is a decreasing sequence of sets in R then $\cap_{j=1}^{\infty} A_j \in R$;

    (iii) if $(A_j)_{j=1}^{\infty}$ is a sequence of sets in R and if there exists $B \in R$ such that $\cup_{j=1}^{\infty} A_j \subseteq B$ then $\cup_{j=1}^{\infty} A_j \in R$.

28.1.4 Prove Proposition 28.1.6.

28.2 Borel sets

Suppose that (X, τ) is a topological space. The σ-field $\mathcal{B}$ generated by the collection of open sets is called the *Borel σ-field*, and its elements are called *Borel sets*. Since the Lebesgue measurable subsets of $\mathbf{R}$ form a σ-algebra which contains the open subsets of $\mathbf{R}$, the Borel σ-field $\mathcal{B}$ is contained in the σ-field $\mathcal{L}$ of Lebesgue measurable sets. The restriction of Lebesgue measure to the bounded Borel sets of $\mathbf{R}$ is called *Borel measure*. The collection of open subsets of $\mathbf{R}$ is closed under countable unions, but not under countable intersections. Recall that a G_δ is a countable intersection of open sets. The collection of open subsets is contained in the collection of G_δ sets, which is closed under countable intersections. This, in turn, is not closed under countable unions. We can therefore consider the collection of $G_{\delta\sigma}$ sets (countable unions of G_δ sets), $G_{\delta\sigma\delta}$ sets, and so on. But it happens that this is a strictly increasing sequence, and that if we consider countable unions and countable intersections of all such sets, the resulting collection is not closed under countable unions or countable intersections. Thus there is no simple way to describe what a typical Borel set looks like: it is often necessary to proceed indirectly.

Proposition 28.2.1   *Consider the following collections of subsets of* $\mathbf{R}$.

$$\mathcal{F}_1 = \{(r, \infty) : r \in \mathbf{Q}\}, \quad \mathcal{F}_2 = \{(-\infty, r] : r \in \mathbf{Q}\},$$
$$\mathcal{F}_3 = \{[r, \infty) : r \in \mathbf{Q}\}, \quad \mathcal{F}_4 = \{(-\infty, r) : r \in \mathbf{Q}\},$$
$$\mathcal{F}_5 = \{(c, \infty) : c \in \mathbf{R}\}, \quad \mathcal{F}_6 = \{(-\infty, c] : c \in \mathbf{R}\},$$
$$\mathcal{F}_7 = \{[c, \infty) : c \in \mathbf{R}\}, \quad \mathcal{F}_8 = \{(-\infty, c) : c \in \mathbf{R}\},$$
$$\mathcal{F}_9 = \{U : U \text{ open}\}, \quad \mathcal{F}_{10} = \{K : K \text{ compact}\},$$
$$\mathcal{F}_{11} = \{A : A \text{ a } G_\delta \text{ set}\}, \quad \mathcal{F}_{12} = \{B : B \text{ an } F_\sigma \text{ set}\}.$$

Then $\mathcal{B} = \sigma(\mathcal{F}_i)$ *for* $1 \leq i \leq 12$.

Proof   Since $\mathcal{B} = \sigma(\mathcal{F}_9)$, it is enough to show that $\mathcal{F}_i \subseteq \sigma(\mathcal{F}_j)$ for $i \neq j$. Taking complements, $\mathcal{F}_1 = \mathcal{F}_2$, $\mathcal{F}_3 = \mathcal{F}_4$, $\mathcal{F}_5 = \mathcal{F}_6$, $\mathcal{F}_7 = \mathcal{F}_8$, $\mathcal{F}_9 \supseteq \mathcal{F}_{10}$ and $\mathcal{F}_{11} = \mathcal{F}_{12}$. Since an open set is the countable union of compact sets,

$\mathcal{F}_9 \supseteq \mathcal{F}_{10}$. Since

$$(r, \infty) = \cup_{n=1}^{\infty}[r + 1/n, \infty) \text{ and } [r, \infty) = \cap_{n=1}^{\infty}(r - 1/n, \infty),$$

$\mathcal{F}_1 = \mathcal{F}_3$, and similarly $\mathcal{F}_5 = \mathcal{F}_7$. Since

$$(c, \infty) = \cup\{(r, \infty) : r \in \mathbf{Q}, r > c\},$$

and since this is a countable union, $\mathcal{F}_1 = \mathcal{F}_7$. If $c < d$ then

$$(c, d) = (c, \infty) \cap (-\infty, d) \in \mathcal{F}_7.$$

Since an open set is the countable union of open intervals, it follows that $\mathcal{F}_7 = \mathcal{F}_9$. Finally it is clear that $\mathcal{F}_9 = \mathcal{F}_{11}$. □

Proposition 28.2.2 *Suppose that (X_1, τ_1) and (X_2, τ_2) are topological spaces, equipped with their Borel σ-fields $\mathcal{B}_1$ and $\mathcal{B}_2$. If $f : X_1 \to X_2$ is a continuous mapping, then f is measurable.*

Proof This is an immediate consequence of Corollary 28.1.9. □

Lebesgue asserted, mistakenly, that the continuous image of a Borel set is a Borel set. The Russian mathematician Suslin showed that this was not so; indeed, there exists a Borel set B in $\mathbf{R}^2$ such that $\pi_1(B) = \{x \in \mathbf{R} : (x, y) \in B \text{ for some } y \in \mathbf{R}\}$ is not a Borel set. In fact, Suslin's results opened up a large new theory, descriptive set theory, which is far too difficult to describe here.[1]

Exercise

28.2.1 Let F be the subset of the Cantor set C defined in Exercise 27.6.1. Show that F is not a Borel set. Deduce that the Borel σ-field in $\mathbf{R}$ is a proper sub-σ-field of the Lebesgue σ-field $\mathcal{L}$.

(This result can also be proved by showing that the Borel σ-field $\mathcal{B}$ has the same cardinality as $\mathbf{R}$: there is a bijection of $\mathcal{B}$ onto $\mathbf{R}$. On the other hand, since every subset of the Cantor set C is Lebesgue measurable, $\mathcal{L}$ has cardinality at least as big as the cardinality of $P(C)$. But there is a bijection of C onto $\mathbf{R}$, and so there is a bijection of $\mathcal{L}$ onto $P(\mathbf{R})$. By Cantor's theorem (Volume I, Theorem 1.6.3), the inclusion mapping $\mathcal{B} \to \mathcal{L}$ cannot be surjective.)

---

[1] See A. Kechris, *Classical Descriptive Set Theory*, Springer, 1995.

28.3 Measurable real-valued functions

Throughout this section, we shall consider a measurable space (X, Σ) and real-valued functions defined on X.

Let us describe some notation that we shall use. We shall, for example, consider sets of the form $\{x : |f(x) - g(x)| < 1/k\}$. When the context is clear, we shall denote this by $(|f - g| < 1/k)$, and use similar notation for other such sets. Thus

$$(f \in A) = f^{-1}(A) \text{ and } (f = c) = f^{-1}(\{c\}).$$

Similarly, if a set such as $(f \in A)$ is a measurable subset of X, and ϕ is a function on Σ, we write $\phi(f \in A)$ for $\phi((f \in A))$.

We say that a real-valued function f on X is Σ-*measurable* (or simply *measurable*, if it is clear what Σ is) if it is a measurable mapping of (X, Σ) into $(\mathbf{R}, \mathcal{B})$; that is, $f^{-1}(A) \in \Sigma$ for every Borel set A in $\mathbf{R}$. If X is an interval I (finite or infinite) in $\mathbf{R}$ and Σ is the σ-field of Lebesgue measurable sets, then we say that f is *Lebesgue measurable*; if Σ is the σ-field of Borel sets, then we say that f is *Borel measurable*.

It follows from Corollary 28.1.9 that a real-valued function f on X is Σ-measurable if and only if $(f \in A) \in \Sigma$ for all A in some $\mathcal{F}_j$, where $\mathcal{F}_j$ is one of the collections of subsets of $\mathbf{R}$ defined in Proposition 28.2.1. In many cases, as in the next proposition, it is convenient to use the collection $\mathcal{F}_5 = \{(c, \infty) : c \in \mathbf{R}\}$. Thus f is measurable if and only if $(f > c) \in \Sigma$ for each $c \in \mathbf{R}$.

Proposition 28.3.1   (i) *If $A \subseteq X$, then the indicator function I_A of A is Σ-measurable if and only if $A \in \Sigma$.*
(ii) *A simple function $f = \sum_{j=1}^{n} \alpha_j I_{A_j}$ (where $\alpha_i < \alpha_j$ and $A_i \cap A_j = \emptyset$ for $i < j$) is Σ-measurable if and only if $A_j \in \Sigma$ for $1 \leq j \leq n$.*

Proof

(i) For

$$(I_A > c) = \begin{cases} X & \text{if } c < 0, \\ A & \text{if } 0 \leq c < 1, \\ \emptyset & \text{if } c \geq 1. \end{cases}$$

(ii) We can suppose that $X = \cup_{i=1}^{n} A_i$.

$$\text{If } c < a_1 \text{ then } (f > c) = X = \cup_{i=1}^{n} A_i,$$

$$\text{if } a_j \leq c < a_{j+1} \text{ then } (f > c) = \cup_{i=j+1}^{n} A_i,$$

$$\text{and if } c \geq a_n \text{ then } (f > c) = \emptyset.$$

$\square$

Recall (Volume II, Exercise 13.1.6) that a real-valued function f on a topological space (X, τ) is *upper semi-continuous* if for each $x \in X$ and each $\epsilon > 0$, there is a neighbourhood $N(x)$ of x such that $f(y) < f(x) + \epsilon$ for $y \in N(x)$, or equivalently, if $(f < c)$ is open, for each $c \in \mathbf{R}$. *Lower semi-continuity* is defined similarly.

Corollary 28.3.2 *If (X, τ) is a topological space and Σ is the σ-field of Borel measurable sets, then upper semi-continuous and lower semi-continuous functions on X are measurable functions.*

Proposition 28.3.3 *If f is a measurable function on X and ϕ is a Borel measurable function on $\mathbf{R}$, then $\phi \circ f$ is a measurable function on X.*

Proof If A is a Borel subset of $\mathbf{R}$ then $\phi^{-1}(A)$ is a Borel subset of $\mathbf{R}$, and so $(\phi \circ f)^{-1}(A) = f^{-1}(\phi^{-1}(A)) \in \Sigma$. $\square$

Theorem 28.3.4 *Suppose that f and g are real-valued Σ-measurable functions on X. Then each of the functions*

$$-f, \ f^+, \ f^-, \ |f|, \ f + g, \ fg, \ f \vee g, \ \text{and } f \wedge g$$

is Σ-measurable. The sets $(f < g)$, $(f \leq g)$ and $(f = g)$ are in Σ.

Proof We consider some continuous real-valued functions on $\mathbf{R}$. Let

$$\phi_1(x) = -x, \ \phi_2(x) = x^+, \ \phi_3(x) = x^-, \ \phi_4(x) = |x| \text{ and } \phi_5(x) = x^2.$$

Then $\phi_i \circ f$ is Σ-measurable, for $1 \leq i \leq 5$: the functions $-f$, f^+, f^-, $|f|$ and f^2 are Σ-measurable.

Addition and multiplication are a little bit more complicated. Since

$$(f + g > c) = \cup\{(f > r) \cap (g > c - r) : r \in \mathbf{Q}\},$$

and since $\mathbf{Q}$ is countable, $f + g$ is Σ-measurable.

Since $fg = \frac{1}{4}((f + g)^2 - (f - g)^2)$, fg is Σ-measurable.

Next, $f \vee g = \frac{1}{2}(f + g + |f - g|)$ and $f \wedge g = \frac{1}{2}(f + g - |f - g|)$, and so they are both Σ-measurable.

Finally, $f(x) < g(x)$ if and only if there exists $r \in \mathbf{Q}$ such that $f(x) < r < g(x)$, so that $(f < g) = \cup\{(f < r) \cap (g > r) : r \in \mathbf{Q}\}$, which is in Σ, since $\mathbf{Q}$ is countable. Similarly, $(f \le g) = X \setminus (g < f) \in \Sigma$, and $(f = g) = (f \le g) \setminus (f < g) \in \Sigma$. $\qquad\square$

Thus the set $\mathcal{L}^0 = \mathcal{L}^0(X, \Sigma, \mu)$ of real-valued measurable functions on X is a real vector space, when addition and scalar multiplication are defined pointwise.

We now consider sequences of measurable functions, suprema, infima and limits. These may be infinite, and so we need to consider extended real-valued functions, functions taking values in $\overline{\mathbf{R}} = \{-\infty\} \cup \mathbf{R} \cup \{\infty\}$. We say that such a function f is Σ-measurable if $(f \in A) \in \Sigma$ for each Borel set A in $\mathbf{R}$ and if both $(f = -\infty)$ and $(f = \infty)$ are in Σ.

Theorem 28.3.5 *Suppose that $(f_n)_{n=1}^{\infty}$ is a sequence of extended real-valued Σ-measurable functions on X. Then each of the extended real-valued functions*

$$\sup_{n \in \mathbf{N}} f_n, \ \inf_{n \in \mathbf{N}} f_n, \ \limsup_{n \to \infty} f_n, \ and \ \liminf_{n \to \infty} f_n$$

is Σ-measurable.

Proof Since

$$\left(\sup_{n \in \mathbf{N}} f_n > c\right) = \cup_{n \in \mathbf{N}}(f_n > c),$$

$$\left(\sup_{n \in \mathbf{N}} f_n = \infty\right) = \cap_{k \in \mathbf{N}}(\cup_{n \in \mathbf{N}}(f_n > k)),$$

$$\left(\sup_{n \in \mathbf{N}} f_n = -\infty\right) = \cap_{n \in \mathbf{N}}(f_n = -\infty),$$

the function $\sup_{n \in \mathbf{N}} f_n$ is Σ-measurable. The proof for $\inf_{n \in \mathbf{N}} f_n$ is exactly similar.

Since

$$\limsup_{n \to \infty} f_n = \inf_n (\sup_{k \ge n} f_k),$$

the function $\limsup_{n \to \infty} f_n$ is Σ-measurable, and so, similarly is $\liminf_{n \to \infty} f_n$. $\qquad\square$

Corollary 28.3.6 *The set*

$$C^* = \{x : f_n(x) \ converges \ in \ \overline{\mathbf{R}} \ as \ n \to \infty\}$$

is in Σ. Let $f(x) = \lim_{n \to \infty} f_n(x)$ if $x \in C^*$, and let $f(x) = 0$ otherwise. Then f is Σ-measurable.

Proof   For

$$C^* = (\limsup_{n \to \infty} f_n = \liminf_{n \to \infty} f_n) \text{ and } f = (\limsup_{n \to \infty} f_n).I_{C^*}.$$

□

Corollary 28.3.7   *If each f_n is real-valued, then the set*

$$C = \{x : f_n(x) \text{ converges in } \mathbf{R} \text{ as } n \to \infty\}$$

is in Σ. Let $f(x) = \lim_{n \to \infty} f_n(x)$ if $x \in C$, and let $f(x) = 0$ otherwise. Then f is Σ-measurable.

Proof   For $C = C^* \setminus ((\liminf_{n \to \infty} f_n = +\infty) \cup (\limsup_{n \to \infty} f_n = -\infty))$.

□

28.4 Measure spaces

Many of the results about Lebesgue measure extend to a more general setting. A *finite measure space* is a triple (X, Σ, μ), where (X, Σ) is a measurable space and μ is a *countably additive*, or *σ-additive*, mapping of Σ into $\mathbf{R}^+$: if (A_n) is a sequence of disjoint elements of Σ then $\mu(\cup_{n=1}^{\infty} A_n) = \sum_{n=1}^{\infty} \mu(A_n)$. (Note that, since all the summands are non-negative, the sum does not depend upon the order of summation.) The function μ is called a *measure*. (The adjective 'finite' is included, because $\mu(X)$ is a real number, and so is finite.) Thus if I is a bounded interval in $\mathbf{R}$ then $(I, \mathcal{L}(I), \lambda)$, where $\mathcal{L}(I)$ is the σ-field of Lebesgue measurable subsets of I and λ is Lebesgue measure, is an example of a finite measure space. Similarly, if $\mathcal{B}(I)$ is the σ-field of Borel measurable subsets of I and λ is the restriction of λ to $\mathcal{B}$, then $(I, \mathcal{B}(I), \lambda)$ is a finite measure space; in this case, λ is called *Borel measure* on I.

The construction of Lebesgue measure was quite complicated, and the same is true of other measure spaces. We shall defer the construction of other measure spaces until Chapter 30.

Theorem 28.4.1   *Suppose that (X, Σ, μ) is a finite measure space. Suppose that $A, B \in \Sigma$, and that $(A_n)_{n=1}^{\infty}$ is a sequence in Σ.*

(i) $\mu(A \cup B) + \mu(A \cap B) = \mu(A) + \mu(B)$.
(ii) If $B \subseteq A$ then $\mu(B) \leq \mu(A)$.
(iii) (Upwards continuity) If $(A_n)_{n=1}^{\infty}$ is an increasing sequence then

$$\mu(\cup_{n=1}^{\infty} A_n) = \sup_{n \in \mathbf{N}} \mu(A_n).$$

(iv) (Downwards continuity) If $(A_n)_{n=1}^{\infty}$ is a decreasing sequence then

$$\mu(\cap_{n=1}^{\infty} A_n) = \inf_{n \in \mathbf{N}} \mu(A_n).$$

(v) $\sup_{n \in \mathbf{N}} \mu(A_n) \leq \mu(\cup_{n=1}^{\infty} A_n) \leq \sum_{n=1}^{\infty} \mu(A_n).$
(vi) $\mu(\cap_{n=1}^{\infty} A_n) \leq \inf_{n \in \mathbf{N}} \mu(A_n).$

Proof   The proof is left as an exercise for the reader.   □

Corollary 28.4.2   *If f is a measurable real-valued function on X, then $\mu(|f| > n) \to 0$ as $n \to \infty$.*

Proof   For the sequence $(|f| > n)_{n=1}^{\infty}$ decreases to the empty set.   □

Suppose that (X, Σ, μ) is a finite measure space and that $(A_n)_{n=1}^{\infty}$ is a sequence in Σ. We define

$$\limsup_{n \to \infty} A_n = \cap_{n=1}^{\infty} \left(\cup_{j=n}^{\infty} A_j \right) \text{ and } \liminf_{n \to \infty} A_n = \cup_{n=1}^{\infty} \left(\cap_{j=n}^{\infty} A_j \right).$$

Thus $x \in \limsup_{n \to \infty} A_n$ if and only if x is frequently in A_n: for each $n \in \mathbf{N}$ there exists $m \geq n$ such that $x \in A_m$, or, equivalently, $x \in A_n$ for infinitely many n. Similarly, $x \in \liminf_{n \to \infty} A_n$ if and only if x is eventually in A_n: there exists $n \in \mathbf{N}$ such that $x \in A_m$ for $m \geq n$.

Corollary 28.4.3 (The first Borel–Cantelli lemma)   *If $\sum_{n=1}^{\infty} \mu(A_n) < \infty$ then $\mu(\limsup_{n \to \infty} A_n) = 0$.*

Proof   Suppose that $\epsilon > 0$. There exists $N \in \mathbf{N}$ such that $\sum_{n=N+1}^{\infty} \mu(A_n) < \epsilon$. Then

$$\mu(\limsup_{n \to \infty} A_n) \leq \mu(\cup_{n=N+1}^{\infty} A_n) \leq \sum_{n=N+1}^{\infty} \mu(A_n) < \epsilon.$$

Since ϵ is arbitrary, the result follows.   □

Measure spaces provide the natural setting for probability theory. A *probability space* is a finite measure space $(\Omega, \Sigma, \mathbf{P})$ for which $\mathbf{P}(\Omega) = 1$. The elements of Σ are called *events*, $\mathbf{P}$ is called a *probability measure*, and $\mathbf{P}(A)$ is called the *probability* of A. The development of probability theory has contributed greatly to measure theory, but it would take us too far afield to investigate this[2].

---

[2] There are many excellent texts on probability theory; my favourite is R. Durrett, *Probability: Theory and Examples*, Cambridge University Press, 2010.

When we constructed Lebesgue measure on $\mathbf{R}$, we began by restricting attention to bounded sets, and then extending the results to more general sets. We can do the same in the present setting. For example, we could consider an arbitrary set X, consider the σ-field of all subsets of X and define μ to be counting measure on A: $\mu(A)$ is the number of elements in A, if A is finite, and $\mu(A) = \infty$ if A is infinite. We shall however restrict our attention to σ-finite measure spaces. A *σ-finite measure space* is a measurable space (X, Σ), together with a sequence $(I_k)_{k=1}^{\infty}$ of disjoint elements of Σ whose union is X, and a function μ on Σ (a σ-finite measure), taking values in $[0, \infty]$, with the properties that

(i) μ is countably additive; if $(A_n)_{n=1}^{\infty}$ is a sequence of disjoint elements of Σ then $\mu(\cup_{n \in \mathbf{N}} A_n) = \sum_{n=1}^{\infty} \mu(A_n)$, and
(ii) $\mu(I_k) < \infty$ for $k \in \mathbf{N}$.

Thus $A \in \Sigma$ if and only if $A \cap I_k \in \Sigma$ for $k \in \mathbf{N}$, and then $\mu(A) = \sum_{k=1}^{\infty} \mu(A \cap I_k)$. The results of Theorem 27.5.1 extend easily to σ-finite measure spaces.

We can also define σ-finite measures in terms of increasing sequences. Let $(J_k)_{k=1}^{\infty}$ be an increasing sequence in Σ whose union is X. Then $A \in \Sigma$ if and only if $A \cap J_k$ in Σ for each $k \in \mathbf{N}$. Then μ is a σ-finite measure on Σ if it is countably additive and if $\mu(J_k)$ is finite, for each $k \in \mathbf{N}$. Then $\mu(A) = \lim_{k \to \infty} \mu(A \cap J_k)$. This definition is clearly equivalent to the one above.

In future we shall use the term 'measure' to mean either a finite measure or a σ-finite measure.

Suppose that (X, Σ, μ) is a finite or σ-finite measure space. An element N of Σ is a *null set* if $\mu(N) = 0$. If $M \in \Sigma$ and $M \subset N$, where N is a null set, then M is a null set. If $(N_n)_{n=1}^{\infty}$ is a sequence of null sets, then $\mu(\cup_{n=1}^{\infty} N_n) \leq \sum_{n=1}^{\infty} \mu(N_n) = 0$, so that $\cup_{n=1}^{\infty} N_n$ is a null set. Thus the collection of null sets is a σ-ring contained in Σ.

A measure space (X, Σ, μ) is *complete* if every subset of a null set is measurable, and is therefore a null set. Lebesgue measure is complete, since if $M \subseteq N$, where N is a null set, then

$$0 \leq \lambda_*(M) \leq \lambda^*(M) \leq \lambda(N) = 0,$$

so that $\lambda_*(M) = \lambda^*(M) = 0$.

It is quite easy to complete a measure space.

Theorem 28.4.4 *Suppose that (X, Σ, μ) is a finite or σ-finite measure space. Let*

$$\mathcal{N} = \{M \subseteq X : \text{ there exists a null set } N \text{ such that } M \subseteq N\}.$$

Let $\hat{\Sigma} = \{A \cup N : A \in \Sigma, N \in \mathcal{N}\}$. Then $\hat{\Sigma}$ is a σ-field containing Σ. If $B = A \cup M \in \hat{\Sigma}$, let $\hat{\mu}(B) = \mu(A)$. Then $\hat{\mu}$ is well defined, and $(X, \hat{\Sigma}, \hat{\mu})$ is a complete measure space. If $B \in \hat{\Sigma}$ there exists A, C in Σ with $A \subseteq B \subseteq C$ and $\mu(A) = \hat{\mu}(B) = \mu(C)$.

Proof   If $B = A \cup M = A' \cup M' \in \hat{\Sigma}$, there exists a null set N such that $M \cup M' \subseteq N$. Since $A \triangle A' \subseteq M \cup M' \subseteq N$, $\mu(A) = \mu(A')$, and so $\hat{\mu}$ is well defined. If $(B_n)_{n=1}^{\infty} = (A_n \cup M_n)_{n=1}^{\infty}$ is a disjoint sequence in $\hat{\Sigma}$, then $\cup_{n=1}^{\infty} B_n = (\cup_{n=1}^{\infty} A_n) \cup (\cup_{n=1}^{\infty} M_n)$. Since $\cup_{n=1}^{\infty} M_n \in \mathcal{N}$,

$$\hat{\mu}(\cup_{n=1}^{\infty} B_n) = \mu(\cup_{n=1}^{\infty} A_n) = \sum_{n=1}^{\infty} \mu(A_n) = \sum_{n=1}^{\infty} \hat{\mu}(B_n),$$

so that $\hat{\mu}$ is a measure which extends μ.

If $B = A \cup M \in \hat{\Sigma}$, there exists a null set N such that $M \subseteq N$. Let $C = A \cup N$. Then $C \in \Sigma$, $A \subseteq B \subseteq C$ and $\mu(A) = \hat{\mu}(B) = \mu(C)$.   □

Exercises

28.4.1 Suppose that Σ is a σ-field of subsets of a set X, and that $\mu : \Sigma \to \mathbf{R}^+$ is a function. Show that the following are equivalent.
   (i) μ is a measure.
   (ii) *(Upwards continuity)* If $(A_n)_{n=1}^{\infty}$ is an increasing sequence in Σ and $A = \cup_{n=1}^{\infty} A_n$, then $\mu(A) = \sup_{n \in \mathbf{N}} \mu(A_n)$.
   (iii) *(Downwards continuity)* If $(A_n)_{n=1}^{\infty}$ is a decreasing sequence and $A = \cap_{n=1}^{\infty} A_n$, then $\mu(A) = \inf_{n \in \mathbf{N}} \mu(A_n)$.
   (iv) *(Downwards continuity at $\emptyset$)* If $(A_n)_{n=1}^{\infty}$ is a decreasing sequence and $\cap_{n=1}^{\infty} A_n = \emptyset$, then $\mu(A_n) \to 0$ as $n \to \infty$.
28.4.2 Suppose that (X, τ) is a compact Hausdorff space. Show that the collection R of subsets of X which are both open and closed is a field. Suppose that $\phi : R \to \mathbf{R}^+$ is finitely additive. Show that it is countably additive.
28.4.3 Suppose that (X, Σ) is a measurable space and that $(I_\gamma)_{\gamma \in \Gamma}$ is an uncountable family of disjoint elements of Σ whose union is X, and a function μ on Σ, taking values in $[0, \infty]$, with the properties that
   (i) μ is countably additive; if $(A_n)_{n=1}^{\infty}$ is a sequence of disjoint elements of Σ then $\mu(\cup_{n \in \mathbf{N}} A_n) = \sum_{n=1}^{\infty} \mu(A_n)$, and

(ii) $0 < \mu(I_\gamma) < \infty$ for $\gamma \in \Gamma$.

Show that if $A \in \Sigma$ and $\mu(A) < \infty$ then $\{\gamma : \mu(A \cap I_\gamma) < \infty\}$ is countable.

28.4.4 Does Corollary 28.4.2 hold for σ-finite measure spaces?

28.4.5 Does the first Borel–Cantelli lemma hold for σ-finite measure spaces?

28.5 Null sets and Borel sets

Let us now consider the null sets of $(\mathbf{R}, \mathcal{L}, \lambda)$.

Proposition 28.5.1 *A subset A of $\mathbf{R}$ is a λ-null set if and only if $\lambda^*(A) = 0$; that is, given $\epsilon > 0$ there exists a sequence $(I_n)_{n=1}^\infty$ of open intervals which cover A for which $\sum_{n=1}^\infty l(I_n) < \epsilon$.*

Proof   Immediate.   □

Corollary 28.5.2 $(\mathbf{R}, \mathcal{L}, \lambda)$ *is a complete measure space.*

As an example, a countable set, such as the set $\mathbf{Q}$ of rationals, is a null set, since it is a countable union of singleton sets, which are null sets. Thus there is a decreasing sequence $(U_n)_{n=1}^\infty$ of open sets containing $\mathbf{Q}$ with $\lambda(U_n) \to 0$ as $n \to \infty$.

On the other hand, Cantor's ternary set is a null set which is not countable.

Recall that a subset of $\mathbf{R}$ is a G_δ set if it is the intersection of a sequence of open sets, and is a K_σ set if it is the union of a sequence of compact sets. It follows from Theorem 27.5.1 that G_δ sets, K_σ sets and F_σ sets are Lebesgue measurable.

Theorem 28.5.3 *Suppose that A is a subset of $\mathbf{R}$. The following are equivalent.*

(i) A is Lebesgue measurable.

(ii) There exist a K_σ set C and a null set M disjoint from C such that $A = C \cup M$.

(iii) There exist a G_δ set B and a null set N contained in B such that $A = B \setminus N$.

Proof   Clearly (ii) implies (i). Suppose that A is Lebesgue measurable. For each $k \in \mathbf{Z}$, $A \cap (k, k+1]$ is Lebesgue measurable, and so for each $n \in \mathbf{N}$ there exists a compact set $K_{k,n}$ such that $K_{k,n} \subseteq A \cap (k, k+1]$, and

$\lambda(K_{k,n}) > \lambda(A) - 1/n$. Let $L_k = \cup_{n=1}^{\infty} K_{k,n}$, and let $N_k = (A \cap (k, k+1]) \setminus L_k$. Then

$$\lambda(N_k) = \lambda(A \cap (k, k+1]) - \lambda(L_k) \leq \lambda(A) - \lambda(K_{k,n}) \leq 1/n,$$

for each $n \in \mathbf{N}$, so that N_k is a null set. Then

$$A = C \cup N, \text{ where } C = \cup \{K_{k,n} : k \in \mathbf{Z}, n \in \mathbf{N}\} \text{ and } N = \cup_{k \in \mathbf{Z}} N_k.$$

C is a K_σ set, N is a null set and $C \cap N = \emptyset$. Thus (i) implies (ii). Finally, it follows by taking complements that (ii) and (iii) are equivalent. $\square$

Thus there exists a G_δ set B containing $\mathbf{Q}$ which is a null set. But $B \neq \mathbf{Q}$. Suppose that $B = \cap_{n=1}^{\infty} U_n$ (with each U_n open). Let $(r_n)_{n=1}^{\infty}$ be an enumeration of $\mathbf{Q}$, and let $V_n = U_n \setminus \{r_n\}$. Then each V_n is a dense open subset of $\mathbf{R}$, so that $\cap_{n=1}^{\infty} V_n = B \setminus \mathbf{Q}$ is a dense subset of $\mathbf{R}$, by Baire's category theorem.

Corollary 28.5.4 *Suppose that A is a subset of $\mathbf{R}$. The following are equivalent.*

(i) *A is Lebesgue measurable.*
(ii) *There exists a Borel set B and a null set N contained in B such that $A = B \setminus N$.*
(iii) *There exists a Borel set C and a null set M disjoint from C such that $A = C \cup M$.*

Proof Immediate. $\square$

Exercise

28.5.1 Let $g : [0, 1] \to [0, 2)$ be the mapping of Exercise 27.6.1. If A is a Borel subset of $[0, 1]$, is $g(A)$ a Borel subset of $[0, 2]$?

28.6 Almost sure convergence

In measure theory, the first Borel–Cantelli lemma is frequently used to show that a property holds on a measure space, except possibly on a set of measure 0. If so, we say that it holds *almost everywhere* or *almost surely*. Thus if $(f_n)_{n=1}^{\infty}$ is a sequence in $\mathcal{L}^0(X, \Sigma, \mu)$ and if $f \in \mathcal{L}^0(X, \Sigma, \mu)$ then we say that $f_n \to f$ almost surely, as $n \to \infty$, if there exists a null set N such that $f_n(x) \to f(x)$ for all $x \in X \setminus N$. Note that we do not require that $f_n(x)$ does not converge to $f(x)$ for $x \in N$; we are content to remain ignorant about what happens on N.

Proposition 28.6.1 *Suppose that (X, Σ, μ) is a measure space, that $(f_n)_{n=1}^{\infty}$ and $(g_n)_{n=1}^{\infty}$ are sequences in $\mathcal{L}^0(X, \Sigma, \mu)$ and that $f, g \in \mathcal{L}^0(X, \Sigma, \mu)$. If $f_n \to f$ and $g_n \to g$ almost everywhere, as $n \to \infty$, then $f_n + g_n \to f + g$ and $f_n g_n \to fg$ almost everywhere, as $n \to \infty$.*

Proof   For the union of two null sets is a null set.   □

Suppose that (X, Σ, μ) is a finite measure space, that $(f_n)_{n=1}^{\infty}$ is a sequence in $\mathcal{L}^0(X, \Sigma, \mu)$. We say that $f_n \to f$ *almost uniformly* if for each $\epsilon > 0$ there exists $A \in \Sigma$ with $\mu(A) < \epsilon$ such that $f_n \to f$ uniformly on $X \setminus A$ as $n \to \infty$. This terminology is standard, but is unfortunate, since 'almost' usually involves sets of measure 0: 'nearly uniform convergent' would be better.

How are these notions of convergence related? The next result is rather remarkable.

Theorem 28.6.2 (Egorov's theorem) *Suppose that (X, Σ, μ) is a finite measure space, that $(f_n)_{n=1}^{\infty}$ is a sequence in $\mathcal{L}^0(X, \Sigma, \mu)$ and that $f \in \mathcal{L}^0(X, \Sigma, \mu)$. Then $f_n \to f$ almost everywhere as $n \to \infty$ if and only if $f_n \to f$ almost uniformly as $n \to \infty$.*

Proof   Suppose first that $f_n \to f$ almost everywhere as $n \to \infty$ and that $\epsilon > 0$. Let

$$C = \{x : f_n(x) \text{ converges in } \mathbf{R} \text{ as } n \to \infty\},$$

and let

$$B_{n,k} = (|f_n - f| > 1/k) \text{ and } A_{n,k} = \cup_{m \geq n} B_{m,k}, \text{ for } n, k \in \mathbf{N}.$$

First, keep k fixed. Then $(A_{n,k})_{n=1}^{\infty}$ is a decreasing sequence in Σ, and $\cap_{n=1}^{\infty} A_{n,k} \subseteq (\limsup_{n \to \infty} |f_n - f| \geq 1/k) \subseteq X \setminus C$, so that, since $\mu(X \setminus C) = 0$, $\mu(A_{n,k}) \to 0$ as $n \to \infty$, by lower continuity. Thus there exists n_k such that $\mu(A_{n_k,k}) < \epsilon/2^k$.

Now let $A = \cup_{k=1}^{\infty} A_{n_k,k}$. Then $\mu(A) \leq \sum_{k=1}^{\infty} \mu(A_{n_k,k}) < \epsilon$. If $x \notin A$ and $k \in \mathbf{N}$, then $x \notin A_{n_k,k}$, and so $x \notin B_{n,k}$ for $n \geq n_k$. Thus $|f_n(x) - f(x)| \leq 1/k$ for $n \geq n_k$; $f_n \to f$ almost uniformly.

Conversely, suppose that $f_n \to f$ almost uniformly as $n \to \infty$. For each $k \in \mathbf{N}$ there exists $A_k \in \Sigma$ with $\mu(A_k) < 1/k$ such that $f_n \to f$ uniformly on

$X \setminus A_k$ as $n \to \infty$. Let $A = \cap_{k=1}^{\infty} A_k$. Then $\mu(A) = 0$ and $f_n \to f$ pointwise on $X \setminus A$. Thus $f_n \to f$ almost everywhere as $n \to \infty$. □

Let us also establish an easy, but important, approximation result that we shall need later, when we consider integration. Here the convergence is pointwise.

Theorem 28.6.3 *Suppose that f is a real-valued measurable function on a measure space (X, Σ, μ). There exists a sequence $(D_n)_{n=1}^{\infty}$ of simple measurable functions which converges pointwise to f. If f is non-negative, then the sequence $(D_n)_{n=1}^{\infty}$ can be taken to be a pointwise increasing sequence of non-negative functions.*

Proof   We use a simple construction. Suppose that $f \in \mathcal{L}^0(X, \Sigma, \mu)$, and that $n \in \mathbf{N}$. We divide the interval $[-n, n)$ into $2n.2^n$ disjoint intervals, each of length $1/2^n$; we set $I_{j,n} = ((j-1)/2^n, j/2^n]$ for $-n2^n + 1 \le j \le n.2^n$. Let $A_{j,n} = (f \in I_{j,n})$; then $A_{j,n}$ is measurable. We set

$$D_n(f) = \sum_{j=-n2^n+1}^{n.2^n} ((j-1)/2^n) I_{A_{j,n}}.$$

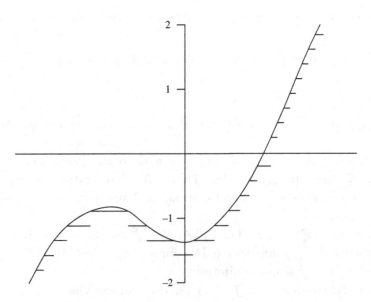

Figure 28.6.

Thus $D_n(f)$ is a simple measurable function, and if $n > |f(x)|$ then $D_n(f)(x) < f(x) \leq D_n(f)(x) + 1/2^n$. Consequently, $D_n(f) \to f$ pointwise. If $f \geq 0$, then the functions D_n are non-negative, and the sequence $(D_n)_{n=1}^{\infty}$ increases pointwise to f. $\qquad\square$

29

Integration

29.1 Integrating non-negative functions

After so much consideration of measure and of measurable functions, we are now in a position to develop the theory of integration. We begin by considering the integral of a non-negative extended-real-valued measurable function f defined on a finite or σ-finite measure space (X, Σ, μ). We define the *tail distribution function* λ_f to be

$$\lambda_f(t) = \mu(f > t), \text{ for } t \in [0, \infty).$$

If $\lambda_f(t) = \infty$ for some $t > 0$ (so that $\lambda_f(s) = \infty$ for $0 < s \le t$) we define $\int_X f \, d\mu = \infty$. Otherwise, the function λ_f is a decreasing real-valued function on $[0, \infty)$. It is continuous on the right, since if $t_n \searrow t$ as $n \to \infty$ then $(f > t_n) \nearrow (f > t)$, so that $\mu(f_n > t) \nearrow \mu(f > t)$. If $\mu(X) = \infty$, it may happen that $\lambda_f(t) \to \infty$ as $t \searrow 0$. We define

$$\int_X f \, d\mu = \int_0^\infty \lambda_f(t) \, dt.$$

Here, the integral on the right is an improper Riemann integral, taking values in $[0, \infty]$; this integral exists, since λ_f is a decreasing function. Note that $\lambda_f(t) \to \mu(f = \infty)$ as $t \to \infty$, so that if $\mu(f = \infty) > 0$ then $\int_X f \, d\mu = \infty$. Note also that $\int_X f \, d\mu = 0$ if and only if $f = 0$ almost everywhere. Figure 29.1 shows why this definition is made: the 'area under the curve' is shifted to the left, and then evaluated.

Note that if $0 \le f \le g$ almost everywhere then $\lambda_f \le \lambda_g$, and so $\int_X f \, d\mu \le \int_X g \, d\mu$. In particular, if $f = g$ almost everywhere then $\int_X f \, d\mu = \int_X g \, d\mu$, and if $f = 0$ almost everywhere then $\int_X f \, d\mu = 0$.

As an important but easy example, let us consider the integral of a non-negative simple measurable function f. We can suppose that $f = \sum_{j=1}^k \alpha_j I_{A_j}$, where $A_1, \ldots, A_k$ are disjoint measurable sets of finite

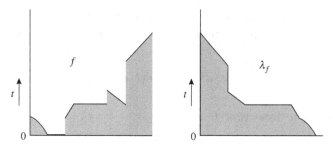

Figure 29.1.

measure, and $0 < \alpha_1 < \cdots < \alpha_k$. Let $\alpha_0 = 0$ and let $B_j = \cup_{i=j}^{k} A_i$. Then $B_j = (f > \alpha_{j-1})$, so that

$$\lambda_f(t) = \begin{cases} \mu(B_j) & \text{for } \alpha_{j-1} \le t < \alpha_j, \text{ for } 1 \le j \le k, \\ 0 & \text{for } \alpha_k \le t. \end{cases}$$

Thus

$$\int_X f \, d\mu = \sum_{j=1}^{k} (\alpha_j - \alpha_{j-1}) \mu(B_j)$$

$$= \sum_{j=1}^{k} (\alpha_j - \alpha_{j-1}) \left(\sum_{i=j}^{k} \mu(A_i) \right)$$

$$= \sum_{i=1}^{k} \left(\sum_{j=1}^{i} (\alpha_j - \alpha_{j-1}) \right) \mu(A_i) = \sum_{i=1}^{k} \alpha_i \mu(A_i).$$

It follows by elementary arguments that if $f = \sum_{j=1}^{k} \alpha_j I_{A_j}$, where the α_j are non-negative, but not necessarily distinct, and the A_j are not necessarily disjoint, then $\int_X f \, d\mu = \sum_{j=1}^{k} \alpha_j \mu(A_j)$.

We need the following easy, but fundamentally important, result concerning improper integrals of decreasing functions.

Theorem 29.1.1 *Suppose that $(g_n)_{n=1}^{\infty}$ is a sequence of decreasing non-negative functions on $(0, \infty)$, which increases pointwise to a function g. Then*

$$\int_0^{\infty} g(t) \, dt = \lim_{n \to \infty} \int_0^{\infty} g_n(t) \, dt.$$

Proof The function g is a decreasing non-negative function, so that the improper Riemann integral exists. We consider the case where $\int_0^{\infty} g(t) \, dt$ is finite; the proof when it is infinite is proved in exactly the same way.

The sequence $(\int_0^\infty g_n(t)\,dt)_{n=1}^\infty$ is increasing, and is bounded above by $\int_0^\infty g(t)\,dt$. Suppose that $\epsilon > 0$. There exist $0 < a < b < \infty$ such that

$$\int_a^b g(t)\,dt > \int_0^\infty g(t)\,dt - \epsilon/3,$$

and there exists a dissection $D = (a = t_0 < \cdots < t_k = b)$ of $[a,b]$ such that

$$s_D(g) = \sum_{j=1}^k g(t_j)(t_j - t_{j-1}) > \int_a^b g(t)\,dt - \epsilon/3.$$

Since $g_n \to g$ pointwise, there exists $N \in \mathbf{N}$ such that $g_n(t_j) > g(t_j) - \epsilon/3(b-a)$, for $1 \le j \le k$ and $n \ge N$. If $n \ge N$, then

$$\int_0^\infty g_n(t)\,dt \ge \int_a^b g_n(t)\,dt \ge s_D(g_n) = \sum_{j=1}^k g_n(t_j)(t_j - t_{j-1})$$

$$= \sum_{j=1}^k g(t_j)(t_j - t_{j-1}) - \sum_{j=1}^k (g(t_j) - g_n(t_j))(t_j - t_{j-1})$$

$$\ge s_D(g) - \epsilon/3 \ge \int_0^\infty g(t)\,dt - \epsilon,$$

which establishes the theorem. $\qquad\qquad\qquad\qquad\qquad\qquad\qquad\qquad$ $\square$

This enables us to prove the fundamental theorem of integration theory.

Theorem 29.1.2 (The monotone convergence theorem) *Suppose that $(f_n)_{n=1}^\infty$ is an increasing sequence of non-negative measurable functions on a finite or σ-finite measure space (X, Σ, μ) which converges pointwise almost everywhere to a function f. Then*

$$\int_X f_n\,d\mu \to \int_X f\,d\mu \text{ as } n \to \infty.$$

Proof   The sequence $(\int_X f_n\,d\mu)_{n=1}^\infty$ is increasing, and converges to a limit less than or equal to $\int_X f\,d\mu$. The importance of the result is that equality holds.

If $\lambda_{f_N}(t) = \infty$ for some $N \in \mathbf{N}$ and some $t > 0$ then $\int_X f_n\,d\mu = \infty$ for $n \ge N$, and the result holds trivially. Otherwise, if $t > 0$ then the sequence $(f_n > t)_{n=1}^\infty$ of sets in Σ increases to $(f > t)$, so that $\lambda_{f_n}(t) \to \lambda_f(t)$ as $n \to \infty$. If $\int_0^\infty \lambda_f(t)\,dt < \infty$ then the result follows from Theorem 29.1.1. If $\int_0^\infty \lambda_f(t)\,dt = \infty$, then given $M > 0$ there exists T such that $\int_0^T \lambda_f(t)\,dt >$

M. But then $\lim_{n\to\infty} \int_0^T \lambda_{f_n}(t)\,dt > M$, and so $\lim_{n\to\infty} \int_0^\infty \lambda_{f_n}(t)\,dt > M$. Since this holds for all $M > 0$, $\int_X f_n\,d\mu \to \infty$ as $n \to \infty$. □

The following example shows how powerful and useful this theorem is; it provides a direct proof of Theorem 26.7.5.

Example 29.1.3 If $x > 0$ then

$$\Gamma(x) = \lim_{n\to\infty} \frac{n!n^x}{x(x+1)\dots(x+n)}.$$

The functions $t^{x-1}(1 - t/n)^n I_{[0,n]}$ increase pointwise to the function $t^{x-1}e^{-t}$ on $[0,\infty)$ as $n \to \infty$, so that

$$\Gamma(x) = \int_0^\infty t^{x-1}e^{-t}\,dt = \lim_{n\to\infty} \int_0^n t^{x-1}(1 - \frac{t}{n})^n\,dt.$$

Making the change of variables $s = t/n$,

$$\int_0^n t^{x-1}(1 - \frac{t}{n})^n\,dt = n^x \int_0^1 s^{x-1}(1 - s)^n\,ds = n^x B(x, n+1),$$

where B is the beta function. As in Volume 1, Section 10.3, if $x > 0$ and $y > 0$ then, integrating by parts, $B(x, y+1) = (y/(x+y))B(x, y)$. Using this repeatedly,

$$\int_0^n t^{x-1}(1 - \frac{t}{n})^n\,dt = \frac{n!n^x}{x(x+1)\dots(x+n)}.$$

We use the monotone convergence theorem to prove a useful inequality.

Theorem 29.1.4 (Fatou's lemma) *Suppose that $(f_n)_{n=1}^\infty$ is a sequence of non-negative measurable functions on a finite or σ-finite measure space (X, Σ, μ). Then*

$$\int_X \liminf_{n\to\infty} f_n\,d\mu \leq \liminf_{n\to\infty} \int f_n\,d\mu.$$

In particular, if $f_n \to f$ almost everywhere then

$$\int_X f\,d\mu \leq \liminf_{n\to\infty} \int f_n\,d\mu.$$

Proof   Let $g_n = \inf_{j \geq n} f_j$. Then $g_n \leq f_n$ and g_n increases pointwise to $\lim\inf_{n \to \infty} f_n$, so that

$$\int_X \lim_{n \to \infty} \inf f_n \, d\mu = \lim_{n \to \infty} \int_X g_n \, dx \leq \lim_{n \to \infty} \inf \int f_n \, d\mu.$$

$\square$

Equality need not hold: for example if $f_n = nI_{(0,1/n]}$ or if $f_n = (1/n)I_{[0,n]}$ then $f_n \to 0$ pointwise, but $\int_{\mathbf{R}} f_n \, d\lambda = 1$, for $n \in \mathbf{N}$.

The integral respects algebraic operations.

Theorem 29.1.5   *Suppose that f and g are non-negative measurable functions on a finite or σ-finite measure space (X, Σ, μ), and that α and β are non-negative numbers. Then*

$$\int_X (\alpha f + \beta g) \, d\mu = \alpha \int_X f \, d\mu + \beta \int_X g \, d\mu.$$

Proof   First suppose that $f = \sum_{j=1}^k \alpha_j I_{A_j}$ and $g = \sum_{m=1}^n \beta_m I_{B_m}$ are simple measurable functions, with $\cup_{j=1}^k A_j = \cup_{m=1}^n B_m = X$. Then

$$\alpha f + \beta g = \sum_{j=1}^k \sum_{m=1}^n (\alpha \alpha_j + \beta \beta_m) I_{A_j \cap B_m},$$

so that

$$\int_X (\alpha f + \beta g) \, d\mu = \sum_{j=1}^k \sum_{m=1}^n (\alpha \alpha_j + \beta \beta_m) \mu(A_j \cap B_m)$$

$$= \alpha \sum_{j=1}^k \alpha_j \left(\sum_{m=1}^n \mu(A_j \cap B_m) \right)$$

$$+ \beta \sum_{m=1}^n \beta_m \left(\sum_{j=1}^k \mu(A_j \cap B_m) \right)$$

$$= \alpha \sum_{j=1}^k \alpha_j \mu(A_j) + \beta \sum_{m=1}^n \beta_m \mu(B_m)$$

$$= \alpha \int_X f \, d\mu + \beta \int_X f \, d\mu.$$

If f and g are measurable then, by Theorem 28.6.3, there exist increasing sequences $(f_n)_{n=1}^\infty$ and $(g_n)_{n=1}^\infty$ of non-negative simple measurable functions which converge pointwise to f and g respectively. Then the sequence $(\alpha f_n + \beta g_n)_{n=1}^{\infty}$ of simple functions increases pointwise to $\alpha f + \beta g$. Applying the theorem of monotone convergence,

$$\int_X (\alpha f + \beta g)\, d\mu = \lim_{n\to\infty} \int_X (\alpha f_n + \beta g_n)\, d\mu$$

$$= \lim_{n\to\infty} \left(\alpha \int_X f_n\, d\mu + \beta \int_X g_n\, d\mu \right)$$

$$= \alpha \int_X f\, d\mu + \beta \int_X f\, d\mu.$$

$\square$

Suppose that f is a non-negative measurable function on a finite or σ-finite measure space (X, Σ, μ), We can also integrate f over measurable sets. If $A \in \Sigma$, we set $\int_A f\, d\mu = \int_X f I_A, d\mu$, where I_A is the indicator function of A. Alternatively, let $\Sigma_A = \{B \in \Sigma, B \subseteq A\}$. Then Σ_A is a σ-field of subsets of A, the restriction μ_A of μ to Σ_A is a measure on Σ_A, the restriction f_A of f to A is Σ_A-measurable, and $\int_A f\, d\mu = \int_A f_A\, d\mu_A$.

Exercise

29.1.1 Suppose that f is a non-negative measurable function on a measure space (X, Σ, μ) and that $f(x) > 0$ for almost all x. Show that $\int_X f\, d\mu = 0$ if and only if $\mu(X) = 0$.

29.2 Integrable functions

We next consider the problem of integrating a real-valued measurable function f on a finite or σ-finite measure space (X, Σ, μ), when f is not necessarily non-negative. This is done by integrating the positive and negative parts of f separately, and trying to combine the integrals. Thus we make the following definitions:

- $\int_X f\, d\mu = \int_X f^+\, d\mu - \int_X f^-\, d\mu$ if $\int_X f^+\, d\mu < \infty$ and $\int_X f^-\, d\mu < \infty$;
- $\int_X f\, d\mu = \infty$ if $\int_X f^+\, d\mu = \infty$ and $\int_X f^-\, d\mu < \infty$;
- $\int_X f\, d\mu = -\infty$ if $\int_X f^+\, d\mu < \infty$ and $\int_X f^-\, d\mu = \infty$.
- $\int_X f\, d\mu$ is not defined if $\int_X f^+\, d\mu = \infty$ and $\int_X f^-\, d\mu = \infty$.

The function f is said to be *integrable* if $\int_X f^+\, d\mu < \infty$ and $\int_X f^-\, d\mu < \infty$; this is clearly the case if and only if $\int_X |f|\, d\mu < \infty$.

Proposition 29.2.1 *If f and g are integrable, if h is a bounded measurable function and if $\alpha \in \mathbf{R}$ then hf, αf and $f + g$ are integrable, and*

$$\int_X \alpha f \, d\mu = \alpha \int_X f \, d\mu \quad \text{and} \quad \int_X (f + g) \, d\mu = \int_X f \, d\mu + \int_X g \, d\mu.$$

Proof   Let $M = \sup_{x \in X} |h(x)|$. Since $(hf)^+ \le M|f|$ and $(hf)^- \le M|f|$, the function hf is integrable, and so therefore is αf. If $\alpha \ge 0$ then

$$\int_X \alpha f \, d\mu = \int_X (\alpha f)^+ \, d\mu - \int_X (\alpha f)^- \, d\mu = \int_X \alpha f^+ \, d\mu - \int_X \alpha f^- \, d\mu$$

$$= \alpha \left(\int_X f^+ \, d\mu - \int_X f^- \, d\mu \right) = \alpha \int_X f \, d\mu,$$

and if $\alpha < 0$ then

$$\int_X \alpha f \, d\mu = \int_X (\alpha f)^+ \, d\mu - \int_X (\alpha f)^- \, d\mu = \int_X |\alpha| f^- \, d\mu - \int_X |\alpha| f^+ \, d\mu$$

$$= |\alpha| \left(\int_X f^- \, d\mu - \int_X f^+ \, d\mu \right) = -|\alpha| \int_X f \, d\mu = \alpha \int_X f \, d\mu.$$

Since $|f + g| \le |f| + |g|$, the function $f + g$ is integrable.
Since $(f + g)^+ + f^- + g^- = (f + g)^- + f^+ + g^+$,

$$\int_X (f+g)^+ \, d\mu + \int_X f^- \, d\mu + \int_X g^- \, d\mu = \int_X (f+g)^- \, d\mu + \int_X f^+ \, d\mu + \int_X g^+ \, d\mu.$$

Rearranging,

$$\int_X (f + g) \, d\mu = \int_X f \, d\mu + \int_X g \, d\mu.$$

$\square$

We denote the set of integrable functions on (X, Σ, μ) by $\mathcal{L}^1_{\mathbf{R}}(X, \Sigma, \mu)$. The proposition shows that $\mathcal{L}^1_{\mathbf{R}}(X, \Sigma, \mu)$ is a vector space and the mapping $f \to \int_X f \, d\mu$ is a linear functional on it.

We have the following simple consequence of the monotone convergence theorem.

Proposition 29.2.2 (Beppo Levi's theorem)   *If $(f_n)_{n=1}^{\infty}$ is a sequence of integrable functions increasing pointwise to f, then*

$$\int_X f_n \, d\mu \to \int_X f \, d\mu \text{ as } n \to \infty.$$

Proof   For $\int_X (f_n - f_1)\, d\mu \to \int_X (f - f_1)\, d\mu$ as $n \to \infty$, by the monotone convergence theorem, and so

$$\int_X f\, d\mu = \int_X (f - f_1)\, d\mu + \int_X f_1\, d\mu$$

$$= \lim_{n \to \infty} \left(\int_X (f_n - f_1)\, d\mu \right) + \int_X f_1\, d\mu = \lim_{n \to \infty} \left(\int_X f_n\, d\mu \right).$$

□

Here is an application.

Example 29.2.3   Euler's number γ is given by the formula

$$\gamma = \int_0^1 \frac{1 - e^{-t}}{t}\, dt - \int_1^\infty \frac{e^{-t}}{t}\, dt.$$

In Volume 1, Exercise 8.8.9, it was shown that

$$-\gamma = \lim_{n \to \infty} \left(\int_0^1 \frac{(1 - t/n)^n - 1}{t}\, dt + \int_1^n \frac{(1 - t/n)^n}{t}\, dt \right);$$

the first integrand increases to $(e^{-t} - 1)/t$ and the second to e^{-t}/t.

A much more important result follows from Fatou's lemma.

Theorem 29.2.4 (The dominated convergence theorem)   *Suppose that* $(f_n)_{n=1}^\infty$ *is a sequence of measurable functions which converges pointwise almost everywhere to* f, *and that* g *is an integrable function such that* $|f_n| \le |g|$, *for each* n. *Then*

$$\int_X f_n\, d\mu \to \int_X f\, d\mu \text{ as } n \to \infty.$$

Proof   The functions $|g| + f_n$ and $|g| + f$ are non-negative. By Fatou's lemma,

$$\int_X (|g| + f)\, d\mu \le \lim_{n \to \infty} \inf \int_X (|g| + f_n)\, d\mu = \int_X |g|\, d\mu + \lim_{n \to \infty} \inf \int_X f_n\, d\mu,$$

so that f is integrable, and $\int_X f\, d\mu \le \liminf_{n \to \infty} \int_X f_n\, d\mu$. Similarly, the functions $|g| - f_n$ and $|g| - f$ are non-negative, so that $\int_X (-f)\, d\mu \le \liminf_{n \to \infty} \int_X (-f_n)\, d\mu$; thus

$$\int_X f\, d\mu \ge \lim_{n \to \infty} \sup \int_X f_n\, d\mu \ge \lim_{n \to \infty} \inf \int_X f_n\, d\mu \ge \int_X f\, d\mu.$$

Consequently, all the terms are equal, and the result follows.   □

Corollary 29.2.5   *Further, $\int_X |f_n - f| \, d\mu \to 0$ as $n \to \infty$.*

Proof   For $|f_n - f| \leq 2g$ almost everywhere, and $|f_n - f| \to 0$ almost everywhere.   □

Corollary 29.2.6 (The bounded convergence theorem)   *Suppose that $(f_n)_{n=1}^\infty$ is a uniformly bounded sequence of measurable functions on a finite measure space (X, Σ, μ) which converges pointwise almost everywhere to f. Then*

$$\int_X f_n \, d\mu \to \int_X f \, d\mu \ \text{ and } \ \int_X |f_n - f| \, d\mu \to 0 \text{ as } n \to \infty.$$

Proof   Take g to be the constant function taking the value $M = \sup\{|f_n(x)| : n \in \mathbf{N}, x \in X\}$.   □

The dominated convergence theorem and the bounded convergence theorem can be applied to infinite series, as the exercises show.

What is the relation between the Riemann integral and the Lebesgue integral? In order to answer this, we need to introduce the notions of the upper and lower envelopes of a bounded function. Let f be *any* bounded real-valued function on an interval I. We define the *upper envelope* $M(f)$ and the *lower envelope* $m(f)$ as

$$M(f)(x) = \inf_{\delta > 0} \left(\sup\{f(y) : y \in I, |x - y| < \delta\}\right),$$
$$m(f)(x) = \sup_{\delta > 0} \left(\inf\{f(y) : y \in I, |x - y| < \delta\}\right).$$

Proposition 29.2.7   *If $M(f)$ is the upper envelope of a bounded real-valued function f on an interval I and $m(f)$ is the lower envelope of f, then $M(f)$ is upper semi-continuous and $m(f)$ is lower semi-continuous.*

Proof   Suppose that $x \in I$ and $\epsilon > 0$. There exists $\delta > 0$ such that if $y \in (x - \delta, x + \delta) \cap I$ then $f(y) < M(f)(x) + \epsilon$. For such y, there exists $\eta > 0$ such that $(y - \eta, y + \eta) \subseteq (x - \delta, x + \delta)$, and so $M(f)(y) < M(f)(x) + \epsilon$. The lower semi-continuity of $m(f)$ is proved similarly.   □

Thus $M(f)$ and $m(f)$ are measurable, even though f need not be. Note that $m(f)(x) \leq f(x) \leq M(f)(x)$, and that $m(f)(x) = M(f)(x)$ if and only if f is continuous at x. The function $\Omega(f) = M(f) - m(f)$ is the *oscillation* of f.

Theorem 29.2.8 *Suppose that f is a bounded real-valued function on a closed interval $I = [a, b]$. Then*

$$\int_I M(f) \, d\lambda = \overline{\int_a^b} f(x) \, dx, \text{ the upper Riemann integral of } f,$$

$$\int_I m(f) \, d\lambda = \underline{\int_a^b} f(x) \, dx, \text{ the lower Riemann integral of } f.$$

Proof Suppose that $\epsilon > 0$. There exists a step function $v \geq f$ such that $\int_a^b v(x) \, dx \leq \overline{\int_a^b} f(x) \, dx + \epsilon$. Then $M(f)(x) \leq v(x)$ except possibly at the finite set of points of discontinuity of v, and so $\int_I M(f) \, d\mu \leq \int_a^b v(x) \, dx$. Since ϵ is arbitrary, it follows that $\int_I M(f) \, d\lambda \leq \overline{\int_a^b} f(x) \, dx$.

Let $M_n(f)(x) = \sup\{M(f)(y) : y \in [x - 1/n, x + 1/n] \cap I\}$, for $n \in \mathbf{N}$. Since $M(f)$ is upper semi-continuous, $M_n(f)$ decreases pointwise to $M(f)$. Thus $\int_I M_n(f) \, d\lambda \to \int_I M(f) \, d\lambda$ as $n \to \infty$, by the theorem of bounded convergence. Hence there exists N such that $\int_I M_N(f) \, d\lambda < \int_I M(f) \, d\lambda + \epsilon$. Let

$$D = (a = x_0 < \cdots < x_k = b)$$

be a dissection of $[a, b]$ with mesh size less than $1/N$. If

$$K_j = \sup\{f(x) : x \in [x_{j-1}, x_j]\},$$

then $K_j \leq M_N(f)(x)$ for $x \in [x_{j-1}, x_j]$, and so

$$\overline{\int_a^b} f(x) \, dx \leq \sum_{j=1}^k K_j(x_j - x_{j-1}) \leq \int_I M_N(f) \, d\lambda < \int_I M(f) \, d\lambda + \epsilon.$$

Since ϵ is arbitrary, it follows that $\overline{\int_a^b} f(x) \, dx \leq \int_I M(f) \, d\lambda$.

The other equality is proved similarly. $\square$

Corollary 29.2.9 *The function f is Riemann integrable if and only if it is continuous almost everywhere. If so, then the Riemann integral of f and the Lebesgue integral of f are equal.*

Proof The theorem implies that f is Riemann integrable if and only if $\int_I (M(f) - m(f)) \, d\lambda = 0$. Since $M(f) \geq f \geq m(f)$, this happens if and only if $M(f) = f = m(f)$ almost everywhere; that is, if and only if f is continuous almost everywhere.

If f is Riemann integrable, then $f = M(f)$ almost everywhere, so that f is Lebesgue measurable and

$$\int_a^b f(x)\, dx = \int_I M(f)\, d\lambda = \int_I f\, d\lambda.$$

$\square$

Complex-valued integrable functions can also be defined. If f is a complex-valued measurable function on (X, Σ), and $f = g + ih$, where g and h are real-valued functions, then f is integrable if g and h are, and the integral is defined as

$$\int_X f\, d\mu = \int_X g\, d\mu + i \int_X h\, d\mu.$$

The set of complex-valued integrable functions is denoted by $\mathcal{L}_{\mathbb{C}}^1(X, \Sigma, \mu)$. It is readily verified that $\mathcal{L}_{\mathbb{C}}^1(X, \Sigma, \mu)$ is a complex vector space, and that the mapping $f \to \int_X f\, d\mu$ is a complex linear functional on it. The reader should verify that the dominated convergence theorem and the bounded convergence theorem also hold for complex-valued functions.

Exercises

29.2.1 Suppose that ϕ is a non-negative finitely additive function on the Borel sets of $[0, 1]$, and that $\phi(A) = \sup\{\phi(K) : K \text{ compact}, K \subseteq A\}$ for each Borel set A. Show that ϕ is countably additive.

29.2.2 By making the substitution $x = ny$, calculate

$$\lim_{n \to \infty} n \int_0^1 (1 - y)^n \cos ay\, dy.$$

29.2.3 Prove the following form of Dirichlet's test. Suppose that $(f_n)_{n=1}^\infty$ is a decreasing sequence of integrable functions which converges to 0 almost everywhere and that $(g_n)_{n=1}^\infty$ is a sequence of bounded measurable functions for which the sequence of partial sums $(\sum_{j=1}^n g_j)_{n=1}^\infty$ is uniformly bounded. Show that $\sum_{n=1}^\infty f_n g_n$ converges almost everywhere to an integrable function s, and that

$$\int_X s\, d\mu = \sum_{n=1}^\infty \int_X f_n g_n\, d\mu.$$

29.2.4 Let f be the indicator function of a fat Cantor set. Show that there exists no function equal to f almost everywhere which is Riemann integrable.

The remaining exercises provide examples of the use of the theorem of dominated convergence, and the theorem of bounded convergence..

29.2.5 Use the thorem of dominated convergence to prove Kronecker's lemma:

$$\text{If } a_j \geq 0 \text{ and } \sum_{j=1}^{\infty} \frac{a_j}{j} < \infty \text{ then } \frac{1}{n} \sum_{j=1}^{n} a_j \to 0 \text{ as } n \to \infty.$$

29.2.6 Suppose that μ is a finite Borel measure on the Euclidean space $\mathbf{R}^d$. If $y \in \mathbf{R}^d$, let

$$\hat{\mu}(y) = \int_{\mathbf{R}^d} e^{-i\langle x,y \rangle} \, d\mu(x);$$

$\hat{\mu}$ is the *Fourier transform* of μ. Show that $\hat{\mu}$ is a bounded continuous function on $\mathbf{R}^d$.

29.2.7 Suppose that $1 < \alpha < 2$. Let $f_n(x) = n^\alpha x/(1 + n^2 x^2)$, for $x \in [0, 1]$. Show that $f_n \to 0$ pointwise, but not uniformly. Use the theorem of dominated convergence to show that $\int_0^1 f_n(x) \, dx \to 0$ as $n \to \infty$. Show that this can also be proved by making a change of variables.

29.2.8 Suppose that $0 < x \leq \pi$. Let

$$s_n(x) = \sum_{j=1}^{n} \frac{\sin jx}{j} \text{ and let } t_n(x) = \sum_{j=1}^{n} \cos jx.$$

Show that

$$s_n(x) = \int_0^x t_n(t) \, dt = \int_0^{x/2} \frac{\sin(2n + 1)t}{\sin t} \, dt - \frac{x}{2}.$$

29.2.9 Show that if $0 < a < b \leq \pi/2$ then

$$\int_a^b \frac{\sin(2n + 1)t}{\sin t} \, dt \to 0 \text{ as } n \to \infty.$$

Deduce that if $0 < x < \pi$ then $s_n(x) \to (\pi - x)/2$ as $n \to \infty$.

29.2.10 Show that

$$0 < \int_0^{x/2} \frac{\sin(2n + 1)t}{\sin t} \, dt \leq \int_0^{\pi/(2n+1)} \frac{\sin(2n + 1)t}{\sin t} \, dt$$

$$= \int_0^{\pi} \frac{\sin t}{(2n + 1)\sin(t/(2n + 1))} \, dt < \frac{\pi}{2} \int_0^{\pi} \frac{\sin t}{t} \, dt.$$

Deduce that there exists K such that

$$\sup_{n \in \mathbf{N}} |s_n(x)| \leq K, \text{ for } x \in [0, \pi].$$

29.2.11 Let

$$u_n(x) = \sum_{j=1}^{n} \frac{(-1)^{j+1} \sin jx}{j} \text{ and let } v_n(x) = \sum_{j=1}^{n} \frac{\sin(2j-1)x}{2j-1}.$$

By making a change of variables, show that $u_n(x) \to x/2$ as $n \to \infty$ and that $\sup_{n \in \mathbf{N}} |u_n(x)| \leq K$ (the constant in the previous exercise), for $x \in [0, \pi)$. Deduce that $v_n(x) \to \pi/2$ as $n \to \infty$ and that $\sup_{n \in \mathbf{N}} |v_n(x)| \leq 2K$, for $x \in (0, \pi)$.

29.2.12 Suppose that f is a Lebesgue integrable function on $[0, \pi]$. Show that

$$\frac{2}{\pi} \sum_{n=1}^{\infty} \frac{1}{2n-1} \int_0^{\pi} f(t) \sin(2n-1)t \, dt = \int_0^{\pi} f(t) \, dt.$$

29.3 Changing measures and changing variables

In this section, we establish various results involving changes of measure and changes of variable.

Proposition 29.3.1 *Suppose that ϕ is a measurable mapping from a finite measure space (X, Σ, μ) into a measurable space (Y, T). If $A \in T$, let $\phi_* \mu(A) = \mu(\phi^{-1}(A))$. Then $\phi_* \mu$ is a finite measure on T.*

Proof If $(A_n)_{n=1}^{\infty}$ is a sequence of disjoint sets in T then $(\phi^{-1}(A_n))_{n=1}^{\infty}$ is a sequence of disjoint sets in Σ, and so

$$\phi_* \mu(\cup_{n \in \mathbf{N}} A_n) = \mu(\cup_{n \in \mathbf{N}} \phi^{-1}(A_n)) = \sum_{n=1}^{\infty} \mu(\phi^{-1}(A_n)) = \sum_{n=1}^{\infty} \phi_* \mu(A_n).$$

$\square$

Some care is needed when (X, Σ, μ) is σ-finite. For example, if $\phi : \mathbf{R} \to \mathbf{R}$ is defined by $\phi(x) = \sin x$ and if A is a Borel subset of $\mathbf{R}$ then $\lambda(\phi^{-1}(A)) = 0$ or ∞.

Proposition 29.3.2 *Suppose that ϕ is a measurable mapping from a σ-finite measure space (X, Σ, μ) into a measurable space (Y, T). Suppose also that there exists an increasing sequence $(A_n)_{n=1}^{\infty}$ in T with union Y such* *that $\mu(\phi^{-1}(A_n)) < \infty$ for each $n \in \mathbf{N}$. If $A \in T$, let $\phi_*\mu(A) = \mu(\phi^{-1}(A))$. Then $\phi_*\mu$ is a σ-finite measure on T.*

Proof　Suppose that $(B_j)_{j=1}^{\infty}$ is an increasing sequence in T, with union B. Then

$$\phi_*(\mu)(B) = \mu(\phi^{-1}(B)) = \lim_{n\to\infty} \mu(\phi^{-1}(B \cap A_n))$$

$$= \lim_{n\to\infty} \lim_{j\to\infty} \mu(\phi^{-1}(B_j \cap A_n)) = \lim_{j\to\infty} \lim_{n\to\infty} \mu(\phi^{-1}(B_j \cap A_n))$$

$$= \lim_{j\to\infty} \mu(\phi^{-1}(B_j)) = \lim_{j\to\infty} \phi_*\mu(B_j),$$

so that $\phi_*\mu$ is a measure on T. Since $\phi_*\mu(A_n) < \infty$, it is σ-finite. ☐

The measure $\phi_*\mu$ is called the *image measure*, or *push-forward measure*. If ϕ is real-valued, then $\phi_*\mu$ is called the *distribution* of ϕ. In the case where μ is a probability measure, it is also called the *law* of ϕ.

For example, if $(X, \Sigma, \mu) = ((0, 2\pi], \mathcal{L}, \lambda)$ and $e(t) = e^{it}$, then the image measure is Lebesgue measure on $\mathbf{T}$, and if we replace $\mathcal{L}$ by $\mathcal{B}$, then the image measure is Borel measure on $\mathbf{T}$. On the other hand, if $(X, \Sigma, \mu) = ((0, 1], \mathcal{L}, \lambda)$ and $h(t) = e^{2\pi it}$, then the image measure is called *Haar measure* on $\mathbf{T}$.

Proposition 29.3.3　*Suppose that ϕ is a measurable mapping from a measure space (X, Σ, μ) into a measurable space (Y, T), and that $\phi_*\mu$ is finite or σ-finite. Suppose that f is a measurable function on Y. Then f is an integrable function on $(Y, T, \phi_*\mu)$ if and only if $f \circ \phi$ is an integrable function on X; if so, then*

$$\int_X f \circ \phi \, d\mu = \int_Y f \, d(\phi_*\mu).$$

Proof　Suppose that $f \geq 0$. Then

$$\int_Y f \, d(\phi_*\mu) = \int_0^{\infty} \phi_*\mu(f > t) \, dt = \int_0^{\infty} \mu(f \circ \phi > t) \, dt = \int_X f \circ \phi \, d\mu.$$

Finally the result follows for general f by considering f^+ and f^-, and subtracting. ☐

Corollary 29.3.4   *If g is an integrable real-valued function on (X, Σ, μ) then*

$$\int_X g \, d\mu = \int_{\mathbf{R}} t \, d(g_* \mu)(t).$$

Proof   Take $\phi = g$ and f the identity mapping on $\mathbf{R}$.   □

We can also construct new measures by multiplying by integrable functions.

Proposition 29.3.5   *Suppose that g is a non-negative integrable function on a measure space (X, Σ, μ). If $A \in \Sigma$, let $(g.d\mu)(A) = \int_A g \, d\mu$. Then $g.d\mu$ is a finite measure on (X, Σ), and if f is a measurable function for which fg is μ integrable, then $\int_X f \, d(g.d\mu) = \int_X fg \, d\mu$.*

Proof   Suppose that $(A_n)_{n=1}^\infty$ is a sequence of disjoint elements of Σ, with union A. Let $B_n = \cup_{j=1}^n A_j$, for $n \in \mathbf{N}$. Since gI_{B_n} increases pointwise to gI_A, it follows from monotone convergence that

$$(g \cdot d\mu)(A) = \lim_{n \to \infty} (g \cdot d\mu)(B_n) = \lim_{n \to \infty} \sum_{j=1}^n (g \cdot d\mu)(A_n) = \sum_{j=1}^\infty (g \cdot d\mu)(A_n),$$

and so $g.d\mu$ is a finite measure.

If $f = I_A$ then $\int_X f \, d(g \cdot d\mu) = \int_X fg \, d\mu$, and the result also holds for simple functions, by linearity. If $f \geq 0$ then there exists an increasing sequence $(f_n)_{n=1}^\infty$ of simple functions increasing pointwise to f. Then $f_n g$ increases pointwise to fg, and so by monotone convergence,

$$\int_X f \, d(g \cdot d\mu) = \lim_{n \to \infty} \int_X f_n \, d(g \cdot d\mu) = \lim_{n \to \infty} \int_X f_n g \, d\mu = \int_X fg \, d\mu.$$

Finally the result follows for integrable f by considering f^+ and f^-, and subtracting.   □

29.4 Convergence in measure

In order to go further, we need to introduce another mode of convergence of measurable functions, namely *convergence in measure*. Suppose that (X, Σ, μ) is a finite measure space, and that $\mathcal{L}^0 = \mathcal{L}^0(X, \Sigma, \mu)$ is the vector space of of real-valued measurable functions on X. Suppose that $(f_n)_{n=1}^\infty$ is a sequence in $\mathcal{L}^0$ and that $f \in \mathcal{L}^0$. We say that $f_n \to f$ *in measure* if, for each $c > 0$, $\mu(|f_n - f| > c) \to 0$ as $n \to \infty$. In the case where $(X, \Sigma, \mathbf{P})$ is a probability space, probabilists call 'convergence in measure' *convergence in probability*.

Proposition 29.4.1 *Suppose that (X, Σ, μ) is a finite measure space, that $(f_n)_{n=1}^{\infty}$ is a sequence in $\mathcal{L}^0(X, \Sigma, \mu)$ and that $f \in \mathcal{L}^0$. If $f_n \to f$ almost everywhere, then $f_n \to f$ in measure.*

Proof   By Egorov's theorem, $f_n \to f$ almost uniformly. Thus if $c > 0$ there exists $A \in \Sigma$ with $\mu(A) < c$ such that $f_n \to f$ uniformly on $X \setminus A$; hence $f_n \to f$ in measure.   □

The following example shows that the converse of this proposition is not true. Let $(X, \Sigma, \mu) = ((0, 1], \mathcal{L}, \lambda)$. If $n = 2^k + j \in \mathbf{N}$, with $0 \le j < 2^k$, let f_n be the indicator function of the interval $(j/2^k, (j+1)/2^k]$. If $0 < c < 1$ then $\lambda(f_n > c) = 1/2^k$, and if $c \ge 1$ then $\lambda(f_n > c) = 0$. Thus $f_n \to 0$ in measure as $n \to \infty$. On the other hand, if $x \in (0, 1]$ then $f_n(x) = 0$ for infinitely many values of n, and equals 1 for infinitely many values of n, so that f_n does not converge at any point of $(0, 1]$.

Nevertheless, convergence in measure and convergence almost everywhere are closely related.

Theorem 29.4.2 *Suppose that (X, Σ, μ) is a finite measure space, that $(f_n)_{n=1}^{\infty}$ is a sequence in $\mathcal{L}^0(X, \Sigma, \mu)$ and that $f_n \to f$ in measure. Then there exists a subsequence $(f_{n_k})_{k=1}^{\infty}$ which converges almost everywhere to f as $k \to \infty$.*

Proof   We use the first Borel–Cantelli lemma. For each $k \in \mathbf{N}$ there exists n_k such that $\mu(|f_n - f| > 1/k) < 1/2^k$ for $n \ge n_k$. We can clearly suppose that $(n_k)_{k=1}^{\infty}$ is a strictly increasing sequence. Let $B_k = (|f_{n_k} - f| > 1/k)$; then $\sum_{k=1}^{\infty} \mu(B_k) < \infty$, so that $\mu(\limsup_{k\to\infty} B_k) = 0$. If $x \notin \limsup_{k\to\infty} B_k$ then there exists K such that $x \notin \bigcup_{k=K}^{\infty} B_k$, so that $|f_{n_k}(x) - f(x)| \le 1/k$ for $k \ge K$, and $f_{n_k}(x) \to f(x)$ as $k \to \infty$. Thus $f_{n_k} \to f$ almost everywhere.   □

Convergence in measure can be characterized by a pseudometric. We define the function ϕ_0 on $[0, \infty)$ by setting

$$\phi_0(t) = t \text{ for } 0 \le t \le 1, \text{ and } \phi_0(t) = 1 \text{ for } 1 < t < \infty.$$

Since ϕ_0 is bounded and continuous, $\phi_0 \circ |f|$ is integrable, for each $f \in \mathcal{L}^0$.

Theorem 29.4.3 *If $f, g \in \mathcal{L}^0(X, \Sigma, \mu)$, let*

$$\rho(f, g) = \int_X \phi_0(|f - g|) \, d\mu.$$

Then ρ is a pseudometric on $\mathcal{L}^0(X, \sigma, \mu)$, and $\rho(f, g) = 0$ if and only if $f = g$ almost everywhere. If $(f_n)_{n=1}^\infty$ is a sequence in $\mathcal{L}^0(X, \Sigma, \mu)$, then $\rho(f_n, f) \to 0$ if and only if $f_n \to f$ in measure.

Proof   Clearly $\rho(f, g) = \rho(g, f)$. Suppose that $f, g, h \in \mathcal{L}^0(X, \Sigma, \mu)$. Since ϕ_0 is an increasing function and $\phi_0(x + y) \leq \phi_0(x) + \phi_0(y)$ for $x, y \in [0, \infty)$,

$$\rho(f, h) = \int_X \phi_0(|f - h|) \, d\mu \leq \int_X \phi_0(|f - g| + |g - h|) \, d\mu$$

$$\leq \int_X \phi_0(|f - g|) \, d\mu + \int_X \phi_0(|g - h|) \, d\mu = \rho(f, g) + \rho(g, h).$$

Also $\rho(f, g) = 0$ if and only if $\phi_0(|f - g|) = 0$ almost everywhere, and this happens if and only if $\mu(|f - g| > 0) = 0$; that is, if and only if $f = g$ almost everywhere. Thus ρ is a pseudometric on $\mathcal{L}^0(X, \Sigma, \mu)$.

Suppose that $f_n \to f$ in measure, and that $\epsilon > 0$. There exists n_0 such that $\mu(|f_n - f| > \epsilon/2\mu(X)) < \epsilon/2$, for $n \geq n_0$. Then

$$\rho(f_n, f) = \int_{|f_n - f| \leq \epsilon/2\mu(X)} \phi_0(|f_n - f|) \, d\mu + \int_{|f_n - f| > \epsilon/2\mu(X)} \phi_0(|f_n - f|) \, d\mu$$
$$< \epsilon/2 + \epsilon/2 = \epsilon$$

for $n \geq n_0$, and so $\rho(f_n, f) \to 0$ as $n \to \infty$.

Conversely, suppose that $\rho(f_n, f) \to 0$ as $n \to \infty$, and that $0 < c \leq 1$. Then

$$\mu(|f_n - f| > c) \leq \frac{1}{c} \int_{|f_n - f| > c} \phi_0(|f_n - f|) \, d\mu \leq \rho(f_n, f)/c \to 0$$

as $n \to \infty$, so that $f_n \to f$ in measure as $n \to \infty$. $\square$

How does convergence in measure relate to the algebraic structure of $\mathcal{L}^0(X, \Sigma, \mu)$?

Proposition 29.4.4   *Suppose that (X, Σ, μ) is a finite measure space, that $(f_n)_{n=1}^\infty$ and $(g_n)_{n=1}^\infty$ are sequences in $\mathcal{L}^0(X, \Sigma, \mu)$ and that $f, g \in \mathcal{L}^0(X, \Sigma, \mu)$. If $f_n \to f$ and $g_n \to g$ in measure, as $n \to \infty$, then $f_n + g_n \to f + g$ and $f_n g_n \to fg$ in measure, as $n \to \infty$.*

Proof   It is easy to see that $f_n + g_n \to f + g$ in measure, as $n \to \infty$. Let us establish the result for products. First we show that $f_n^2 \to f^2$ in measure as $n \to \infty$. Suppose that $0 < \epsilon < 1$. There exists $k \in \mathbf{N}$ such that if $B = (|f| > k)$ then $\mu(B) < \epsilon/3$, and there exists n_0 such that if $C_n = (|f_n - f| > \epsilon/3(2k+1))$ then $\mu(C_n) < \epsilon/3$, for $n \geq n_0$. Let $A_n = X \setminus (B \cup C_n)$.

If $x \in A_n$ then $|f_n(x) + f(x)| \leq 2k + 1$, so that $\phi_0(|f_n^2 - f^2|) \leq \epsilon/3$. Thus if $n \geq n_0$ then

$$\int_X |\phi_0(f_n^2 - f^2)| \, d\mu$$

$$\leq \int_{A_n} |\phi_0(f_n^2 - f^2)| \, d\mu + \int_B |\phi_0(f_n^2 - f^2)| \, d\mu + \int_{C_n} |\phi_0(f_n^2 - f^2)| \, d\mu$$

$$\leq \epsilon/3 + \mu(B) + \mu(C_n) < \epsilon,$$

and so $f_n^2 \to f^2$ in measure.

The general case follows by polarization. $(f_n + g_n)^2 \to (f + g)^2$ and $(f_n - g_n)^2 \to (f - g)^2$ in measure, as $n \to \infty$, and so

$$f_n g_n = \tfrac{1}{4}((f_n + g_n)^2 - (f_n - g_n)^2) \to \tfrac{1}{4}((f + g)^2 - (f - g)^2) = fg$$

in measure as $n \to \infty$. $\qquad\qquad\qquad\qquad\qquad\qquad\qquad\qquad\qquad\qquad$ □

We now define an equivalence relation $\sim$ on the space $\mathcal{L}^0(X, \Sigma, \mu)$ of real-valued measurable functions on X by setting $f \sim g$ if $f = g$ almost everywhere, and denote the quotient space by $L^0 = L^0(X, \Sigma, \mu)$. Since $f \sim g$ if and only if $\rho(f, g) = 0$, it follows that if we define $d_0([f], [g]) = \rho(f, g)$ then this is well-defined (it does not depend upon the choice of representatives), and d_0 is a metric on $L^0(X, \Sigma, \mu)$ (See Volume II, Section 11.1). Further,

$$[0] = \mathcal{N}^0 = \{f : f = 0 \text{ almost everywhere}\},$$

and L^0 is the quotient vector space $\mathcal{L}^0/\mathcal{N}^0$.

We shall follow the usual custom, and write f both for a measurable function and for its equivalence class. For example, we shall write '$f_n \to f$ in measure if and only if $d_0(f_n, f) \to 0$ as $n \to \infty$'. This practice is so well established that the reader needs to become accustomed to it. In fact, since countable unions of null sets are null sets, the transition between functions and their equivalence classes is usually quite straightforward. Very occasionally it is necessary to argue in a more detailed way.

We now consider properties of the metric space $(L_0(X, \Sigma, \mu), d_0)$. It follows from Proposition 29.4.4 that the mappings $(f, g) \to f + g$ and $(f, g) \to fg$ from $L^0 \times L^0$ into L^0 are continuous. Notice also that a translation mapping is an isometry of L^0.

Theorem 29.4.5 *The space $S(X, \Sigma, \mu)$ of simple measurable functions is a dense linear subspace of $L^0(X, \Sigma, \mu)$.*

Proof Suppose that $f \in L^0$ and that $0 < \epsilon \leq 1$. By Corollary 28.4.2, there exists $n \in \mathbf{N}$ such $\mu(|f| > n) < \epsilon/2$. By the construction in Theorem 28.6.3, there exists a simple measurable function g such that $|g(x) - f(x)| < \epsilon/2\mu(X)$ for $x \in (|f| \le n)$. Then

$$\rho_0(f, g) = \int_{(|f|>n)} \phi_0(|f - g|)\, d\mu + \int_{(|f|\le n)} \phi_0(|f - g|)\, d\mu < \epsilon/2 + \epsilon/2 = \epsilon.$$

$\square$

The next theorem lies at the heart of many of the results in the theory of measure and integration.

Theorem 29.4.6 *If (X, Σ, μ) is a finite measure space then $(L^0(X, \Sigma, \mu), d_0)$ is a complete metric space.*

Proof   Suppose that $(f_n)_{n=0}^\infty$ is a d_0-Cauchy sequence in L^0. We shall show that there is a subsequence $(f_{n_k})_{k=1}^\infty$ which converges almost everywhere to an element f of L^0. Then $f_{n_k} \to f$ in measure, and so $f_n \to f$ in measure.

For each $k \in \mathbf{N}$ there exists n_k such that $\mu(|f_n - f_m| > 1/2^k) < 1/2^k$ for $m, n \ge n_k$. We can clearly suppose that $(n_k)_{k=1}^\infty$ is a strictly increasing sequence. Let $B_k = (|f_{n_k} - f_{n_{k+1}}| > 1/2^k)$, and let $C = \limsup_{k\to\infty} B_k$. Since $\sum_{k=1}^\infty \mu(B_k) < \infty$, $\mu(C) = 0$, by the first Borel–Cantelli lemma. If $x \notin C$ then there exists k_0 such that $x \notin B_k$ for $k \ge k_0$. Thus

$$|f_{n_k}(x) - f_{n_{k+1}}(x)| \le 1/2^k, \text{ for } k \ge k_0,$$

and so $|f_{n_l}(x) - f_{n_m}(x)| < 2/2^k$ for $l > m \ge k_0$. Hence $(f_{n_k}(x))_{k=1}^\infty$ is a real Cauchy sequence, which converges, by the general principle of convergence. Thus $(f_{n_k}(x))_{k=1}^\infty$ converges, to $f(x)$, say. If $x \in C$, set $f(x) = 0$. Then f is measurable, and $f_n \to f$ almost everywhere. $\square$

We have seen that convergence in measure can be characterized very satisfactorily in terms of a metric. Is the same true for convergence almost everywhere? The next result shows that the answer is 'no'.

Proposition 29.4.7   *Suppose that τ is a topology on $L^0(X, \Sigma, \mu)$ with the property that if $(f_n)_{n=1}^\infty$ is a sequence in L^0 which converges almost everywhere to f, then $f_n \to f$ in the topology τ, as $n \to \infty$. It then follows that if $f_n \to f$ in measure then $f_n \to f$ in the topology τ, as $n \to \infty$.*

Proof   Suppose not. Then there exists a neighbourhood N of f and a subsequence $(f_{n_k})_{k=1}^\infty$ such that $f_{n_k} \notin N$ for all $k \in \mathbf{N}$. Let $g_k = f_{n_k}$. Then $g_k \to f$ in measure as $k \to \infty$, and so there is a subsequence $(g_{k_l})_{l=1}^\infty$ such that $g_{k_l} \to f$ almost everywhere, as $l \to \infty$. But $g_{k_l} \notin N$, for all $l \in \mathbf{N}$, giving a contradiction. $\square$

We can extend these results in two ways. First, suppose that (X, Σ, μ) is a σ-finite measure space. Suppose that $(f_n)_{n=1}^{\infty}$ is a sequence in $L^0(X, \Sigma, \mu)$ and that $f \in L^0(X, \Sigma, \mu)$. If $A \in \Sigma$ and $\mu(A) > 0$, let $\pi_A : L^0(X, \Sigma, \mu) \to L^0(A, \Sigma, \mu)$ be the restriction mapping. Then f_n *converges locally in measure* to f if $\pi_A(f_n) \to \pi_A(f)$ in measure, as $n \to \infty$, for each $A \in \Sigma$ with $\mu(A) > 0$.

Proposition 29.4.8 *Suppose that (X, Σ, μ) is a σ-finite measure space, with a sequence $(I_k)_{k=1}^{\infty}$ of disjoint elements of Σ of finite positive measure, whose union is X. Suppose that $(f_n)_{n=1}^{\infty}$ is a sequence in $L^0(X, \Sigma, \mu)$, and that $f \in L^0(X, \Sigma, \mu)$. Then $f_n \to f$ locally in measure if and only if $\pi_{I_k}(f_n) \to \pi_{I_k}(f)$ in measure, as $n \to \infty$, for each $k \in \mathbf{N}$.*

Proof A worthwhile exercise. □

Then $L^0(X, \Sigma, \mu)$ is isomorphic to the product $\prod_{k=1}^{\infty} L^0(I_k, \Sigma, \mu)$, and local convergence in measure is characterized by a complete product metric such as

$$d_0(f, g) = \sum_{k=1}^{\infty} \frac{d_0(\pi_{I_k}(f), \pi_{I_k}(g))}{2^k \mu(I_k)}.$$

Secondly, we can consider the space $L^0_{\mathbf{C}}(X, \Sigma, \mu)$ of (equivalence classes of) complex-valued measurable functions. Results corresponding to those established above are obtained, usually by considering real and imaginary parts.

Exercises

29.4.1 Show that the set of step functions is dense in $L^0([0,1], \mathcal{L}, \lambda)$.

29.4.2 Show that the space $C([0,1])$ of continuous functions is dense in $L^0([0,1], \mathcal{L}, \lambda)$.

29.4.3 Suppose that $(f_n)_{n=1}^{\infty}$ is a sequence of measurable functions on a finite measure space (X, Σ, μ). Show that f_n converges in measure if and only if whenever $(g_j)_{j=1}^{\infty} = (f_{n_j})_{j=1}^{\infty}$ is a subsequence of $(f_n)_{n=1}^{\infty}$ there exists a subsequence $(h_k)_{k=1}^{\infty} = (g_{j_k})_{k=1}^{\infty}$ of $(g_j)_{j=1}^{\infty}$ which converges almost everywhere.

Use this to give another proof of Proposition 29.4.4.

The remaining exercises show how to construct a metric which defines convergence in measure, without using integration. Suppose that (X, Σ, μ) is a finite measure space, and that $f \in \mathcal{L}^0(X, \Sigma, \mu)$ is non-negative. Let λ_f be the tail distribution of f: $\lambda_f(t) = \mu(f > t)$.

29.4.4 Show that λ_f is a decreasing right-continuous function on $[0, \infty)$, and that $\lambda_f(t) \to 0$ as $t \to \infty$.

29.4.5 Let $\phi(f) = \inf\{t : \lambda_f(t) \leq t\}$. If $f, g \in \mathcal{L}^0(X, \Sigma, \mu)$, let $\rho(f, g) = \phi(|f - g|)$. Show that ρ is a pseudometric on $\mathcal{L}^0(X, \Sigma, \mu)$, and that $\rho(f, g) = 0$ if and only if $f = g$ almost everywhere.

29.4.6 Let d be the corresponding metric on $L^0(X, \Sigma, \mu)$. Show that d is uniformly equivalent to the metric d_0 defined above.

29.5 The spaces $L^1_{\mathbf{R}}(X, \Sigma, \mu)$ and $L^1_{\mathbf{C}}(X, \Sigma, \mu)$

We now consider metric properties of the real vector space $\mathcal{L}^1_{\mathbf{R}}(X, \Sigma, \mu)$ of real-valued integrable functions, and the complex vector space $\mathcal{L}^1_{\mathbf{C}}(X, \Sigma, \mu)$ of complex-valued integrable functions on a finite or σ-finite measure space (X, Σ, μ). We shall concentrate on the complex case: the corresponding results in the real case are easier, and the details are left to the reader.

Proposition 29.5.1 *The function $\rho_1(f) = \int_X |f| \, d\mu$ is a seminorm on $\mathcal{L}^1_{\mathbf{C}}(X, \Sigma, \mu)$. $\rho_1(f) = 0$ if and only if $f = 0$ almost everywhere.*

Proof

$$\rho_1(f + g) = \int_X |f + g| \, d\mu \leq \int_X (|f| + |g|) \, d\mu$$

$$= \int_X |f| \, d\mu + \int_X |g| \, d\mu = \rho_1(f) + \rho_1(g)$$

and

$$\rho_1(\alpha f) = \int_X |\alpha f| \, d\mu = \int_X |\alpha| . |f| \, d\mu = |\alpha| \int_X |f| \, d\mu = |\alpha| \rho_1(f).$$

Further, $\rho_1(f) = 0$ if and only if $\int |f| \, d\mu = 0$, if and only if $|f| = 0$ almost everywhere, if and only if $f = 0$ almost everywhere.   □

Thus the quotient space $L^1_{\mathbf{C}}(X, \Sigma, \mu) = \mathcal{L}^1_{\mathbf{C}}(X, \Sigma, \mu)/\mathcal{N}$ becomes a normed space when we set $\|[f]\|_1 = \rho_1(f)$. Again, we write f both for an integrable function and for its equivalence class in $L^1_{\mathbf{C}}(X, \Sigma, \mu)$.

Theorem 29.5.2 *$L^1_{\mathbf{C}}(X, \Sigma, \mu)$ is a linear subspace of $L^0_{\mathbf{C}}(X, \Sigma, \mu)$. The inclusion mapping*

$$j : (L^1_{\mathbf{C}}(X, \Sigma, \mu), \|.\|_1) \to (L^0(X, \Sigma, \mu), d_0)$$

is uniformly continuous.

Proof   We use Markov's inequality.

Lemma 29.5.3 (Markov's inequality) *If* $f \in L^1_{\mathbf{C}}(X, \Sigma, \mu)$ *and* $\alpha > 0$ *then* $\mu(|f| \geq \alpha) \leq \|f\|_1 / \alpha$.

Proof For $\alpha I_{(|f| \geq \alpha)} \leq |f|$, so that

$$\alpha \mu(|f| \geq \alpha) = \int_X \alpha I_{(|f| \geq \alpha)} \, d\mu \leq \int_X |f| \, d\mu = \|f\|_1 \, .$$

$\square$

    Thus if $\epsilon > 0$ and $\|f - g\|_1 < \epsilon^2$ then $\mu(|f - g| > \epsilon) \leq \epsilon$, so that $d_0(f, g) \leq \epsilon$. $\square$

Theorem 29.5.4 *The normed space* $(L^1_{\mathbf{C}}(X, \Sigma, \mu), \|.\|_1)$ *is complete.*

Proof We use Proposition 12.1.9 of Volume II, which implies that $(L^1(X, \Sigma, \mu), \|.\|_1)$ is complete if $j(M_\epsilon(f))$ is d_0-closed in $(L^0(X, \Sigma, \mu), d_0)$, for each $f \in L^1(X, \Sigma, \mu)$ and each $\epsilon > 0$ (where $M_\epsilon(f) = \{g : \|f - g\|_1 \leq \epsilon\}$).
    Suppose that $g_n \in M_\epsilon(f)$, and that $d_0(g_n, g) \to 0$ as $n \to \infty$. There exists a subsequence $(g_{n_k})_{k=1}^\infty$ such that $g_{n_k} \to g$ almost everywhere, as $k \to \infty$, and so $|g_{n_k} - f| \to |g - f|$ almost everywhere, as $k \to \infty$. By Fatou's lemma,

$$\int_X |g - f| \, d\mu \leq \lim_{k \to \infty} \inf \int_X |g_{n_k} - f| \, d\mu \leq \epsilon,$$

so that $g \in M_\epsilon(f)$. $\square$

Proposition 29.5.5 *The vector space* $S_{\mathbf{R}}(X, \Sigma, \mu)$ *of simple real-valued measurable functions is dense in* $(L^1_{\mathbf{R}}(X, \Sigma, \mu), \|.\|_1)$.

Proof First, suppose that f is a non-negative function in $L^1_{\mathbf{R}}(X, \Sigma, \mu)$. There exists an increasing sequence $(f_n)_{n=1}^\infty$ of non-negative functions in $S(X, \Sigma, \mu)$ which converges pointwise to f, and so $\int_X f_n \, d\mu \to \int_X f \, d\mu$, by monotone convergence. Thus

$$\int_X (f - f_n) \, d\mu = \int_X f \, d\mu - \int_X f_n \, d\mu \to 0 \text{ as } n \to \infty.$$

Thus f is in the closure of $S(X, \Sigma, \mu)$. If $f \in L^1(X, \Sigma, \mu)$ then f^+ and f^- are both in the closure of $S(X, \Sigma, \mu)$, and so therefore is f. $\square$

Corollary 29.5.6 *The vector space* $S_{\mathbf{C}}(X, \Sigma, \mu)$ *of simple complex-valued measurable functions is dense in* $(L^1_{\mathbf{C}}(X, \Sigma, \mu), \|.\|_1)$.

Proof Consider real and imaginary parts. $\square$

Exercises

29.5.1 Suppose that (X, Σ, μ) is a finite measure space.

(i) Let $\Phi(A, B) = \mu(A \Delta B) = \mu(A \setminus B) + \mu(B \setminus A)$, for $A, B \in \Sigma$. Show that Φ is a pseudometric on Σ.

(ii) Let $(M_\mu(X, \Sigma), d)$ be the quotient space, with the quotient metric. $(M_\mu(X, \Sigma), d)$ is the *measure algebra* of (X, Σ, μ). Show that the mapping $[A] \to [I_A]$ from $(M_\mu(X, \Sigma), d)$ to $(L^1(X, \Sigma, \mu), \|.\|_1)$ is an isometry.

(iii) Show that $(M_\mu(X, \Sigma), d)$ is a complete metric space.

29.5.2 Suppose that $(f_n)_{n=0}^\infty$ and $(g_n)_{n=0}^\infty$ are sequences in $L^1(X, \Sigma.\mu)$, and that $|f_n| \le g_n$ for $n \in \mathbf{Z}^+$. Suppose that $f_n \to f_0$ almost everywhere, that $g_n \to g_0$ almost everywhere and that $\|g_n\|_1 \to \|g_0\|_1$ as $n \to \infty$. Use Fatou's lemma to show that $\int_X f_n \, d\mu \to \int_X f_0 \, d\mu$ as $n \to \infty$. Deduce that $f_n \to f_0$ in norm as $n \to \infty$.

29.5.3 Suppose that (X, Σ, μ) is a finite measure space. Let

$$P = \{f \in L^1_{\mathbf{R}}(X, \Sigma, \mu) : f \ge 0, \int_X f \, d\mu = 1\}.$$

Is P a closed subset of $L^1_{\mathbf{R}}(X, \Sigma, \mu)$?

29.5.4 Show that the vector space of step functions is dense in $L^1([0, 1], \mathcal{L}, \lambda)$.

29.5.5 Show that the vector space $C([0, 1])$ of continuous functions on $[0, 1]$ is dense in $L^1([0, 1], \mathcal{L}, \lambda)$, and that the vector space of continuous functions of compact support is dense in $L^1(\mathbf{R}, \mathcal{L}, \lambda)$.

29.5.6 Suppose that $f \in L^1([-1, 1], \lambda)$. Set $f(x) = 0$ for $|x| > 1$. Show that

$$\int_{-1}^{1} |f(x + h) - f(x)| \, dx \to 0 \text{ as } h \to 0.$$

29.5.7 Suppose that $f \in L^1(\mathbf{R}, \lambda)$, Let $\hat{f}(y) = \int_{-\infty}^{\infty} f(x)e^{-ixy} \, dx$, for $y \in \mathbf{R}$. By considering suitable approximations, show that $\hat{f}$ is a bounded continuous function on $\mathbf{R}$ and that $|\hat{f}(y)| \to 0$ as $|y| \to \infty$.

29.6 The spaces $L^p_{\mathbf{R}}(X, \Sigma, \mu)$ and $L^p_{\mathbf{C}}(X, \Sigma, \mu)$, for $0 < p < \infty$

We now introduce some further spaces of functions, and of equivalence classes of functions. Suppose that $0 < p < \infty$. We define $\mathcal{L}^p_{\mathbf{R}} = \mathcal{L}^p_{\mathbf{R}}(X, \Sigma, \mu)$ to be the collection of those real-valued measurable functions for which

$\int_X |f|^p \, d\mu < \infty$, and define $\mathcal{L}^p_C = \mathcal{L}^p_C(X, \Sigma, \mu)$ to be the collection of those complex-valued measurable functions for which $\int_X |f|^p \, d\mu < \infty$. We shall establish results in the complex case, and again leave it to the reader to verify that corresponding results hold in the real case.

If $f \in \mathcal{L}^p_C$ and α is a scalar, then $\alpha f \in \mathcal{L}^p_C$. Since

$$|a + b|^p \le 2^p \max(|a|^p, |b|^p) \le 2^p(|a|^p + |b|^p),$$

$f + g \in \mathcal{L}^p_C$ if $f, g \in \mathcal{L}^p_C$. Thus $\mathcal{L}^p_C$ is a vector space.

Theorem 29.6.1   *(i) If $1 \le p < \infty$ then $\phi_p(f) = (\int_X |f|^p \, d\mu)^{1/p}$ is a semi-norm on $\mathcal{L}^p_C$, and $\phi_p(f) = 0$ if and only if $f = 0$ almost everywhere.*

(ii) If $0 < p < 1$ then $\rho_p(f, g) = \int_X |f - g|^p \, d\mu$ is a pseudometric on $\mathcal{L}^p_C$, and $\rho_p(f, g) = 0$ if and only if $f = g$ almost everywhere.

Proof   The proof depends on the facts that the function t^p is convex on $[0, \infty)$ for $1 \le p < \infty$ and is concave for $0 < p < 1$.

(i) As in Proposition 29.5.1, $\phi_p(\alpha f) = |\alpha| \phi_p(f)$, and $\phi_p(f) = 0$ if and only if $f = 0$ almost everywhere. If f or g is zero almost everywhere then trivially $\phi_p(f + g) = \phi_p(f) + \phi_p(g)$. Otherwise, let $F = f / \phi_p(f)$ and let $G = g / \phi_p(g)$, so that $\phi_p(F) = \phi_p(G) = 1$. Let $\lambda = \phi_p(g)/(\phi_p(f) + \phi_p(g))$, so that $0 < \lambda < 1$. Then

$$f = (\phi_p(f) + \phi_p(g))(1 - \lambda)F \text{ and } g = (\phi_p(f) + \phi_p(g))\lambda G,$$

so that

$$\begin{aligned}
|f + g|^p &= (\phi_p(f) + \phi_p(g)^p \, |(1 - \lambda)F + \lambda G|^p \\
&\le (\phi_p(f) + \phi_p(g))^p \, ((1 - \lambda)|F| + \lambda|G|)^p \\
&\le (\phi_p(f) + \phi_p(g))^p \, ((1 - \lambda)|F|^p + \lambda|G|^p),
\end{aligned}$$

since the function t^p is convex, for $1 \le p < \infty$. Integrating,

$$\int_x |f + g|^p \, d\mu \le (\phi_p(f) + \phi_p(g))^p \left((1 - \lambda) \int_x |F|^p \, d\mu + \lambda \int_X |G|^p \, d\mu \right)$$

$$= (\phi_p(f) + \phi_p(g))^p.$$

Thus we have established *Minkowski's inequality*

$$\left(\int_X |f + g|^p \, d\mu \right)^{1/p} \le \left(\int_X |f|^p \, d\mu \right)^{1/p} + \left(\int_X |g|^p \, d\mu \right)^{1/p},$$

and shown that ϕ_p is a semi-norm.

(ii) If $0 < p < 1$, the function t^{p-1} is decreasing on $(0, \infty)$, so that if a and b are non-negative and $a + b > 0$, then

$$(a + b)^p = a(a + b)^{p-1} + b(a + b)^{p-1} \leq a^p + b^p.$$

Integrating,

$$\int_X |f + g|^p \, d\mu \leq \int_X (|f| + |g|)^p \, d\mu \leq \int_X |f|^p \, d\mu + \int_X |g|^p \, d\mu;$$

this is enough to show that ρ_p is a pseudometric.   □

As with $L^1_{\mathbf{C}}$, we define $L^p_{\mathbf{C}}(X, \Sigma, \mu)$ to be the quotient space $\mathcal{L}^p_{\mathbf{C}}(X, \Sigma, \mu)/\mathcal{N}$, and set

$$\|[f]\|_p = \phi_p(f) \text{ for } p \geq 1, \text{ and } d_p([f], [g]) = \rho_p(f, g) \text{ for } 0 < p < 1;$$

$\|.\|_p$ is a norm, and d_p is a metric. We again write f both for a function in $\mathcal{L}^p_{\mathbf{C}}(X, \Sigma, \mu)$ and for its equivalence class in $L^p_{\mathbf{C}}(X, \Sigma, \mu)$.

Theorem 29.6.2   *If $0 < p < \infty$ then the inclusion mapping*

$$j : (L^p_{\mathbf{C}}(X, \Sigma, \mu), \|.\|_p) \to (L^0_{\mathbf{C}}(X, \Sigma, \mu), d_0)$$

is uniformly continuous.

Proof   Markov's inequality shows that if $f \in L^p_{\mathbf{C}}(X, \Sigma, \mu)$ and $\alpha > 0$ then $\mu(|f| \geq \alpha) \leq (\int_X |f|^p \, d\mu)/\alpha^p$.

If $p \geq 1$, $\epsilon > 0$ and $\|f - g\|_p < \epsilon^{1+1/p}$ then $\mu(|f - g| > \epsilon) \leq \epsilon$, so that $d_0(f, g) \leq \epsilon$.

If $0 < p < 1$, $\epsilon > 0$ and $d_p(f, g) < \epsilon^{p+1}$ then $\mu(|f - g| > \epsilon) \leq \epsilon$, so that $d_0(f, g) \leq \epsilon$.   □

Theorem 29.6.3 $(L^p_{\mathbf{C}}, \|.\|_p)$ *is a Banach space for $1 \leq p < \infty$ and $(L^p_{\mathbf{C}}, d_p)$ is a complete metric space for $0 < p < 1$.*

Proof   The proof is essentially the same as the proof of Theorem 29.5.4, inserting an exponent p when this is required.   □

Proposition 29.6.4   *If $0 < p < 1$ then the vector space $S_{\mathbf{R}}(X, \Sigma, \mu)$ of simple real-valued measurable functions is dense in $(L^p_{\mathbf{R}}(X, \Sigma, \mu), d_p)$ and the vector space $S_{\mathbf{C}}(X, \Sigma, \mu)$ of simple complex-valued measurable functions is dense in $(L^p_{\mathbf{C}}(X, \Sigma, \mu), d_p)$.*

If $1 \leq p < \infty$ then the vector space $S_{\mathbf{R}}(X, \Sigma, \mu)$ of simple real-valued measurable functions is dense in $(L^p_{\mathbf{R}}(X, \Sigma, \mu), \|.\|_p)$ and the vector space

$S_{\mathbf{C}}(X, \Sigma, \mu)$ of simple complex-valued measurable functions is dense in $(L_{\mathbf{C}}^p(X, \Sigma, \mu), \|.\|_p)$.

Proof   Once again, by considering real and imaginary parts, and positive and negative parts, it is enough to approximate a non-negative function f by simple measurable functions. There exists an increasing sequence $(f_n)_{n=1}^\infty$ of non-negative functions in $S(X, \Sigma, \mu)$ which converges pointwise to f. Then $(f - f_n)^p \to 0$ pointwise as $n \to \infty$, and $(f - f_n)^p \leq f^p$, and so by dominated convergence $\int_X (f - f_n)^p \, d\mu \to 0$ as $n \to \infty$; this gives the result. □

In metric space terms, these results are the most important results of integration theory. When $(X, \Sigma, \mu) = (\mathbf{R}, \mathcal{L}, \lambda)$, the vector space of step functions, and the vector space of continuous functions of compact support, are dense in $L^p(X, \Sigma, \mu)$ (Exercise 29.6.2). The results show that $L_{\mathbf{C}}^p(X, \Sigma, \mu)$ can be thought of as the completion of $S_{\mathbf{C}}(X, \Sigma, \mu)$, when $S_{\mathbf{C}}(X, \Sigma, \mu)$ is given an appropriate norm, or metric, and that $L_{\mathbf{C}}^p(\mathbf{R}, \mathcal{L}, \lambda)$ can be thought of as the completion of step functions, or the vector space of continuous functions of compact support, when it is given an appropriate norm, or metric.

The Banach space $L_{\mathbf{C}}^2(X, \Sigma, \mu)$ is particularly important.

Theorem 29.6.5   *If $f, g \in L_{\mathbf{C}}^2(X, \Sigma, \mu)$ then $f\bar{g} \in L_{\mathbf{C}}^1(X, \Sigma, \mu)$. The function $\langle f, g \rangle = \int_X f\bar{g} \, d\mu$ is an inner product on $L_{\mathbf{C}}^2(X, \Sigma, \mu)$, which defines the norm, so that $L_{\mathbf{C}}^2(X, \Sigma, \mu)$ is a Hilbert space.*

Proof   Since $f\bar{g} = \frac{1}{2}((f + \bar{g})^2 - f^2 - \bar{g}^2)$, $f\bar{g} \in L_{\mathbf{C}}^1(X, \Sigma, \mu)$. It then follows that $\langle f, g \rangle = \int_X f\bar{g} \, d\mu$ is an inner product on $L_{\mathbf{C}}^2(X, \Sigma, \mu)$ which defines the norm on $L_{\mathbf{C}}^2(X, \Sigma, \mu)$. □

We can also establish Hölder's inequality. Recall that if $1 < p < \infty$ and $1/p + 1/p' = 1$, and if a, b are non-negative, then

$$ab \leq \frac{a^p}{p} + \frac{b^{p'}}{p'},$$

with equality if and only if $a^p = b^{p'}$.

If $z = re^{i\theta}$ is a non-zero complex number in polar form, we define the *signum* sgn (z) to be $e^{i\theta}$. We define sgn $(0) = 0$.

Theorem 29.6.6 (Hölder's inequality)   *Suppose that $1 < p < \infty$, that $1/p + 1/p' = 1$ and that $f \in L_{\mathbf{C}}^p(X, \Sigma, \mu)$ and $g \in L_{\mathbf{C}}^{p'}(X, \Sigma, \mu)$. Then $fg \in L^1(X, \Sigma, \mu)$, and*

$$\left| \int_X fg \, d\mu \right| \leq \int_X |fg| \, d\mu \leq \|f\|_p \, \|g\|_{p'}.$$

Equality holds throughout if and only if either $\|f\|_p \|g\|_{p'} = 0$, *or* $g = \lambda.\overline{\mathrm{sgn}\,(f)}.|f|^{p-1}$ *almost everywhere, where* $\lambda \neq 0$.

Proof   The result is trivial if either f or g is zero. Otherwise, by scaling, it is enough to consider the case where $\|f\|_p = \|g\|_{p'} = 1$. Then, by the inequality above, $|fg| \leq |f|^p/p + |g|^{p'}/p'$; integrating,

$$\int_X |fg|\,d\mu \leq \int_X \frac{|f|^p}{p}\,d\mu + \int_X \frac{|g|^{p'}}{p'}\,d\mu = 1/p + 1/p' = 1 = \|f\|_p \|g\|_{p'}.$$

Thus $fg \in L^1_{\mathbf{C}}(X,\Sigma,\mu)$ and so $|\int_X fg\,d\mu| \leq \int |fg|\,d\mu \leq \|f\|_p \|g\|'_p$.

If $g = \overline{\mathrm{sgn}\,(f)}.|f|^{p-1}$ almost everywhere, then $fg = |fg| = |f|^p = |g|^{p'}$ almost everywhere, so that

$$\left|\int fg\,d\mu\right| = \int_X |fg|\,d\mu = \|f\|_p^p = \|g\|_{p'}^{p'} = \|f\|_p \|g\|_{p'}.$$

By scaling, the result holds if $g = \lambda.\overline{\mathrm{sgn}\,(f)}.|f|^{p-1}$.

Conversely, suppose that

$$\left|\int_X fg\,d\mu\right| = \int_X |fg|\,d\mu = \|f\|_p \|g\|_{p'}.$$

Then, again by scaling, we need only consider the case where $\|f\|_p = \|g\|_{p'} = 1$. Since $|\int_X fg\,d\mu| = \int_X |fg|\,d\mu$, $fg = |fg|$ almost everywhere, so that either $f(x)g(x) = 0$ or $\overline{\mathrm{sgn}\,(f(x))} = \mathrm{sgn}\,(g(x))$, for almost all x. Since

$$\int_X |fg|\,d\mu = 1 = \int_X |f|^p/p\,d\mu + \int |g|^{p'}/p'\,d\mu \text{ and } |f|^p/p + |g|^{p'}/p' \geq |fg|,$$

$|fg| = |f|^p/p + |g|^{p'}/p'$ almost everywhere, and so $|f|^p = |g|^{p'}$ almost everywhere. Thus $|g| = |f|^{p/p'} = |f|^{p-1}$ almost everywhere, and $g = \mathrm{sgn}\,(g)|g| = \overline{\mathrm{sgn}\,(f)}|f|^{p-1}$ almost everywhere. $\qquad\square$

Hölder's inequality shows that there is a natural scale of inclusions for the L^p spaces, when the underlying space has finite measure.

Corollary 29.6.7   *Suppose that (X,Σ,μ) is a measure space, that $\mu(X) < \infty$ and that $0 < p < q < \infty$. Then $L^q_{\mathbf{C}}(X,\Sigma,\mu) \subseteq L^p_{\mathbf{C}}(X,\Sigma,\mu)$.*

If $1 \leq p < q < \infty$ and $f \in L^q_{\mathbf{C}}(X,\Sigma,\mu)$ then $\|f\|_p \leq \mu(X)^{1/p-1/q}\|f\|_q$.

Proof   Suppose that $f \in L^q_{\mathbf{C}}(X,\Sigma,\mu)$, where $q < \infty$. Let $r = q/(q-p)$, so that $p/q + 1/r = 1$ and $1/rp = 1/p - 1/q$. We apply Hölder's inequality to the functions I_X and $|f|^p$, using exponents r and q/p:

$$\int_X |f|^p \, d\mu \leq (\mu(X))^{1/r} \left(\int_X |f|^q \, d\mu \right)^{p/q},$$

so that if $p \geq 1$ then

$$\|f\|_p \leq (\mu(X))^{1/rp} \left(\int_X |f|^q \, d\mu \right)^{1/q} = \mu(X)^{1/p - 1/q} \|f\|_q.$$

$\square$

We leave it as an exercise for the reader to establish the corresponding inequalities when $0 < p < 1 \leq q < \infty$ and when $0 < p < q < 1$.

Suppose that $(\Omega, \Sigma, \mathbf{P})$ is a probability space. It follows from this corollary that $L_{\mathbf{C}}^2(\Omega, \Sigma, \mathbf{P}) \subseteq L_{\mathbf{C}}^1(\Omega, \Sigma, \mathbf{P})$ and that if $f \in L_{\mathbf{C}}^2(\Omega, \Sigma, \mathbf{P})$ then

$$\left| \int_\Omega f \, d\mathbf{P} \right| \leq \int_\Omega |f| \, d\mathbf{P} \leq \|f\|_2.$$

If $f \in L_{\mathbf{C}}^1(\Omega, \Sigma, \mathbf{P})$, the quantity $\int_\Omega f \, d\mathbf{P}$ is called the *expectation* or *mean* of f, and is denoted by $\mathbf{E}(f)$. If $f \in L_{\mathbf{C}}^2(\Omega, \Sigma, \mathbf{P})$, the quantity $\sigma_f^2 = \int_\Omega |f - \mathbf{E}(f)|^2 \, d\mathbf{P}$ is called the *variance* of f. Note that

$$\sigma_f^2 = \int_\Omega (f - \mathbf{E}(f))(\bar{f} - \mathbf{E}(\bar{f})) \, d\mathbf{P}$$

$$= \int_\Omega f\bar{f} \, d\mathbf{P} - 2\mathbf{E}(f)\mathbf{E}(\bar{f}) + \mathbf{E}(f)\mathbf{E}(\bar{f}) = \|f\|_2^2 - |\mathbf{E}(f)|^2.$$

Proposition 29.6.8 (Chebyshev's inequality) *If* $f \in L_{\mathbf{C}}^2(\Omega, \Sigma, \mathbf{P})$ *and* $t > 0$ *then*

$$\mathbf{P}(|f - \mathbf{E}(f)| \geq t) \leq \sigma_f^2 / t^2.$$

Proof By Markov's inequality,

$$\mathbf{P}(|f - \mathbf{E}(f)| \geq t) = \mathbf{P}(|f - \mathbf{E}(f)|^2 \geq t^2) \leq \sigma_f^2 / t^2.$$

$\square$

Here is an application.

Proposition 29.6.9 (The second Borel–Cantelli lemma) *Suppose that $(A_n)_{n=1}^{\infty}$ is a sequence of events in a probability space $(\Omega, \Sigma, \mathbf{P})$ for which*

$$\sum_{n=1}^{\infty} \mathbf{P}(A_n) = \infty \ and\ \mathbf{P}(A_i \cap A_j) \leq \mathbf{P}(A_i).\mathbf{P}(A_j)\ for\ i \neq j.$$

Then

$$\mathbf{P}(\liminf_{n \to \infty} A_n) = \mathbf{P}(\{x : x \in A_n\ infinitely\ often\}) = 1.$$

Proof   Let $p_j = \mathbf{P}(A_j)$, let $s_n = \sum_{j=1}^{n} p_j$ and let $N_n = \sum_{j=1}^{n} I_{A_j}$. $(N_n)_{n=1}^{\infty}$ is an increasing sequence of functions. We apply Chebyshev's inequality to N_n. $\mathbf{E}(N_n) = s_n$, and

$$\sigma_{N_n}^2 = \int_{\Omega} \left(\sum_{j=1}^{n} (I_{A_j} - p_j) \right)^2 d\mathbf{P}$$

$$= \sum_{j=1}^{n} \int_{\Omega} (I_{A_j} - p_j)^2 d\mathbf{P} + 2 \sum_{1 \leq i < j \leq n} \int_{\Omega} (I_{A_i} - p_i)(I_{A_j} - p_j) d\mathbf{P}$$

$$= \sum_{j=1}^{n} p_j(1 - p_j) + 2 \sum_{1 \leq i < j \leq n} (\mathbf{P}(A_i \cap A_j) - p_i p_j) \leq s_n.$$

If $k \in \mathbf{N}$ and $s_n \geq k$ then

$$\mathbf{P}(N_n \leq k) \leq \mathbf{P}(|N_n - s_n| \geq s_n - k) \leq \frac{s_n}{(s_n - k)^2},$$

by Chebyshev's inequality. Thus

$$\mathbf{P}(\lim_{n \to \infty} N_n \leq k) = \lim_{n \to \infty} \mathbf{P}(N_n \leq k) = 0.$$

Hence $\mathbf{P}(\lim_{n \to \infty} N_n > k) = 1$, and so $N_n \to \infty$ almost everywhere. This clearly implies the result. □

The second Borel–Cantelli lemma is frequently used when the events A_n are pairwise independent ($\mathbf{P}(A_i \cap A_j) = \mathbf{P}(A_i).\mathbf{P}(A_j)$ for $i \neq j$) or independent ($\mathbf{P}(\cap_{j=1}^{k} A_{i_j}) = \prod_{j=1}^{k} \mathbf{P}(A_{i_j})$ for $i_1 < \cdots < i_k$).

Exercises

29.6.1 Show that the set of step functions is dense in $L^p([0,1], \mathcal{L}, \lambda)$, for $0 < p < \infty$.

29.6.2 Show that $C([0, 1])$ of continuous functions is dense in $L^p([0, 1], \mathcal{L}, \lambda)$, and that the vector space of continuous functions of compact support is dense in $L^p(\mathbf{R}, \mathcal{L}, \lambda)$, for $0 < p < \infty$.

29.6.3 Suppose in Corollary 29.6.7 that $0 < p < 1 \leq q < \infty$. What is the corresponding inequality relating $d_p(f, 0)$ and $\|g\|_q$? Suppose that $0 < p < q < 1$. What is the corresponding inequality relating $d_p(f, 0)$ and $d_q(f, 0)$?

29.6.4 Suppose that (X, Σ, μ) is a finite measure space. Show that if f is a non-negative measurable function and $p > 0$ then

$$\int_X f^p \, d\mu = p \int_0^\infty t^{p-1} \lambda_f(t) \, dt.$$

Deduce that if $p < q$ and $\lambda_f(t) = O(t^{-q})$ then $f \in L^p(X, \Sigma, \mu)$.

29.6.5 Suppose that $0 < p < q < \infty$. Show that

$$t^{-1/q} I_{[0,1]}(t) \in L_{\mathbf{R}}^p(\mathbf{R}, \mathcal{L}, \lambda) \setminus L_{\mathbf{R}}^q(\mathbf{R}, \mathcal{L}, \lambda),$$

and $t^{-1/p} I_{[1,\infty)}(t) \in L_{\mathbf{R}}^q(\mathbf{R}, \mathcal{L}, \lambda) \setminus L_{\mathbf{R}}^p(\mathbf{R}, \mathcal{L}, \lambda)$.

In this case, there are no natural inclusions.

29.6.6 Suppose that $(\Omega, \Sigma, \mathbf{P})$ is a probability space. Suppose that $0 < h < p$ and that $f \in L^{p+h}(\Omega, \Sigma, \mathbf{P})$. Show that

$$\left(\int_\Omega |f|^p \, d\mathbf{P} \right)^2 \leq \left(\int_\Omega |f|^{p+h} \, d\mathbf{P} \right) \left(\int_\Omega |f|^{p-h} \, d\mathbf{P} \right).$$

29.7 The spaces $L_{\mathbf{R}}^\infty(X, \Sigma, \mu)$ and $L_{\mathbf{C}}^\infty(X, \Sigma, \mu)$

Suppose that (X, Σ, μ) is a finite or σ-finite measure space. A function f in $\mathcal{L}_{\mathbf{R}}^0(X, \Sigma, \mu)$ is *essentially bounded above* if there exists M such that $f < M$ almost everywhere; that is, there exists M such that $\mu(f > M) = 0$. The essential supremum $\operatorname{ess\,sup}(f)$ is then defined to be $\inf\{t : \mu(f > t) = 0\}$. If $(t_n)_{n=1}^\infty$ is a decreasing sequence for which $t_n \to \operatorname{ess\,sup}(f)$, then $\mu(f > \operatorname{ess\,sup}(f)) = \lim_{n \to \infty} \mu(f > t_n) = 0$, while if $t < \operatorname{ess\,sup}(f)$ then $\mu(f > t) > 0$. f is *essentially bounded below* if $-f$ is essentially bounded above, and f is *essentially bounded* if it is essentially bounded above and below.

We define $\mathcal{L}_{\mathbf{R}}^\infty = \mathcal{L}_{\mathbf{R}}^\infty(X, \Sigma, \mu)$ to be $\{f \in \mathcal{L}_{\mathbf{R}}^0 : f \text{ is essentially bounded}\}$. $\mathcal{L}_{\mathbf{R}}^\infty$ is a linear subspace of $\mathcal{L}_{\mathbf{R}}^0$.

Theorem 29.7.1   *The function* $p(f) = \operatorname{ess\,sup}(f)$ *is a seminorm on* $\mathcal{L}^\infty(X, \Sigma, \mu)$, *and*

$$\{f : p(f) = 0\} = \mathcal{N}_{\mathbf{R}}^\infty(X, \Sigma, \mu) = \{f \in \mathcal{L}_{\mathbf{R}}^\infty : f = 0 \text{ almost everywhere}\}.$$

Let $\|.\|_\infty$ be the corresponding norm on the quotient space $L^\infty(X, \Sigma, \mu)$. Then $(L^\infty(X, \Sigma, \mu), \|.\|_\infty)$ is a Banach space.

Proof   The first statement follows easily from the definitions. If $f \in \mathcal{L}^\infty$, let $B = (|f| > \operatorname{ess\,sup}(|f|))$, and let $f' = fI_{X\setminus B}$. Then $\operatorname{ess\,sup}(|f'|) = \sup(|f'|)$, and $f' - f \in \mathcal{N}_\mathbf{R}^\infty(X, \Sigma, \mu)$, so that $[f] = [f']$; this idea is always useful in considering convergence in the norm $\|.\|_\infty$. In order to prove completeness, we use Proposition 14.2.5 of Volume II; it is enough to show that if $\sum_{n=1}^\infty \|[f_n]\|_\infty < \infty$, then $\sum_{n=1}^\infty [f_n]$ converges in norm. We can pick representatives f_n' in $\mathcal{L}_\mathbf{R}^\infty$ for which $\operatorname{ess\,sup}(|f_n'|) = \sup(|f_n'|)$. Then $\sum_{n=1}^\infty (\sup |f_n'|) < \infty$, and so the sum $\sum_{n=1}^\infty f_n(x)$ converges uniformly on X to a bounded measurable function f on X. Consequently $\sum_{n=1}^\infty [f_n]$ converges in norm to $[f]$.   □

Norm convergence in $L^\infty(X, \Sigma, \mu)$ is called *uniform convergence almost everywhere*.

We can also consider the space $\mathcal{L}_\mathbf{C}^\infty(X, \Sigma, \mu)$ of essentially bounded complex-valued functions (measurable functions f for which $|f|$ is essentially bounded), and the corresponding Banach space $(L_\mathbf{C}^\infty(X, \Sigma, \mu), \|.\|_\infty)$.

Exercises

29.7.1 If (X, Σ, μ) is a finite measure space, show that the set of simple measurable is dense in $(L_\mathbf{R}^\infty(X, \Sigma, \mu), \|.\|_\infty)$.

29.7.2 Show that $C_\mathbf{R}([0, 1])$ is not dense in $L_\mathbf{R}^\infty([0, 1], \mathcal{L}, \lambda)$.

29.7.3 Give an example of an element of $L_\mathbf{R}^\infty([0, 1], \mathcal{L}, \lambda)$ which cannot be approximated by step functions in the $\|.\|_\infty$ norm.

30

Constructing measures

30.1 Outer measures

We used outer measure to define Lebesgue measure. Can we do the same in a more general situation?

An *outer measure* on a non-empty set X is a mapping μ^* from the set $P(X)$ of subsets of X into $\mathbf{R}^+$ which satisfies

(a) $\mu^*(\emptyset) = 0$,
(b) if $E \subseteq F$ then $\mu^*(E) \leq \mu^*(F)$, and
(c) if $(E_n)_{n=1}^{\infty}$ is a sequence in $P(X)$, then $\mu^*(\cup_{n=1}^{\infty} E_n) \leq \sum_{n=1}^{\infty} \mu^*(E_n)$.

The function μ_* defined by $\mu_*(E) = \mu^*(X) - \mu^*(X \setminus E)$ is the corresponding inner measure. By (c), $\mu^*(X) \leq \mu^*(E) + \mu^*(X \setminus E)$, and so $\mu_*(E) \leq \mu^*(E)$, for all $E \in P(X)$.

Thus Lebesgue outer measure λ^* on a finite interval is an example of an outer measure.

First we show that if (X, Σ, μ) is a complete finite measure, then it defines an outer measure, and the resulting outer measure determines the measure space (X, Σ, μ).

Theorem 30.1.1 *Suppose that (X, Σ, μ) is a complete finite measure space. If $E \in P(X)$, let $\mu^*(E) = \inf\{\mu(A) : A \in \Sigma, E \subseteq A\}$.*

(i) *If $E \subseteq X$, there exist sets A and B in Σ such that $A \subseteq E \subseteq B$, $\mu(A) = \mu_*(E)$ and $\mu(B) = \mu^*(E)$.*
(ii) *μ^* is an outer measure on X.*
(iii) *if $E \subseteq X$ then $E \in \Sigma$ if and only if $\mu^*(F) = \mu^*(F \cap E) + \mu^*(F \setminus E)$ for all $F \subseteq X$.*

Proof   (i) For each $n \in \mathbf{N}$ there exists $B_n \in \Sigma$ with $E \subseteq B_n$ and $\mu(B_n) < \mu^*(E) + 1/n$. Let $B = \cap \{B_n : n \in \mathbf{N}\}$. Then $B \in \Sigma$ and $E \subseteq B$, so that

$$\mu^*(E) \leq \mu(B) \leq \mu(B_n) < \mu^*(E) + 1/n, \text{ for all } n \in \mathbf{N}.$$

Thus $\mu^*(E) = \mu(B)$. Applying this result to $X \setminus E$, it follows that there exists $C \in \Sigma$ with $X \setminus E \subseteq C$ and $\mu^*(X \setminus E) = \mu(C)$. Let $A = X \setminus C$, so that $A \subseteq E$. Since $\mu^*(X) = \mu(X)$, it follows that

$$\mu_*(E) = \mu^*(X) - \mu^*(X \setminus E) = \mu(X) - \mu(C) = \mu(A).$$

(ii) Conditions (a) and (b) are clearly satisfied. Suppose that $(E_n)_{n=1}^{\infty}$ is a sequence of subsets of X. For each $n \in \mathbf{N}$ there exists $B_n \in \Sigma$ with $E_n \subseteq B_n$ and $\mu^*(E_n) = \mu(B_n)$. Then $B = \cup_{n=1}^{\infty} B_n \in \Sigma$ and

$$\mu^*(\cup_{n=1}^{\infty} E_n) \leq \mu(B) \leq \sum_{n=1}^{\infty} \mu(B_n) = \sum_{n=1}^{\infty} \mu^*(E_n).$$

(iii) Suppose that the condition is satisfied. There exist sets B and C in Σ such that $E \subseteq B$, $X \setminus E \subseteq C$, $\mu^*(E) = \mu(B)$ and $\mu^*(X \setminus E) = \mu(C)$. Then $B \cup C = X$ and

$$\mu(B) + \mu(C) = \mu^*(E) + \mu^*(X \setminus E) = \mu^*(X) = \mu(X),$$

so that $\mu(B \cap C) = \mu(B \cup C) - \mu(B) - \mu(C) = 0$. Since the measure μ is complete, $E \cap (B \cap C) \in \Sigma$. Since $E = (X \setminus C) \cup (E \cap (B \cap C))$, $E \in \Sigma$.

Conversely, suppose that $E \in \Sigma$ and that F is a subset of X. By condition (c), $\mu^*(F \cap E) + \mu^*(F \setminus E) \geq \mu^*(F)$; we need to prove the converse inequality. There exists $A \in \Sigma$ such that $F \subseteq A$ and $\mu^*(F) = \mu(A)$. Let $B = E \cap A$ and let $C = A \setminus E$, so that A is the disjoint union of B and C. Since $E \cap F \subseteq B$ and $F \setminus E \subseteq C$,

$$\mu^*(F) = \mu(A) = \mu(B) + \mu(C) \geq \mu^*(F \cap E) + \mu^*(F \setminus E). \qquad \square$$

This result helps explain the definition of Σ in the next theorem.

Theorem 30.1.2   *Suppose that μ^* is an outer measure on X. Let*

$$\Sigma = \{A \subseteq X : \mu^*(E \cap A) + \mu^*(E \setminus A) = \mu^*(E) \text{ for all } E \subseteq X\}.$$

Then Σ is a σ-field, and if μ is the restriction of μ^* to Σ, then μ is a finite measure, and (X, Σ, μ) is a complete measure space.

Proof By condition (c), $\mu^*(E \cap A) + \mu^*(E \setminus A) \geq \mu^*(E)$ for all A and E, and so

$$\Sigma = \{A \subseteq X : \mu^*(E \cap A) + \mu^*(E \setminus A) \leq \mu^*(E) \text{ for all } E \subseteq X\}.$$

The proof comprises six separate steps.

First, we show that Σ is a field. Certainly $X \in \Sigma$. Since $A \cap E = E \setminus (X \setminus A)$ and $E \setminus A = E \cap (X \setminus A)$, it follows that $A \in \Sigma$ if and only if $X \setminus A \in \Sigma$. Thus it is sufficient to show that if A and B are in Σ, then so is $A \cap B$. If $E \subseteq X$, then

$$\begin{aligned}
\mu^*(E) &= \mu^*(E \cap A) + \mu^*(E \setminus A) \\
&= [\mu^*(E \cap A \cap B) + \mu^*((E \cap A) \setminus B)] \\
&\quad + [\mu^*((E \setminus A) \cap B) + \mu^*((E \setminus A) \setminus B)] \\
&\geq \mu^*(E \cap A \cap B) \\
&\quad + \mu^*(((E \cap A) \setminus B) \cup ((E \setminus A) \cap B) \cup ((E \setminus A) \setminus B)) \\
&= \mu^*(E \cap (A \cap B)) + \mu^*(E \setminus (A \cap B)).
\end{aligned}$$

Secondly, we show that μ^* is additive on Σ. If A and B are disjoint elements of Σ, then

$$\mu^*(A \cup B) = \mu^*((A \cup B) \cap A) + \mu^*((A \cup B) \setminus A) = \mu^*(A) + \mu^*(B),$$

so that μ^* is additive on Σ.

Thirdly, suppose that $(A_n)_{n=1}^{\infty}$ is a sequence of disjoint elements in Σ, that $B_n = \cup_{j=1}^{n} A_j$ and that $A = \cup_{n=1}^{\infty} A_n$. We show that if $E \subseteq X$ then $\mu^*(E \cap B_n) = \sum_{j=1}^{n} \mu^*(E \cap A_j)$. We prove this by induction on n. It is trivially true when $n = 1$. Suppose that it is true for n. Since $B_n \in \Sigma$,

$$\begin{aligned}
\mu^*(E \cap B_{n+1}) &= \mu^*((E \cap B_{n+1}) \cap B_n) + \mu^*((E \cap B_{n+1}) \setminus B_n) \\
&= \mu^*(E \cap B_n) + \mu^*(E \cap A_{n+1}) = \sum_{j=1}^{n+1} \mu^*(E \cap A_j),
\end{aligned}$$

which establishes the induction.

Fourthly, we show that $\mu^*(A) = \sum_{n=1}^{\infty} \mu^*(A_n)$. By the previous step, $\sum_{j=1}^{n} \mu^*(A_j) = \mu^*(B_n) \leq \mu^*(A)$. Since this holds for all $n \in \mathbf{N}$, $\mu^*(A) \geq \sum_{n=1}^{\infty} \mu^*(A_n)$: the converse inequality follows from the definition of outer measure.

Fifthly, we show that $A \in \Sigma$. If $E \subseteq X$ and $n \in \mathbf{N}$, then

$$\mu^*(E) = \mu^*(E \cap B_n) + \mu^*(E \setminus B_n)$$

$$= \sum_{j=1}^{n} \mu^*(E \cap A_j) + \mu^*(E \setminus B_n) \geq \sum_{j=1}^{n} \mu^*(E \cap A_j) + \mu^*(E \setminus A),$$

so that

$$\mu^*(E) \geq \sum_{j=1}^{\infty} \mu^*(E \cap A_j) + \mu^*(E \setminus A) \geq \mu^*(E \cap A) + \mu^*(E \setminus A),$$

and $A \in \Sigma$.

Consequently, Σ is a σ-field, and the restriction μ of μ^* to Σ is a finite measure.

Sixthly, we show that Σ is a complete measure. Suppose that $A \in \Sigma$, that $\mu(A) = 0$ and that $F \subseteq A$. Then $\mu^*(F) = 0$, so that if $E \subseteq X$ then

$$\mu^*(E \cap F) + \mu^*(E \setminus F) = \mu^*(E \setminus F) \leq \mu^*(E).$$

Hence $F \in \Sigma$.   □

If we start with an outer measure μ^*, consider the measure μ of this theorem, and use μ to construct an outer measure $\mu^\vee$, as in Theorem 30.1.1, then it does not necessarily follow that $\mu^* = \mu^\vee$. Nor does it necessarily follow that if $\mu^*(A) = \mu_*(A)$ then $A \in \Sigma$, as Exercise 30.1.2 shows.

Exercises

30.1.1 Let μ^* be an outer measure on a set X, let μ be the measure which it defines, and let $\mu^\vee$ be the outer measure defined by μ. Show that $\mu^\vee(E) \geq \mu^*(E)$, for $E \subseteq X$.

30.1.2 Let X be a set with four elements. Let $\mu^*(\emptyset) = 0$, let $\mu^*(E) = 1/3$ if E is a singleton set, let $\mu^*(E) = 1/2$ if E has two points, let $\mu^*(E) = 2/3$ if E has three points and let $\mu^*(X) = 1$. Show that μ^* is an outer measure. What is the corresponding σ-field Σ? What is the corresponding measure μ? What is the outer measure $\mu^\vee$ defined by μ? Which subsets E of X satisfy $\mu^*(E) = \mu_*(E)$?

30.2 Caratheodory's extension theorem

We are now in a position to prove a fundamental extension theorem, which allows us to construct many interesting measure spaces. We need another definition. A collection S of subsets of a set X is a *semi-ring* if

(a) $\emptyset \in S$,

(b) if $A, B \in S$ then $A \cap B \in S$, and

(c) if $A, B \in S$ and $A \subset B$ then there exists a finite sequence $(C_1, \ldots, C_k)$ of disjoint elements of S such that $B \setminus A = \cup_{j=1}^{k} C_j$.

Here are some examples.

- The collection S of all subsets of $\mathbf{R}$ of the form $(a, b]$, $(-\infty, b]$, (a, ∞) or $\mathbf{R}$ is a semi-ring.
- If R_1 is a ring of subsets of a set X_1 and R_2 is a ring of subsets of a set X_2 then the collection of sets of the form $A_1 \times A_2$, with $A_1 \in R_1$ and $A_2 \in R_2$ is a semi-ring of subsets of $X_1 \times X_2$.
- Recall that the Bernoulli sequence space $\Omega(\mathbf{N})$ is the infinite product $\prod_{j=1}^{\infty} \{0, 1\}_j$, and that a j-cylinder set is a set of the form

$$\{x \in \Omega : x_i = a_i \text{ for } 1 \leq i \leq j\}, \text{ where } (a_1, \ldots, a_j) \in \{0, 1\}^j.$$

($\Omega(\mathbf{N})$ was introduced in Volume II, Section 13.2, and cylinder sets in Volume II, Section 15.4.) The collection of cylinder sets in Ω is a semi-ring.

A non-negative real-valued function m on a semi-ring S is a *pre-measure* if $m(\emptyset) = 0$ and if it is σ-additive: if $(A_n)_{n=1}^{\infty}$ is a sequence of disjoint elements of S whose union is in S, then $m(\cup_{n=1}^{\infty} A_n) = \sum_{n=1}^{\infty} m(A_n)$.

Theorem 30.2.1 (The Caratheodory extension theorem) *Suppose that m is a pre-measure on a semi-ring S of subsets of a set X, and that $X \in S$. Then there exists a complete finite measure μ on a σ-field Σ containing S, for which $\mu(A) = m(A)$ for $A \in S$.*

Proof   If $E \subseteq X$, let

$$\mu^*(E) = \inf \left\{ \sum_{n=1}^{\infty} m(A_n) : A_n \in S, E \subseteq \cup_{n=1}^{\infty} A_n \right\}.$$

We show that μ^* is an outer measure, and that if Σ is the σ-field and μ the measure given by Theorem 30.1.2, then Σ and μ have the required properties. Note that, if $A \in S$, then $\mu^*(A) = m(A)$, since m is a pre-measure.

Clearly conditions (a) and (b) of the preceding section are satisfied. Suppose that $(E_n)_{n=1}^{\infty}$ is a sequence of subsets of X, and that $\epsilon > 0$. For each n, there exists a sequence $(A_{nk})_{k=1}^{\infty}$ in S such that $E_n \subseteq \cup_{k=1}^{\infty} A_{nk}$

$\sum_{k=1}^{\infty} m(A_{nk}) < \mu^*(E_n) + \epsilon/2^n$. Then $\cup_{n=1}^{\infty} E_n \subseteq \cup_{n=1}^{\infty} \cup_{k=1}^{\infty} A_{nk}$, and

$$\mu^*(\cup_{n=1}^{\infty} E_n) \le \sum_{n=1}^{\infty} \sum_{k=1}^{\infty} m(A_{nk}) < \sum_{n=1}^{\infty} \mu^*(E_n) + \epsilon.$$

Since ϵ is arbitrary, condition (c) is satisfied; μ^* is an outer measure.

Next, we show that $S \subseteq \Sigma$. Suppose that $B \in S$, that $E \subseteq X$ and that $\epsilon > 0$. There exists a sequence $(A_n)_{n=1}^{\infty}$ in S such that $E \subseteq \cup_{n=1}^{\infty} A_n$ and $\sum_{n=1}^{\infty} m(A_n) \le \mu^*(E) + \epsilon$. Then $E \cap B \subset \cup_{n=1}^{\infty}(A_n \cap B)$; since $A_n \cap B \in S$ for all $n \in \mathbf{N}$, $\mu^*(E \cap B) \le \sum_{n=1}^{\infty} m(A_n \cap B)$. Similarly, $\mu^*(E \setminus B) \le \sum_{n=1}^{\infty} m(A_n \setminus B)$. Since $m(A_n) = m(A_n \cap B) + m(A_n \setminus B)$, it follows that

$$\mu^*(E \cap B) + \mu^*(E \setminus B) \le \sum_{n=1}^{\infty} m(A_n \cap B) + \sum_{n=1}^{\infty} m(A_n \setminus B)$$

$$= \sum_{n=1}^{\infty} m(A_n) < \mu^*(E) + \epsilon.$$

Since ϵ is arbitrary, $\mu^*(E \cap B) + \mu^*(E \setminus B) \le \mu^*(E)$, so that $B \in \Sigma$.

Finally, if $A \in \Sigma$ then $\mu(A) = \mu^*(A) = m(A)$.   □

Corollary 30.2.2   *If $E \subseteq X$, there exists $B \in \Sigma$ such that $E \subseteq B$ and $\mu^*(E) = \mu(B)$. $E \in \Sigma$ if and only if $\mu^*(E) = \mu_*(E)$.*

Proof   If $k \in \mathbf{N}$ there exists a sequence $(A_{nk})_{n=1}^{\infty}$ in S such that $E \subseteq \cup_{n=1}^{\infty} A_{nk}$ and $\sum_{n=1}^{\infty} m(A_{nk}) < \mu^*(E) + 1/k$. Let $B_k = \cup_{n=1}^{\infty} A_{nk}$. Then $B_k \in \Sigma$, and $\mu(B_k) \le \sum_{n=1}^{\infty} m(A_{nk}) < \mu^*(E) + 1/k$. Let $B = \cap_{k=1}^{\infty} B_k$. Then $B \in \Sigma$, $E \subseteq B$ and $\mu(B) \le \mu(B_k) < \mu^*(E) + 1/k$ for all $k \in \mathbf{N}$, so that $\mu(B) = \mu^*(E)$.

If $E \in \Sigma$, then certainly $\mu(E) = \mu^*(E) = \mu_*(E)$. Suppose conversely that $\mu^*(E) = \mu_*(E)$. Then there exist sets A and B in Σ such that $A \subseteq E \subseteq B$ and $\mu(A) = \mu_*(E) = \mu^*(E) = \mu(B)$. Thus $\mu^*(E \setminus A) \le \mu(B \setminus A) = 0$, so that, since the measure space is complete, $E \setminus A \in \Sigma$. Consequently, $E = A \cup (E \setminus A) \in \Sigma$.   □

Example 30.2.3   Finite Borel measures on the Bernoulli sequence space $\Omega(\mathbf{N})$.

$\Omega(\mathbf{N})$ is a compact metrizable space, when it is given the product topology, and the cylinder sets are open and closed. If C is a j-cylinder set, C is the union of the two $(j+1)$-cylinder sets

$$C^{(0)} = \{x \in C : x_{j+1} = 0\} \text{ and } C^{(1)} = \{x \in C : x_{j+1} = 1\}.$$

If μ is a finite Borel measure on $\Omega(\mathbf{N})$, then $\mu(C) = \mu(C^{(0)}) + \mu(C^{(1)})$, for every cylinder set C. Conversely, suppose that m is a non-negative real-valued function on the semi-ring of cylinder sets, which satisfies $m(C) = m(C^{(0)}) + m(C^{(1)})$, for every cylinder set C. Then it is an easy exercise to show that m is additive. But it is then trivially a pre-measure, since if $(C_n)_{n=1}^\infty$ is a disjoint sequence of cylinder sets whose union C is a cylinder set, then all but finitely many sets C_n must be empty, since C is compact and the sets C_n are open. It therefore follows from Caratheodory's extension theorem that there is a measure μ on a σ-field containing the cylinder sets, which extends m. But the cylinder sets generate the Borel σ-field, so that the restriction of μ to the Borel σ-field is a finite Borel measure on $\Omega(\mathbf{N})$.

30.3 Uniqueness

Is the extension provided by Caratheodory's extension theorem uniquely determined? There are two closely related results which show that the answer to this question, and other similar questions, is 'yes'. We need a definition. A collection M of subsets of a set X is a *monotone class* if whenever $(A_n)_{n=1}^\infty$ is an increasing sequence in M then $\cup_{n=1}^\infty A_n \in M$ and whenever $(A_n)_{n=1}^\infty$ is a decreasing sequence in M then $\cap_{n=1}^\infty A_n \in M$.

Theorem 30.3.1 (The monotone class theorem) *If R is a field of subsets of a set X and if M is a monotone class containing R, then M contains the σ-field $\sigma(R)$ generated by R.*

Proof   Since the intersection of monotone classes is clearly a monotone class, there is a smallest monotone class M_0 such that $R \subseteq M_0 \subseteq M$. Since $\sigma(R)$ is a monotone class, $M_0 \subseteq \sigma(R)$. Let

$$M_1 = \{A \in M_0 : E \setminus A \in M_0 \text{ for all } E \in R\}.$$

Then M_1 is a monotone class containing R, and so $M_1 = M_0$. If $A \in M_0$, let

$$M_A = \{E \subseteq X : E \cap A \in M_0\}.$$

M_A is a monotone class. If $B \in R$, then $R \subseteq M_B$, and so $M_0 \subseteq M_B$. This means that if $A \in M_0$ then $A \cap B \in M_0$. This in turn means that $B \in M_A$. But this holds for all $B \in R$, and so $R \subseteq M_A$. Thus $M_0 \subseteq M_A$. Consequently if $A' \in M_0$ then $A' \cap A \in M_0$. Thus M_0 is a ring: since it is also a monotone class, and since $X \in M_0$, it is a σ-field containing R. Thus $\sigma(R) \subseteq M_0 \subseteq M$. □

In the next result, we weaken one condition, and strengthen the other. We need some more definitions. Suppose that X is a set. A π-*system* in

X is a collection Π of subsets of X with the property that if $E, F \in \Pi$, then $E \cap F \in \Pi$. A λ-*system* in X is a collection Λ of subsets of X which satisfies

(i) $X \in \Lambda$,
(ii) if $(A_n)_{n=1}^{\infty}$ is an increasing sequence in Λ, then $\cup_{n=1}^{\infty} A_n \in \Lambda$, and
(iii) If $A, B \in \Lambda$ and $A \subseteq B$ then $B \setminus A \in \Lambda$.

If (A_n) is a decreasing sequence in a λ-system Λ, then

$$\cap_{n=1}^{\infty} A_n = A_1 \setminus \cup_{n=1}^{\infty} (A_1 \setminus A_n),$$

so that a λ-system is a monotone class. Verify that a λ-system which is also a π-system is a σ-field.

Theorem 30.3.2 (Dynkin's π-λ theorem)   *If Π is a π-system of subsets of a set X and Λ is a λ-system containing Π, then Λ contains the σ-field $\sigma(\Pi)$ generated by Π.*

Proof   Let $l(\Pi)$ be the intersection of the λ-systems which contain Π. Then $l(\Pi)$ is a λ-system. We show that $l(\Pi)$ is a π-system, which establishes the result. Suppose that $A \in l(\Pi)$. Let $l_A = \{E \in l(\Pi) : E \cap A \in l(\Pi)\}$. Then $l(A)$ is a λ-system (verify this). Suppose first that $B \in \Pi$. If $C \in \Pi$, then $C \in l_B$, so that $\Pi \subset l_B$. Consequently $l(\Pi) \subseteq l_B$. Now suppose that $A \in l(\Pi)$. If $B \in \Pi$ then $A \in l_B$, and so $B \in l_A$. Thus $\Pi \subseteq l_A$. Consequently $l(\Pi) \subset l_A$. Thus if $A' \in l(\Pi)$ then $A \cap A' \in l(\Pi) : l(\Pi)$ is a π-system.

Note the similarities in the proofs of the two theorems. For many problems, either can be used, but often Dynkin's $\pi - \lambda$ theorem is more convenient. $\square$

Theorem 30.3.3   *Suppose that μ_1 and μ_2 are finite measures on a σ-field Σ, and that Π is a π-system contained in Σ. If $\mu_1(A) = \mu_2(A)$ for all $A \in \Pi$, then $\mu_1(A) = \mu_2(A)$ for all $A \in \sigma(\Pi)$.*

Proof   Let $\Sigma_0 = \{A \in \Sigma : \mu_1(A) = \mu_2(A)\}$. Then Σ_0 is a λ-system containing Π, and so it contains $\sigma(\Pi)$. $\square$

Corollary 30.3.4   *The extension in Caratheodory's extension theorem is unique.*

Proof   For a semi-ring is a π-system. $\square$

Exercise

30.3.1 Use the monotone class theorem to prove Theorem 30.3.3.

30.4 Product measures

Suppose that (X, Σ) and (Y, T) are two measurable spaces. A set of the form $A \times B$, where $A \in \Sigma$ and $B \in T$, is called a *measurable rectangle*. The σ-field generated by the measurable rectangles is called the *product σ-field*, and is denoted by $\Sigma \otimes T$.

Here is an important example.

Example 30.4.1 Suppose that (X, d) and (Y, ρ) are two separable metric spaces, and that Σ is the Borel σ-field of X, T the Borel σ-field of Y. Then $\Sigma \otimes T$ is the Borel σ-field of the product metric space $X \times Y$.

Proof Let $\mathcal{B}$ be the Borel σ-field of $X \times Y$; let X_0 be a countable dense subset of X and let Y_0 be a countable dense subset of Y. Let

$$\mathcal{A} = \{\{(x, y) : d(x, x_0) < 1/n, \rho(y, y_0) < 1/n\} : x_0 \in X_0, \ y_0 \in Y_0, n \in \mathbf{N}\}.$$

Then any open set is a countable union of sets in $\mathcal{A}$, and so $\mathcal{B} = \sigma(\mathcal{A})$. But every element of $\mathcal{A}$ is a measurable rectangle, and so $\mathcal{B} = \sigma(\mathcal{A}) \subseteq \Sigma \otimes T$. On the other hand, if $A \times B$ is a measurable rectangle, and if U is open in Y, then

$$A \times U \in \{C \times U : C \in \Sigma\} \subseteq \mathcal{B},$$

and so $\{A \times D : D \in T\} \subseteq \mathcal{B}$; hence $A \times B \in \mathcal{B}$, and so $\Sigma \otimes T \subseteq \mathcal{B}$. □

Suppose now that μ is a measure on Σ and ν is a measure on T. Can we define a measure $\mu \otimes \nu$ on $\Sigma \otimes T$ in such a way that if $A \times B$ is a measurable rectangle then $(\mu \otimes \nu)(A \times B) = \mu(A).\nu(B)$?

We begin by considering the case where μ and ν are finite measures. If $C \subseteq X \times Y$ and $x \in X$, we define $C^{(x)}$ to be $\{y : (x, y) \in C\}$.

Theorem 30.4.2 *Suppose that $C \in \Sigma \otimes T$. Then $C^{(x)} \in T$ for each $x \in X$, and the function $\nu(C^{(x)})$ is μ-measurable.*

Proof We use Dynkin's π-λ theorem. Let $\mathcal{C}$ be the collection of subsets of $X \times Y$ for which $C^{(x)} \in T$ for each $x \in X$, and the function $\nu(C^{(x)})$ is μ-measurable. If $C, D \in \mathcal{C}$ and $C \subseteq D$ then $(D \setminus C)^{(x)} = D^{(x)} \setminus C^{(x)}$ and $\nu((D \setminus C)^{(x)}) = \nu(D^{(x)}) - \nu(C^{(x)})$, so that $D \setminus C \in \mathcal{C}$. If $(C_n)_{n=1}^{\infty}$ is an increasing sequence in $\mathcal{C}$, with union C, and $x \in X$, then $C^{(x)} = \cup_{n=1}^{\infty}(C_n^{(x)}) \in T$, and $\nu(C^{(x)}) = \lim_{n \to \infty} \nu(C_n^{(x)})$, so that the function $\nu(C^{(x)})$ is μ-measurable. Thus $C \in \mathcal{C}$, and so $\mathcal{C}$ is a λ-system. The set of measurable rectangles is a π-system contained in $\mathcal{C}$. It therefore follows that $\Sigma \otimes T \subseteq \mathcal{C}$. □

We use this to define $\mu \otimes \nu$. If $C \in \Sigma \otimes T$ we set

$$(\mu \otimes \nu)(C) = \int_X \nu(C^{(x)}) \, d\mu(x).$$

Theorem 30.4.3 $\mu \otimes \nu$ *is a finite measure on* $\Sigma \otimes T$, *and*

$$(\mu \otimes \nu)(A \times B) = \mu(A).\nu(B)$$

for every measurable rectangle $A \times B$.

Proof   Certainly $(\mu \otimes \nu)(\emptyset) = 0$, and if $A \times B$ is a measurable rectangle, then $(\mu \otimes \nu)(A \times B) = \mu(A).\nu(B)$. In particular, $(\mu \otimes \nu)(X \times Y) = \mu(X).\nu(Y) < \infty$. If $(C_n)_{n=1}^\infty$ is an increasing sequence in $\Sigma \otimes T$, with union C, then $\nu(C_n^{(x)})$ increases to $\nu(C^{(x)})$ as $n \to \infty$, for each $x \in X$, and so, by the monotone convergence theorem,

$$(\mu \otimes \nu)(C) = \int_X \nu(C^{(x)}) \, d\mu(x) = \lim_{n \to \infty} \int_X \nu(C_n^{(x)}) \, d\mu(x) = \lim_{n \to \infty} (\mu \otimes \nu)(C_n).$$

It therefore follows from Exercise 28.4.1 that $\mu \otimes \nu$ is a finite measure on $\Sigma \otimes T$.   □

We can also define a measure $\mu \widetilde{\otimes} \nu$ on $\Sigma \otimes T$ by reversing the roles of X and Y. If $y \in Y$, let $C_{(y)} = \{x \in X : (x, y) \in C\}$. Then $C_{(y)} \in \Sigma$ for each $y \in Y$, and the function $\mu(C_{(y)})$ is ν-measurable, and we set

$$(\mu \widetilde{\otimes} \nu)(C) = \int_Y \mu(C_{(y)}) \, d\nu(y).$$

Theorem 30.4.4   *The measures* $\mu \otimes \nu$ *and* $\mu \widetilde{\otimes} \nu$ *are the same.*

Proof   The collection $\{C \in \Sigma \otimes T : (\mu \otimes \nu)(C) = (\mu \widetilde{\otimes} \nu)(C)\}$ is a λ-system containing the measurable rectangles, and is therefore equal to $\Sigma \otimes T$.   □

We can extend these results to three or more products. For example, if $((X_i, \Sigma_i, \mu_i))_{i=1}^3$ are three finite measure spaces, then we can construct the measures $(\mu_1 \otimes \mu_2) \otimes \mu_3$ on $(\Sigma_1 \otimes \Sigma_2) \otimes \Sigma_3$ and $\mu_1 \otimes (\mu_2 \otimes \mu_3)$ on $\Sigma_1 \otimes (\Sigma_2 \otimes \Sigma_3)$. Further applications of Dynkin's π -λ theorem show that $(\Sigma_1 \otimes \Sigma_2) \otimes \Sigma_3 = \Sigma_1 \otimes (\Sigma_2 \otimes \Sigma_3)$ and $(\mu_1 \otimes \mu_2) \otimes \mu_3 = \mu_1 \otimes (\mu_2 \otimes \mu_3)$.

We can also consider products of σ-finite measures. Suppose that (X, Σ, μ) and (Y, T, ν) are σ-finite measure spaces, and that $(I_k)_{k=1}^\infty$ is a disjoint sequence of elements of Σ of finite measure whose union is X, and that $(J_l)_{l=1}^\infty$ is a corresponding sequence in T. We construct the product measure $\mu \otimes \nu$ on each set $I_k \times J_l$. If $A \in \Sigma \times T$, we then define $(\mu \otimes \nu)(A) = \sum_{k,l} \mu(A \cap (I_k \times J_l))$. Again,

$$(\mu \otimes \nu)(A) = \int_X \nu(A^{(x)}) \, d\mu(x);$$

but in this case the integrand and the integral can take infinite values.

We can use product measures to illustrate the notion that the integral is the 'area under the curve'. Suppose that f is a non-negative measurable function on a measure space (X, Σ, μ). We consider Borel measure λ on $[0, \infty)$. Let $A_f = \{(x, t) \in X \times [0, \infty) : 0 \le t < f(x)\}$; A_f is the set of points in $X \times [0, \infty)$ which are 'under the curve'. If $x \in X$, then $A_f^{(x)} = \emptyset$ if $f(x) = 0$; otherwise

$$A_f^{(x)} = \{t \in [0, \infty) : 0 \le t < f(x)\} = [0, f(x)).$$

Thus $\mu(A_f^{(x)}) = f(x)$, and $(\mu \otimes \nu)(A_f) = \int_X f(x) \, d\mu(x)$. On the other hand, if $t \in [0, \infty)$ then $A_{f,(t)} = \{x \in X; f(x) > t\}$, so that $\mu(A_{f,(t)}) = \mu(f > t)$ and $(\mu \tilde{\otimes} \nu)(A_f) = \int_0^\infty \mu(f > t) \, dt$. This reveals a certain circularity of argument, but also throws some light on the definition of the integral.

We now consider functions of two variables. (The results extend easily to functions of three or more variables.)

Theorem 30.4.5 (Tonelli's theorem, I) *Suppose that (X, Σ, μ) and (Y, T, ν) are measure spaces, and that f is a non-negative $\Sigma \otimes T$ measurable function on $X \times Y$.*

(i) The function $y \to f(x, y)$ is T-measurable, for each $x \in X$, the extended-real-valued function $x \to \int_Y f(x, y) \, d\nu(y)$ is Σ-measurable, and

$$\int_{X \times Y} f \, d(\mu \otimes \nu) = \int_X \left(\int_Y f(x, y) \, d\nu(y) \right) d\mu(x).$$

(ii) The function $x \to f(x, y)$ is Σ-measurable, for each $y \in Y$, the extended-real-valued function $y \to \int_X f(x, y) \, d\mu(x)$ is T-measurable, and

$$\int_{X \times Y} f \, d(\mu \otimes \nu) = \int_Y \left(\int_X f(x, y) \, d\mu(x) \right) d\nu(y).$$

Proof   (i) Let

$$A_f = \{(x, y, t) \in X \times Y \times [0, \infty) : 0 \le t < f(x, y)\}.$$

For fixed x and t,

$$\{y \in Y : f(x, y) > t\} = \{y \in Y : (x, y, t) \in A_f\},$$

so that $\{y \in Y : f(x,y) > t\}$ is T-measurable; hence the function $y \to f(x,y)$ is T-measurable. Similarly,

$$\int_Y f(x,y)\,d\nu(y) = (\mu \otimes \lambda)(\{(x,t) : (x,y,t) \in A_f\}),$$

so that the extended-real-valued function $x \to \int_Y f(x,y)\,d\nu(y)$ is Σ-measurable. Finally,

$$\int_{X \times Y} f\,d(\mu \otimes \nu) = ((\mu \otimes \nu) \otimes \lambda)(A_f)$$

$$= (\mu \otimes (\nu \otimes \lambda))(A_f)$$

$$= \int_X (\nu \otimes \lambda)(0 \le f(x,y) < t)\,d\mu(x)$$

$$= \int_X \left(\int_Y f(x,y)\,d\nu(y) \right) d\mu(x).$$

The proof of (ii) is exactly similar.   □

The importance of this result is that the integral can be evaluated by repeated integration, and also, and equally important, that we can change the order of integration.

What about functions which may take positive and negative values, but are $\mu \otimes \nu$ integrable?

Theorem 30.4.6 (Fubini's theorem, I) *Suppose that (X, Σ, μ) and (Y, T, ν) are measure spaces, and that f is a $\Sigma \otimes T$ measurable function on $X \times Y$. If f is $\mu \otimes \nu$-integrable, then the function $y \to f(x,y)$ is T-measurable, for every $x \in X$, and is ν-integrable except on a μ-null set N. The function $x \to \int_Y f(x,y)\,d\nu(y)$ is Σ-measurable and μ-integrable on $X \setminus N$, and*

$$\int_{X \times Y} f\,d(\mu \otimes \nu) = \int_{X \setminus N} \left(\int_Y f(x,y)\,d\nu(y) \right) d\mu(x).$$

Conversely, if the function $y \to |f(x,y)|$ is T-measurable except on a μ-null set N and $\int_{X \setminus N} \left(\int_Y |f(x,y)|\,d\nu(y) \right) d\mu(x) < \infty$, then f is $\mu \otimes \nu$-integrable, and

$$\int_{X \times Y} f\,d(\mu \otimes \nu) = \int_{X \setminus N} \left(\int_Y f(x,y)\,d\nu(y) \right) d\mu(x).$$

Further, there exists a ν-null subset M of Y such that

$$\int_{X \times Y} f \, d(\mu \otimes \nu) = \int_{Y \setminus M} \left(\int_X f(x, y) \, d\mu(x) \right) d\nu(y).$$

Proof If f is $\mu \otimes \nu$-integrable, then

$$\int_{X \otimes Y} f^+ \, d(\mu \otimes \nu) < \infty \text{ and } \int_{X \otimes Y} f^- \, d(\mu \otimes \nu) < \infty.$$

It therefore follows that if

$$N^+ = \{x \in X : \int_Y f^+(x, y) \, d\nu(y) = \infty\}$$

$$\text{and } N^- = \{x \in X : \int_Y f^-(x, y) \, d\nu(y) = \infty\},$$

then N^+ and N^- are μ-null sets. Setting $N = N^+ \cup N^-$, the functions $\int_Y f^+(x, y) \, d\nu(y)$ and $\int_Y f^-(x, y) \, d\nu(y)$ are μ-integrable on $X \setminus N$, and so therefore is $\int_Y f(x, y) \, d\nu(y)$. Further,

$$\int_{X \times Y} f \, d(\mu \otimes \nu) = \int_{X \times Y} f^+ \, d(\mu \otimes \nu) - \int_{X \times Y} f^- \, d(\mu \otimes \nu)$$

$$= \int_{X \setminus N} \left(\int_Y f^+(x, y) \, d\nu(y) \right) d\mu(x)$$

$$- \int_{X \setminus N} \left(\int_Y f^-(x, y) \, d\nu(y) \right) d\mu(x)$$

$$= \int_{X \setminus N} \left(\int_Y f(x, y) \, d\nu(y) \right) d\mu(x).$$

Conversely, suppose that (ii) holds. It then follows from Tonelli's theorem that $\int_{X \times Y} |f| \, d(\mu \otimes \nu) < \infty$, so that f is $\mu \otimes \nu$-integrable. □

Fubini's theorem holds because the Lebesgue integral is an absolute integral. The next example illustrates this.

Example 30.4.7 Fubini's theorem for counting measure.

Suppose that $(X, \Sigma, \mu) = (Y, T, \nu) = (\mathbf{N}, P(\mathbf{N}), \mu)$, where μ is counting measure. The $P(\mathbf{N}) \otimes P(\mathbf{N}) = P(\mathbf{N} \times \mathbf{N})$, so that all functions are measurable. Fubini's theorem then states that if f is a function on $\mathbf{N} \times \mathbf{N}$, then $\sum_{i,j} |f(i,j)| < \infty$ if and only if $\sum_{i=1}^{\infty} (\sum_{j=1}^{\infty} |f(i,j)|) < \infty$ If so, then

$\sum_{i=1}^{\infty} f(i,j)$ converges for each j and $\sum_{j=1}^{\infty} f(i,j)$ converges for each i, and

$$\sum_{i,j} f(i,j) = \sum_{j=1}^{\infty} \left(\sum_{i=1}^{\infty} f(i,j) \right) = \sum_{j=1}^{\infty} \left(\sum_{i=1}^{\infty} f(i,j) \right).$$

This is Exercise 4.4.3 of Volume I.

If (X, Σ, μ) and (Y, T, ν) are complete measure spaces, it is natural to consider the completion

$$(X \times Y, \Sigma \hat{\otimes} T, \mu \hat{\otimes} \nu) \text{ of } (X \times Y, \Sigma \otimes T, \mu \otimes \nu).$$

(Note that, unlike $\Sigma \otimes T$, the σ-field $\Sigma \hat{\otimes} T$ depends upon μ and ν.) There are corresponding Tonelli and Fubini theorems.

Theorem 30.4.8 (Tonelli's theorem, II) *Suppose that (X, Σ, μ) and (Y, T, ν) are complete measure spaces, and that f is a non-negative $\Sigma \hat{\otimes} T$ measurable function on $X \times Y$.*

(i) The function $y \to f(x,y)$ is T-measurable, except on a null subset N of X, the extended-real-valued function $x \to \int_Y f(x,y)\,d\nu(y)$ is Σ-measurable on $X \setminus N$, and

$$\int_{X \times Y} f \, d(\mu \hat{\otimes} \nu) = \int_{X \setminus N} \left(\int_Y f(x,y)\,d\nu(y) \right) d\mu(x).$$

(ii) The function $x \to f(x,y)$ is Σ-measurable, except on a null subset M of Y, the extended-real-valued function $y \to \int_X f(x,y)\,d\mu(x)$ is T-measurable on $Y \setminus M$, and

$$\int_{X \times Y} f \, d(\mu \otimes \nu) = \int_{Y \setminus M} \left(\int_X f(x,y)\,d\mu(x) \right) d\nu(y).$$

Proof　There exist $\Sigma \otimes T$-measurable functions g and h on $X \times Y$ such that $0 \le g \le f \le h$ and such that $g = f = h$ $(\mu \otimes \nu)$-almost everywhere. Thus

$$0 = \int_{X \times Y} (h - g) \, d(\mu \otimes \nu) = \int_X \left(\int_Y (h(x,y) - g(x,y))\,d\nu(y) \right) d\mu(x),$$

so that, except on a μ-null subset N of X, $\int_Y (h(x,y) - g(x,y))\,d\nu(y) = 0$. Thus if $x \in X \setminus N$ then $h(x,y) = g(x,y)$ for ν almost all y. Since (Y, T, ν) is complete, if $x \in X \setminus N$ then $h(x,y) = f(x,y) = g(x,y)$ for ν almost all y; hence the function $y \to f(x,y)$ is T measurable, and $\int_Y f(x,y)\,d\nu(y) =$

$\int_Y g(x,y)\, d\nu(y)$. Thus

$$\int_{X \times Y} f\, d(\mu \hat{\otimes} \nu) = \int_{X \times Y} g\, d(\mu \otimes \nu)$$

$$= \int_X \left(\int_Y g(x,y)\, d\nu(y) \right) d\mu(x)$$

$$= \int_{X \setminus N} \left(\int_Y g(x,y)\, d\nu(y) \right) d\mu(x)$$

$$= \int_{X \setminus N} \left(\int_Y f(x,y)\, d\nu(y) \right) d\mu(x).$$

Again, the proof of (ii) is exactly similar. □

Theorem 30.4.9 (Fubini's theorem, II) *Suppose that* (X, Σ, μ) *and* (Y, T, ν) *are complete measure spaces, and that* f *is a* $\Sigma \hat{\otimes} T$ *measurable function on* $X \times Y$. *If* f *is* $\mu \hat{\otimes} \nu$-*integrable, then the function* $y \to f(x,y)$ *is* T-*measurable and* ν-*integrable except on a* μ-*null set* N, *the function* $x \to \int_Y f(x,y)\, d\nu(y)$ *is* Σ-*measurable and* μ-*integrable on* $X \setminus N$, *and*

$$\int_{X \times Y} f\, d(\mu \hat{\otimes} \nu) = \int_{X \setminus N} \left(\int_Y f(x,y)\, d\nu(y) \right) d\mu(x).$$

Conversely, if the function $y \to |f(x,y)|$ *is* T-*measurable except on a* μ-*null set* N *and* $\int_{X \setminus N} \left(\int_Y |f(x,y)|\, d\nu(y) \right) d\mu(x) < \infty$, *then* f *is* $\mu \hat{\otimes} \nu$-*integrable, and*

$$\int_{X \times Y} f\, d(\mu \hat{\otimes} \nu) = \int_{X \setminus N} \left(\int_Y f(x,y)\, d\nu(y) \right) d\mu(x).$$

Further, there exists a ν-*null subset* M *of* Y *such that*

$$\int_{X \times Y} f\, d(\mu \hat{\otimes} \nu) = \int_{Y \setminus M} \left(\int_X f(x,y)\, d\mu(x) \right) d\nu(y).$$

Proof The proof again follows by sandwiching f between two $\Sigma \otimes T$-measurable functions. The details are left to the reader. □

In particular, these last two theorems apply to Lebesgue measurable functions in $\mathbf{R}^d$.

A word of caution about the naming of these results. Many authors simply use 'Fubini's theorem' to refer to any of the theorems that we have called Fubini's theorem or Tonelli's theorem, while some authors also attribute some of the results to E.W. Hobson.

Exercises

30.4.1 Suppose that X is an uncountable set with the discrete metric. Show that the diagonal $\Delta = \{(x,x) : x \in X\}$ is not in $P(X) \otimes P(X)$.

30.4.2 Give the details of the proof of Theorem 30.4.9.

30.4.3 Suppose that $0 < a < b$. Show that the function $f(x,y) = e^{-xy}$ is integrable on $[0,\infty) \times [a,b]$. Use Fubini's theorem to calculate

$$\int_0^\infty \frac{e^{-ax} - e^{-bx}}{x}\, dx.$$

30.4.4 Let $f(0,0) = 0$ and let $f(x,y) = xy/(x^2+y^2)^2$ if $(x,y) \neq (0,0)$. Show that the integrals $\int_{-1}^1 f(x,y)\, d\lambda(x)$ and $\int_{-1}^1 f(x,y)\, d\lambda(y)$ exist and are equal for all $x, y \in [-1,1]$. Is f integrable on $[-1,1] \times [-1,1]$?

30.4.5 Give an example of a Lebesgue measurable function f on the unit square $[0,1] \times [0,1]$ for which $\int_0^1 f(x,y)\, d\lambda(x)$ exists and equals 1 for all y and $\int_0^1 f(x,y)\, d\lambda(y)$ exists and equals 0 for all x. Why does this not contradict Fubini's theorem?

30.4.6 Let $f(x,y) = (x^2 - y^2)/(x^2+y^2)^2$, for $(x,y) \in (0,1) \times (0,1)$. Calculate

$$\int_0^1 \left(\int_0^1 f(x,y)\, dx \right) dy \text{ and } \int_0^1 \left(\int_0^1 f(x,y)\, dy \right) dx.$$

What does this tell you about the integrability of f?

30.5 Borel measures on R, I

There are many other measures defined on the Borel sets $\mathcal{B}$ of $\mathbf{R}$ than Lebesgue measure. We begin by considering finite measures. Let $\mathcal{M}^+(\mathbf{R})$ be the set of finite Borel measures defined on the Borel sets $\mathcal{B}$ of $\mathbf{R}$.

Proposition 30.5.1   *Suppose that $\mu \in \mathcal{M}^+(\mathbf{R})$. Let $F_\mu(t) = \mu((-\infty,t])$. Then F_μ is a non-negative right-continuous increasing function on $\mathbf{R}$, $F_\mu(t) \to 0$ as $t \to -\infty$ and $F_\mu(t) \to \mu(\mathbf{R})$ as $t \to +\infty$.*

Proof   Certainly F_μ is non-negative and increasing. If $t_n \searrow t$ as $n \to \infty$ then $(-\infty, t_n] \searrow (-\infty, t]$, so that $F_\mu(t_n) \searrow F_\mu(t)$, by downwards continuity; thus F_μ is right continuous. Similarly, $(-\infty, -n] \searrow \emptyset$ as $n \to \infty$, so that $F_\mu(t) \to 0$ as $t \to -\infty$, and $(-\infty, n] \nearrow \mathbf{R}$ as $t \to +\infty$, so that $F_\mu(t) \to \mu(\mathbf{R})$ as $t \to \infty$, by upwards continuity. $\square$

The function F_μ is called the *cumulative distribution function* of μ.

Theorem 30.5.2 *Let $\mathcal{F}(\mathbf{R})$ denote the set of non-negative bounded increasing right-continuous functions f on $\mathbf{R}$ for which $f(t) \to 0$ as $t \to -\infty$. Then the mapping $\mu \to F_\mu$ is a bijection of $\mathcal{M}^+(\mathbf{R})$ onto $\mathcal{F}(\mathbf{R})$.*

Proof First we show that the mapping is injective. Suppose that $F_\mu = F_\nu$. The collection of sets Π of the form $(-\infty, b]$ is a π-system, and $\mu(A) = \nu(A)$ for each $A \in \Pi$. Let Λ be the collection of Borel sets B for which $\mu(B) = \nu(B)$. Then Λ is a λ-system containing Π, and so, by Dynkin's π-λ theorem, Λ contains $\sigma(\Pi)$, which is the Borel σ-field. Thus $\mu = \nu$.

In order to show that the mapping is surjective, we use the Caratheodory extension theorem. Suppose that $F \in \mathcal{F}(\mathbf{R})$. Let S be the semi-ring of sets of the form $(a, b]$, together with the empty set. Let $m(\emptyset) = 0$ and let $m((a, b]) = F(b) - F(a)$. We show that m is a pre-measure on S. Suppose that $(a, b]$ is the disjoint union of the sequence $((a_j, b_j])_{j=1}^\infty$. Then

$$\sum_{j=1}^n m((a_j, b_j]) = \sum_{j=1}^n (F(b_j) - F(a_j)) \leq F(b) - F(a) = m((a, b]),$$

and so $\sum_{j=1}^\infty m((a_j, b_j]) \leq m((a, b])$.

We must prove the reverse inequality. Suppose that $\epsilon > 0$. Since F is right continuous, there exists $a < a' < b$ such that $F(a') - F(a) < \epsilon/2$ and for each $j \in \mathbf{N}$ there exists $b_j' > b_j$ such that $F(b_j') < F(b_j) + \epsilon/2^{j+1}$. The open intervals (a_j, b_j') cover the compact set $[a', b]$, and so there exists J such that $[a', b] \subseteq \cup_{j=1}^J (a_j, b_j')$. But this implies that $\sum_{j=1}^J (F(b_j') - F(a_j)) \geq F(b) - F(a')$. Hence

$$\sum_{j=1}^\infty m((a_j, b_j]) = \sum_{j=1}^\infty (F(b_j) - F(a_j)) \geq \sum_{j=1}^\infty (F(b_j') - F(a_j)) - \epsilon/2$$

$$\geq \sum_{j=1}^J (F(b_j') - F(a_j)) - \epsilon/2 \geq (F(b) - F(a')) - \epsilon/2$$

$$> (Fb) - F(a)) - \epsilon = m((a, b]) - \epsilon.$$

Since ϵ is arbitrary, the result follows. By the the Caratheodory extension theorem, there exists a measure μ on a σ-field containing $\sigma(S)$ which extends m. But $\sigma(S)$ is the Borel σ-field, and so F is the image of the restriction of μ to $\sigma(S)$. $\qquad\square$

Suppose that $F \in \mathcal{F}(\mathbf{R})$ and μ_F is the corresponding Borel measure. If $f \in L^1(\mu_F)$, the integral $\int_{\mathbf{R}} f \, d\mu_F$ is frequently written as $\int_X f \, dF$; it is called the *Stieltjes integral* of f with respect to F.

We can also consider the set $\mathcal{R}(\mathbf{R})$ of σ-finite measures on $\mathbf{R}$; here $\mu(\mathbf{R})$ may be infinite, and it may be the case that $\mu(-\infty, t] = \infty$ for all $t \in \mathbf{R}$. The cumulative distribution function is therefore unsuitable. Instead, we consider functions for which $f(0) = 0$. If μ is a σ-finite measure on $\mathbf{R}$, let

$$
J_\mu(t) = \begin{cases} \mu(0, t] & \text{if } t > 0, \\ 0 & \text{if } t = 0, \\ -\mu(t, 0] & \text{if } t < 0, \end{cases}
$$

so that $\mu((a, b]) = J_\mu(b) - J_\mu(a)$.

Theorem 30.5.3 *Let $\mathcal{J}(\mathbf{R})$ denote the set of non-negative increasing right-continuous functions f on $\mathbf{R}$ for which $f(0) = 0$. The mapping $\mu \to J_\mu$ is a bijection of $\mathcal{R}(\mathbf{R})$ onto $\mathcal{J}(\mathbf{R})$.*

Proof   This follows from the previous theorem, for example, by first considering the measure μ_k defined by setting $\mu_k(A) = \mu(A \cap (-k, k])$, and then letting k tend to infinity. The details are left to the reader. □

Another important case concerns σ-finite measures defined on the semi-infinite open interval $(0, \infty)$. If μ is such a measure and $0 < t < \infty$, let $\lambda_\mu(t) = \mu(t, \infty)$: λ_μ is the *tail distribution function* on $(0, \infty)$. This relates to the tail distribution of a non-negative measurable function f on a measure space (X, Σ, μ):

$$
\lambda_f(t) = \mu(f > t) = (f_*\mu)(t, \infty) = \lambda_{f_*\mu}(t).
$$

We are usually only concerned with measures for which $\lambda_\mu(t) < \infty$ for $t > 0$ (although, if μ is not a finite measure, then $\lambda_\mu(t) \to \infty$ as $t \searrow 0$). We denote the set of such measures by $\mathcal{R}_*(0, \infty)$.

Proposition 30.5.4 *If $\mu \in \mathcal{R}_*(0, \infty)$ then λ_μ is a decreasing right-continuous function on $(0, \infty)$ for which $\lambda_\mu(t) \to 0$ as $t \to \infty$.*

Proof   Just like the proof of Proposition 30.5.1. □

Theorem 30.5.5 *Let $\Lambda(\mathbf{R})$ denote the set of decreasing right-continuous functions λ on $(0, \infty)$ for which $f(t) \to 0$ as $t \to \infty$. The mapping $\mu \to \lambda_\mu$ is a bijection of $\mathcal{R}_*(0, \infty)$ onto $\Lambda(\mathbf{R})$.*

Proof   Once again, this is left as an exercise for the reader. □

We shall study Borel measures on $\mathbf{R}$ and their cumulative distribution functions further, in Section 32.4.

Exercises

We can also construct the Borel measure λ_F corresponding to a function F in $\mathcal{F}(\mathbf{R})$, by following the proof of the existence of Lebesgue measure.

30.5.1 If $I = (a, b)$ is an open interval, set $l_F(I) = F(b-) - F(a)$, with similar definitions for semi-infinite and infinite open intervals.

30.5.2 Use this to define $l_F(O)$, for open sets O.

30.5.3 If K is a compact subset of **R**, set $s_F(K) = l(\mathbf{R}) - l(\mathbf{R} \setminus K)$.

30.5.4 If A is a subset of **R**, define the outer measure $(\lambda_F)^*(A)$ and inner measure $(\lambda_F)_*(A)$ as

$$(\lambda_F)^*(A) = \inf\{l_F(U) : U \text{ open}, A \subseteq U\},$$
$$(\lambda_F)_*(A) = \sup\{s_F(K) : K \text{ compact}, K \subseteq A\}.$$

Show that $(\lambda_F)_*(A) \leq (\lambda_F)^*(A)$.

30.5.5 Say that A is λ_F-measurable if equality holds, and then define $\lambda_F(A)$ to be the common value. Show that the set of λ_F-measurable sets is a σ-field Σ_F containing the Borel σ-field.

30.5.6 Show that λ_F is a finite measure on Σ_F.

30.5.7 Show that F is the cumulative distribution function of λ_F.

31

Signed measures and complex measures

31.1 Signed measures

So far we have been concerned with measures which take non-negative values. We now drop this requirement. A *signed measure* σ on a measurable space (X, Σ) is a real-valued function on Σ which is σ-additive: if $(A_n)_{n=1}^{\infty}$ is a sequence of disjoint elements of Σ, then

$$\sigma(\cup_{n=1}^{\infty} A_n) = \sum_{n=1}^{\infty} \sigma(A_n).$$

An important feature of this definition is that infinite values are not allowed. A finite measure is a signed measure; in this setting, we call such a measure a *positive* measure.

Proposition 31.1.1 *Suppose that σ is a signed measure on a measurable space (X, Σ).*

(i) $\sigma(\emptyset) = 0$.
(ii) *If $(A_n)_{n=1}^{\infty}$ is a sequence of disjoint elements of Σ, then $\sum_{n=1}^{\infty} \sigma(A_n)$ converges absolutely.*
(iii) *If $(A_n)_{n=1}^{\infty}$ is an increasing sequence in Σ with union A then $\sigma(A) = \lim_{n\to\infty} \sigma(A_n)$.*
(iv) *If $(B_n)_{n=1}^{\infty}$ is a decreasing sequence in Σ with intersection B then $\sigma(B) = \lim_{n\to\infty} \sigma(B_n)$.*
(v) $\sigma(\Sigma) = \{\sigma(A) : A \in \Sigma\}$ *is a bounded subset of $\mathbf{R}$.*

Proof

(i) Take $A_n = \emptyset$ for $n \in \mathbf{N}$. Then $\sum_{n=1}^{\infty} \sigma(A_n)$ converges, so that $\sigma(\emptyset) = 0$.
(ii) If τ is any permutation of $\mathbf{N}$ then $\sum_{n=1}^{\infty} \sigma(A_{\tau(n)})$ converges, so that the result follows from Theorem 4.5.2 of Volume I.

(iii) Let $C_1 = A_1$ and let $C_n = A_n \setminus A_{n-1}$ for $n > 1$. Then A is the disjoint union of the sequence $(C_n)_{n=1}^{\infty}$, so that

$$\sigma(A) = \sum_{n=1}^{\infty} \sigma(C_n) = \lim_{n \to \infty} \sum_{j=1}^{n} \sigma(C_j) = \lim_{n \to \infty} \sigma(A_n).$$

(iv) Since $(X \setminus B_n)_{n=1}^{\infty}$ increases to $X \setminus B$,

$$\sigma(B) = \sigma(X) - \sigma(X \setminus B) = \sigma(X) - \lim_{n \to \infty} \sigma(X \setminus B_n)$$
$$= \lim_{n \to \infty} (\sigma(X) - \sigma(X \setminus B_n)) = \lim_{n \to \infty} \sigma(B_n).$$

(v) We need a lemma.

Lemma 31.1.2 *Let*

$$\mathcal{H} = \{H \in \Sigma : \{\sigma(C) : C \in \Sigma, C \subseteq H\} \text{ is unbounded}\}.$$

If $H \in \mathcal{H}$, then there exist $H' \in \mathcal{H}$ and $C \in \Sigma$ such that H is the disjoint union $H' \cup C$ and $|\sigma(C)| \geq 1$.

Proof   There exists $D \in \Sigma$ such that $D \subseteq H$ and $|\sigma(D)| \geq |\sigma(H)| + 1$. Then $|\sigma(H \setminus D)| \geq 1$. If $D \in \mathcal{H}$, take $H' = D$ and $C = H \setminus D$. Otherwise, $H \setminus D$ must be in $\mathcal{H}$; take $H' = H \setminus D$ and $C = D$.   □

Suppose that $X \in \mathcal{H}$. Let $A_0 = X$. Applying the lemma repeatedly, there exists a decreasing sequence $(A_n)_{n=1}^{\infty}$ in $\mathcal{H}$, such that if $C_n = A_{n-1} \setminus A_n$ then $|\sigma(C_n)| \geq 1$. But $(C_n)_{n=1}^{\infty}$ is a sequence of disjoint elements of Σ, and so $\sum_{n=1}^{\infty} \sigma(C_n)$ converges. This gives a contradiction. Thus $X \notin \mathcal{H}$, and so $\sigma(\Sigma)$ is bounded.   □

The set $ca_{\mathbf{R}}(X, \Sigma)$ of signed measures on a measure space (X, Σ) contains the finite measures, and is a linear subspace of the vector space space of all real-valued functions on Σ. Thus if π and ν are positive measures then $\pi - \nu$ is a signed measure. We can decompose a signed measure σ as the difference of two positive measures, in a canonical way.

Theorem 31.1.3   *If σ is a signed measure on a measurable space (X, Σ), then there exist disjoint P and N in Σ, with $X = P \cup N$, such that $\sigma(A) \geq 0$ for $A \subseteq P$ and $\sigma(A) \leq 0$ for $A \subseteq N$. Let $\sigma^+(A) = \sigma(A \cap P)$ and let $\sigma^-(A) = -\sigma(A \cap N)$. Then σ^+ and σ^- are positive measures on Σ, and $\sigma = \sigma^+ - \sigma^-$.*

Further, the decomposition is essentially unique; if $X = P' \cup N'$, where P' and N' are disjoint elements of Σ for which $\pi(A) = \sigma(A \cap P') \geq 0$ and $\nu(A) = -\sigma(A \cap N') \geq 0$ for $A \in \Sigma$, then $\pi = \sigma^+$ and $\nu = \sigma^-$.

Proof Say that A is strictly non-negative if $\sigma(B) \geq 0$ for all $B \subseteq A$. First we show that if $A \in \Sigma$ then there exists a strictly non-negative $C \subseteq A$ with $\sigma(C) \geq \sigma(A)$. Suppose not. If $\sigma(A) \leq 0$ then we can take $C = \emptyset$. Suppose that $\sigma(A) > 0$. Let $l_0 = \inf\{\sigma(B) : B \subseteq A\}$: $-\infty < l_0 < 0$. Choose $B_1 \subseteq A$ such that $\sigma(B_1) < l_0/2$, and let $A_1 = A \setminus B_1$. Then $\sigma(A_1) > \sigma(A)$, and if $l_1 = \inf\{\sigma(B) : B \subseteq A_1\}$ then $l_0/2 < l_1 < 0$. Repeating the process, we obtain a decreasing sequence (A_n) such that $\sigma(A_n)$ is increasing, and $l_n = \inf\{\sigma(B) : B \subseteq A_n\} \to 0$. Then $\sigma(\cap_n(A_n)) \geq \sigma(A)$ and $\cap_n(A_n)$ is strictly non-negative.

It follows that

$$M = \sup\{\sigma(A) : A \in \Sigma\} = \sup\{\sigma(A) : A \text{ strictly non-negative}\}.$$

There exist strictly non-negative P_n such that $\sigma(P_n) > M - 1/n$. Then $P = \cup_n P_n$ is strictly non-negative, and $\sigma(P) = M$. It follows that if $A \cap P = \emptyset$ then $\sigma(A) \leq 0$, so that we can take $N = X \setminus P$.

It is then immediate that σ^+ and σ^- are positive measures on (X, Σ), and that $\sigma = \sigma^+ - \sigma^-$.

Finally, suppose that P', N', π and ν satisfy the conditions of the theorem, and that $A \in \Sigma$. If $B \subseteq A \cap P'$ then $\sigma(B) \geq 0$, so that $\sigma^+(A \cap P') = \sigma(A \cap P') = \pi(A)$. Similarly, if $B \subseteq A \cap N'$ then $\sigma(B) \leq 0$, so that $\sigma^+(A \cap N') = 0$. Consequently,

$$\sigma^+(A) = \sigma(A \cap P') = \pi(A \cap P') = \pi(A),$$

so that $\pi = \sigma^+$. Similarly, $\nu = \sigma^-$. □

The decomposition $\sigma = \sigma^+ - \sigma^-$ of this theorem is called the *Jordan decomposition* of σ. We set $|\sigma| = \sigma^+ + \sigma^-$. $|\sigma|$ is a positive measure.

Proposition 31.1.4 *If $\sigma \in ca_{\mathbf{R}}(X, \Sigma)$ and $A \in \Sigma$ then $|\sigma(A)| \leq |\sigma|(A)$ and*

$$|\sigma|(A) = \sup\{|\sigma(B)| + |\sigma(C)| : B, C \in \Sigma, B \cap C = \emptyset, B \cup C = A\}.$$

Proof First,

$$|\sigma(A)| = |\sigma(A \cap P) + \sigma(A \cap N)| \leq |\sigma(A \cap P)| + |\sigma(A \cap N)|$$
$$= |\sigma|(A \cap P) + |\sigma|(A \cap N) = |\sigma|(A).$$

Secondly,

$$|\sigma|(A) = \sigma(A \cap P) + \sigma(A \cap N)$$
$$\leq \sup\{|\sigma(B)| + |\sigma(C)| : B, C \in \Sigma, B \cap C = \emptyset, B \cup C = A\},$$

while if $B, C \in \Sigma$, $B \cap C = \emptyset$ and $B \cup C = A$, then

$$|\sigma(B)| + |\sigma(C)| \leq |\sigma|(B) + |\sigma|(C) = |\sigma|(A). \qquad \square$$

Corollary 31.1.5 *If $(A_n)_{n=1}^\infty$ is a sequence of disjoint elements of Σ with union A then $\sum_{n=1}^\infty |\sigma(A_n)| \leq |\sigma|(A)$.*

Proof For

$$\sum_{n=1}^\infty |\sigma(A_n)| \leq \sum_{n=1}^\infty |\sigma|(A_n) = |\sigma|(A).$$

$\square$

Theorem 31.1.6 *If $\sigma \in ca_{\mathbf{R}}(X, \Sigma)$, let $\|\sigma\|_{ca} = |\sigma|(X)$. Then $\|.\|_{ca}$ is a norm on the vector space $ca_{\mathbf{R}}(X, \Sigma)$ of signed measures on (X, Σ) under which $ca_{\mathbf{R}}(X, \Sigma)$ is complete.*

Proof Let $\sigma = \pi - \nu$ be the Jordan decomposition of σ. If $\lambda \geq 0$ then $\lambda\sigma = \lambda\pi - \lambda\nu$ is the Jordan decomposition of $\lambda\sigma$, so that $\|\lambda\sigma\| = \lambda\|\sigma\|$. If $\lambda < 0$ then $\lambda\sigma = |\lambda|\nu - |\lambda|\pi$ is the Jordan decomposition of $\lambda\sigma$, so that $\|\lambda\sigma\| = |\lambda|\nu(X) + |\lambda|\pi(X) = |\lambda|\|\sigma\|$.

If σ_1, σ_2 are signed measures then

$$\begin{aligned}
\|\sigma_1 + \sigma_2\|_{ca} &= |\sigma_1 + \sigma_2|(X) \\
&= \sup\{|(\sigma_1 + \sigma_2)(A)| + |(\sigma_1 + \sigma_2)(X \setminus A)| : A \in \Sigma\} \\
&\leq \sup\{|(\sigma_1(A)| + |\sigma_2(A)| + |(\sigma_1(X \setminus A)| + |\sigma_2(X \setminus A)| : A \in \Sigma\} \\
&\leq \sup\{|(\sigma_1(A)| + |(\sigma_1(X \setminus A)| : A \in \Sigma\} \\
&\quad + \sup\{|\sigma_2(A)| + |\sigma_2(X \setminus A)| : A \in \Sigma\} \\
&= \|\sigma_1\|_{ca} + \|\sigma_2\|_{ca}.
\end{aligned}$$

Thus $\|.\|_{ca}$ is a norm on $ca_{\mathbf{R}}(X, \Sigma)$.

Suppose that $(\sigma_k)_{k=1}^\infty$ is a Cauchy sequence in $ca_{\mathbf{R}}(X, \Sigma)$. If $A \in \Sigma$ then

$$|\sigma_j(A) - \sigma_k(A)| \leq |\sigma_j - \sigma_k|(A) \leq \|\sigma_j - \sigma_k\|_{ca},$$

so that $(\sigma_k(A))_{k=1}^\infty$ is a Cauchy sequence in $\mathbf{R}$, which converges to $\sigma(A)$, say. We shall show that σ is a signed measure and that $\sigma_k \to \sigma$ in norm as $k \to \infty$.

Suppose that $(A_n)_{n=1}^\infty$ is a sequence of disjoint elements of Σ with union A, and that $\epsilon > 0$. There exists $K \in \mathbf{N}$ such that $\|\sigma_j - \sigma_k\|_{ca} < \epsilon/2$ for $j, k \geq K$. By Corollary 31.1.5,

$$\sum_{n=1}^\infty |\sigma_j(A_n) - \sigma_K(A_n)| \leq |\sigma_j - \sigma_K|(A) \leq \|\sigma_j - \sigma_K\|_{ca} < \epsilon/2,$$

for $j \geq K$.

Letting $j \to \infty$, it follows that $\sum_{n=1}^{\infty} |\sigma(A_n) - \sigma_K(A_n)| \leq \epsilon/2$, and similarly $|\sigma(A) - \sigma_K(A)| \leq \epsilon/2$. Thus

$$\left| \left(\sum_{n=1}^{\infty} \sigma(A_n) \right) - \sigma(A) \right| = \left| \left(\sum_{n=1}^{\infty} (\sigma(A_n) - \sigma_K(A_n)) \right) - (\sigma(A) - \sigma_K(A)) \right|$$

$$\leq \left(\sum_{n=1}^{\infty} (|\sigma(A_n) - \sigma_K(A_n)|) \right) + |\sigma(A) - \sigma_K(A)| \leq \epsilon.$$

Since ϵ is arbitrary, it follows that σ is σ-additive.

Finally, if $A \in \Sigma$ then

$$|\sigma(A) - \sigma_k(A)| + |\sigma(X \setminus A) - \sigma_k(X \setminus A)| < \epsilon$$

for $k \geq K$, so that $\|\sigma - \sigma_k\|_{ca} \leq \epsilon$ for $k \geq K$; $\sigma_k \to \sigma$ as $k \to \infty$. $\qquad \square$

We can use integrable functions to define signed measures.

Theorem 31.1.7 *Suppose that $f \in L^1(X, \Sigma, \mu)$. If $A \in \Sigma$, let $f.d\mu(A) = \int_A f \, d\mu$. Then $f.d\mu$ is a signed measure, and the mapping $f \to f.d\mu$ is an isometric linear mapping of $L^1(X, \Sigma, \mu)$ onto a closed subspace of the space $ca_{\mathbf{R}}(X, \Sigma)$ of signed measures on (X, Σ).*

Proof Suppose that $(B_n)_{n=1}^{\infty}$ is an increasing sequence in Σ, with union B. Then $|f I_{B_n}| \leq |f|$ for each $n \in \mathbf{N}$, and $f I_{B_n} \to f I_B$ pointwise as $n \to \infty$. By the theorem of dominated convergence,

$$f.d\mu(B_n) = \int_{B_n} f \, d\mu \to \int_B f \, d\mu = f.d\mu(B),$$

so that $f.d\mu$ is a signed measure.

Clearly $f.d\mu = f^+.d\mu - f^-.d\mu$, so that

$$\|f.d\mu\|_{ca} = \|f^+.d\mu\|_{ca} + \|f^-.d\mu\|_{ca}$$

$$= \int_X f^+ \, d\mu + \int_X f^- \, d\mu = \int_X |f| \, d\mu = \|f\|_1 \, .$$

Thus the mapping is an isometry. Since $L^1(X, \Sigma, \mu)$ is complete, the image is closed. $\qquad \square$

We return to this topic in Section 32.1.

31.2 Complex measures

We can also consider measures which take complex values. Suppose that (X, Σ) is a measurable space. A *complex measure* σ is a complex-valued function on Σ which is σ-additive: if $(A_n)_{n=1}^{\infty}$ is a sequence of disjoint elements of Σ, then

$$\sigma(\cup_{n=1}^{\infty} A_n) = \sum_{n=1}^{\infty} \sigma(A_n).$$

If σ is a complex-valued measure, then the real and imaginary parts of σ are signed measures. Thus the vector space $ca_{\mathbf{C}}(X, \Sigma)$ of signed measures is the direct sum $ca_{\mathbf{R}}(X, \Sigma) \oplus i.ca_{\mathbf{R}}(X, \Sigma)$, and we can deduce properties of σ from this.

We can give $ca_{\mathbf{C}}(X, \Sigma)$ a complex norm $\|.\|_{ca(\mathbf{C})}$, under which it is a Banach space.

Theorem 31.2.1 *If $\sigma \in ca_{\mathbf{C}}(X, \Sigma)$ and $A \in \Sigma$, let*

$$|\sigma|(A) = \sup\{\sum_{j=1}^{k} |\sigma(A_j)| : A_j \in \Sigma, \{A_1, \dots, A_k\} \text{ a partition of } A\}.$$

Then $|\sigma|$ is a positive measure on (X, Σ). Let $\|\sigma\|_{ca(\mathbf{C})} = |\sigma|(X)$. Then $\|.\|_{ca(\mathbf{C})}$ is a norm on $ca_{\mathbf{C}}(X, \Sigma)$, under which $ca_{\mathbf{C}}(X, \Sigma)$ is complete. The restriction of $\|.\|_{ca(\mathbf{C})}$ to $ca_{\mathbf{R}}(X, \Sigma)$ is the norm $\|.\|_{ca}$ of Theorem 31.1.6.

Proof Suppose that $(B_n)_{n=1}^{\infty}$ is a sequence of disjoint elements of Σ with union B, and that $\{A_1, \dots, A_k\}$ is a partition of B by sets in Σ. Then

$$\sum_{j=1}^{k} |\sigma(A_j)| \leq \sum_{j=1}^{k} \left(\sum_{m=1}^{\infty} |\sigma(A_j \cap B_m)| \right)$$

$$= \sum_{m=1}^{\infty} \left(\sum_{j=1}^{k} |\sigma(A_j \cap B_m)| \right) \leq \sum_{m=1}^{\infty} |\sigma|(B_m);$$

taking the supremum over partitions of B, it follows that $|\sigma|(B) \leq \sum_{m=1}^{\infty} |\sigma|(B_m)$.

Conversely, suppose that $\epsilon > 0$. For each $m \in \mathbf{N}$ there exists a partition $(A_{1,m}, \dots, A_{k_m,m}\}$ of B_m by sets in Σ such that

$$\sum_{j=1}^{k_m} |\sigma(A_{j,m})| > |\sigma|(B_m) - \epsilon/2^m.$$

If $n \in \mathbf{N}$ then $\{A_{j,m} : 1 \le j \le k_m, 1 \le m \le n\} \cup \{X - \cup_{m=1}^n B_m\}$ is a partition of B, so that

$$|\sigma|(B) \ge \sum \{|\sigma(A_{j,m})| : 1 \le j \le k_m, 1 \le m \le n\} + |\sigma(X - \cup_{m=1}^n B_m)|$$

$$\ge \sum_{m=1}^n |\sigma|(B_m) - \epsilon.$$

Since this holds for all $n \in \mathbf{N}$, $|\sigma|(B) \ge \sum_{m=1}^\infty |\sigma|(B_m) - \epsilon$. Since this holds for all $\epsilon > 0$, $|\sigma|(B) \ge \sum_{m=1}^\infty |\sigma|(B_m)$, and so $|\sigma|$ is a measure.

We leave it as an exercise for the reader to show that that $\|.\|_{ca(\mathbf{C})}$ is a norm on $ca_{\mathbf{C}}(X, \Sigma)$.

Suppose that $\sigma \in ca_{\mathbf{R}}(X, \Sigma)$. It follows from the definitions that $\|\sigma\|_{ca} \le \|\sigma\|_{ca(\mathbf{C})}$. Let $\sigma = \pi - \nu$ be the Jordan decomposition of σ, with corresponding dissection $X = P \cup N$. If $\{A_1, \dots, A_k\}$ is a partition of X by sets in Σ,

$$\sum_{j=1}^k |\sigma(A_j)| \le \sum_{j=1}^k (\sigma(A_j \cap P) - \sigma(A_j \cap N)) \le \|\sigma\|_{ca}.$$

Taking the supremum, $\|\sigma\|_{ca(\mathbf{C})} \le \|\sigma\|_{ca}$. Thus $\|\sigma\|_{ca(\mathbf{C})} = \|\sigma\|_{ca}$.

If $\sigma = \sigma_1 + i\sigma_2$, with σ_1, σ_2 signed measures, then

$$\|\sigma\|_{ca(\mathbf{C})} \le \|\sigma_1\|_{ca(\mathbf{C})} + \|\sigma_2\|_{ca(\mathbf{C})} = \|\sigma_1\|_{ca} + \|\sigma_2\|_{ca} \le 2\|\sigma\|_{ca(\mathbf{C})},$$

so that, considered as a real normed space, $(ca_{\mathbf{C}}(X, \Sigma), \|.\|_{ca(\mathbf{C})})$ is isomorphic to $(ca_{\mathbf{R}}(X, \Sigma), \|.\|_{ca}) \oplus (ca_{\mathbf{R}}(X, \Sigma), \|.\|_{ca})$, and is therefore complete. $\square$

The quantity $\|\sigma\|_{ca(\mathbf{C})}$ are called the *total variation* of σ. From now on, we shall denote it by $\|.\|_{ca}$. Theorem 31.2.1 shows that this should cause no confusion.

We also have a complex version of Theorem 31.1.7.

Theorem 31.2.2 *Suppose that $f \in L_{\mathbf{C}}^1(X, \Sigma, \mu)$. If $A \in \Sigma$, let $f.d\mu(A) = \int_A f \, d\mu$. Then $f.d\mu$ is a complex measure, and the mapping $f \to f.d\mu$ is an isometric linear mapping of $L_{\mathbf{C}}^1(X, \Sigma, \mu)$ onto a closed subspace of the space $ca_{\mathbf{C}}(X, \Sigma)$ of complex measures on (X, Σ).*

Proof It follows by considering real and imaginary parts that $f.d\mu$ is a complex measure. If $A_1, \dots, A_n$ are disjoint elements of Σ then

$$\sum_{j=1}^{n} |f.d\mu(A_j)| = \sum_{j=1}^{n} |\int_{A_j} f \, d\mu| \leq \sum_{j=1}^{n} \int_{A_j} |f| \, d\mu \leq \int_X |f| \, d\mu = \|f\|_1 ,$$

and so $\|f.d\mu\|_{ca} \leq \|f\|_1$.

Suppose that $\epsilon > 0$. By Corollary 29.5.6, there exists a simple function $g = \sum_{j=1}^{n} c_j I_{A_j}$ such that $\|f - g\|_1 \leq \epsilon/2$. Then

$$\sum_{j=1}^{n} |f.d\mu(A_j)| = \sum_{j=1}^{n} |\int_{A_j} f \, d\mu|$$

$$\geq \sum_{j=1}^{n} |\int_{A_j} g \, d\mu| - \sum_{j=1}^{n} |\int_{A_j} (f-g) \, d\mu|$$

$$\geq \sum_{j=1}^{n} \int_{A_j} |g| \, d\mu - \sum_{j=1}^{n} \int_{A_j} |f-g| \, d\mu$$

$$\geq \|g\|_1 - \|f-g\|_1 \geq \|f\|_1 - 2\|f-g\|_1 \geq \|f\|_1 - \epsilon,$$

so that $\|f.d\mu\|_{ca} \geq \|f\|_1$. Thus the mapping is an isometry, and again the image is closed. $\qquad \square$

Exercise

31.2.1 Show that $\|.\|_{ca(\mathbf{C})}$ is a norm on $ca_{\mathbf{C}}(X, \Sigma)$.

31.3 Functions of bounded variation

We now consider a signed measure σ on the Borel subsets of $\mathbf{R}$, with Jordan decomposition $\sigma = \pi - \nu$. We define the *cumulative distribution function* of σ in exactly the same way as for positive measures: if $t \in \mathbf{R}$ then $F_\sigma(t) = \sigma((-\infty, t])$. Since $F_\sigma = F_\pi - F_\nu$, F_σ is a bounded right-continuous function on $\mathbf{R}$, $F_\sigma(t) \to 0$ as $t \to -\infty$ and $F_\sigma(t) \to \sigma(\mathbf{R})$ as $t \to +\infty$.

How do we recognize the cumulative distribution function of a signed Borel measure on $\mathbf{R}$? If I is a closed interval in $\mathbf{R}$, we denote the set of all finite strictly increasing sequences $T = (t_0 < t_1 < \cdots < t_k)$ in I by $\mathcal{T}(I)$. Suppose that f is a real-valued function on $\mathbf{R}$. If $T = (t_0 < t_1 < \cdots < t_k) \in \mathcal{T}(I)$, we set

$$v_T^+(f) = \sum_{j=1}^{k} (f(t_j) - f(t_{j-1}))_+,$$

$$v_T^-(f) = \sum_{j=1}^{k} (f(t_j) - f(t_{j-1}))_-$$

$$\text{and } v_T(f) = \sum_{j=1}^{k} |f(t_j) - f(t_{j-1})|.$$

Clearly $v_T(f) = v_T^+(f) + v_T^-(f)$ and $f(t_k) - f(t_0) = v_T^+(f) - v_T^-(f)$. We set

$$v^+(f, I) = \sup_{T \in \mathcal{T}(I)} v_T^+(f),$$

$$v^-(f, I) = \sup_{T \in \mathcal{T}(I)} v_T^-(f)$$

$$\text{and } v(f, I) = \sup_{T \in \mathcal{T}(I)} v_T(f).$$

The quantity $v^+(f, I)$ is the *positive variation* of f on I, $v^-(f, I)$ is the *negative variation* of f on I, and $v(f, I)$ is the *total variation* of f on I. We write $v_f^+(t)$ for $v^+(f, (-\infty, t])$; $v_f^-(t)$ and $v_f(t)$ are defined similarly. A real-valued function f is of *bounded variation* if $v(f, \mathbf{R})$ is finite.

Here are some basic properties of the variations of a function.

Theorem 31.3.1   *Suppose that f, g are real-valued functions on $\mathbf{R}$, and that I is a closed interval in $\mathbf{R}$.*

(i) $v(f, I) = v^+(f, I) + v^-(f, I)$.
(ii) *If $a < b < c$ then $v^+(f, [a, c]) = v^+(f, [a, b]) + v^+(f, [b, c])$, and similar equalities hold for $v^-(f, [a, c])$ and $v(f, [a, c])$.*
(iii) $v^+(f + g, I) \leq v^+(f, I) + v^+(g, I)$, $v^-(f + g, I) \leq v^-(f, I) + v^-(g, I)$ *and $v(f + g, I) \leq v(f, I) + v(g, I)$.*
(iv) $v^+(-f, I) = v^-(f, I)$, $v^-(-f, I) = v^+(f, I)$ *and $v(-f, I) = v(f, I)$.*
(v) *If $\lambda > 0$ then $v^+(\lambda f, I) = \lambda v^+(f, I)$, $v^-(\lambda f, I) = \lambda v^-(f, I)$ and $v(\lambda f, I) = \lambda v(f, I)$.*
(vi) *If $I = [a, b]$ and $v(f, I) < \infty$ then $f(b) - f(a) = v^+(f, I) - v^-(f, I)$.*

Proof   These results follow from the fact that adding extra points to $T \in \mathcal{T}(I)$ does not decrease any of $v^+(f, I)$, $v^-(f, I)$ or $v(f, I)$.

For (i),

$$v^+(f, I) + v^-(f, I) = \sup_{T \in \mathcal{T}(I)} (v_T^+(f) + v_T^-(f)) = \sup_{T \in \mathcal{T}(I)} v_T(f) = v(f, I).$$

(ii) and (iii) are proved similarly, and (iv) and (v) follow from the definitions.

(vi) Given $\epsilon > 0$, there exists $T \in \mathcal{T}(I)$ such that

$$v^+(f, I) - v_T^+(f) < \epsilon/2 \text{ and } v(f, I) - v_T(f) < \epsilon/2,$$

so that

$$|v^+(f, I) - v^-(f, I) - (f(b) - f(a))|$$
$$= |v^+(f, I) - v^-(f, I) - (v_T^+(f) - v_T^-(f))| < \epsilon.$$

Since ϵ is arbitrary, (vi) holds. □

Corollary 31.3.2 *Suppose that f is a function of bounded variation. Then v_f^+, v_f^- and v_f are increasing functions on $\mathbf{R}$ which tend to 0 as $t \to -\infty$. Further $f(t)$ tends to a limit $f(-\infty)$ as $t \to -\infty$, and to a limit $f(+\infty)$ as $t \to +\infty$, and*

$$f(t) = f(-\infty) + v_f^+(t) - v_f^-(t) \text{ for } t \in \mathbf{R}.$$

The set of points of discontinuity of f is countable, and each discontinuity is a jump discontinuity.

Proof The functions v_f^+, v_f^- and v_f are increasing, by (ii). Given $\epsilon > 0$ there exists $T = (t_0 < \cdots < t_k) \in \mathcal{T}(\mathbf{R})$ such that $v_T(f) > v(f, \mathbf{R}) - \epsilon$. If $t < t_0$ then

$$v_f(t) + v(f, [t, t_0]) + v_T(f) \leq v_f(t) + v(f, [t, t_0]) + v(f, [t_0, t_k]) = v_f(t_k)$$
$$\leq v(f, \mathbf{R}) < v_T(f) + \epsilon,$$

so that $v_f(t) \to 0$ as $t \to -\infty$. Consequently $v_f^+(t) \to 0$ and $v_f^-(t) \to 0$ as $t \to -\infty$.

If $s < t$ then

$$f(t) - f(s) = (v_f^+(t) - v_f^-(t)) - (v_f^+(s) - v_f^-(s)) \to 0 \text{ as } s, t \to -\infty,$$

so that $f(t)$ tends to a limit $f(-\infty)$ as $t \to -\infty$. Similarly, $f(t)$ tends to a limit $f(+\infty)$ as $t \to +\infty$. Further

$$f(t) = f(s) + (v_f^+(t) - v_f^-(t)) - (v_f^+(s) - v_f^-(s))$$
$$\to f(-\infty) + (v_f^+(t) - v_f^-(t)) \text{ as } s \to -\infty,$$

so that $f(t) = f(-\infty) + v_f^+(t) - v_f^-(t)$.

The final result follows from the fact that v_f^+ and v_f^- are increasing, and so their sets of points of discontinuity are countable, and each discontinuity is a jump discontinuity. □

We denote by $bv_0(\mathbf{R})$ the vector space of right-continuous functions f on $\mathbf{R}$ of bounded variation for which $f(t) \to 0$ as $t \to -\infty$.

Proposition 31.3.3 *If $f \in bv_0(\mathbf{R})$ then v_f^+, v_f^- and v_f are in $bv_0(\mathbf{R})$.*

Proof   We need to show that each of the functions is right-continuous. Since $v_f^+ = \frac{1}{2}(v_f + f)$ and $v_f^- = \frac{1}{2}(v_f - f)$, it is enough to show that v_f is right-continuous. Suppose that $t \in \mathbf{R}$ and that $\epsilon > 0$. There exists $\delta > 0$ such that $|f(s) - f(t)| < \epsilon/2$ for $t < s < t + \delta$. Choose $t < r < t + \delta$. There exists $T = (t = t_0 < t_1 < \cdots < t_k = r) \in \mathcal{T}([t, r])$ for which $v_T(f) > v(f, [t, r]) - \epsilon/2$. Then

$$v_f(t_1) - v_f(t) = v(f, [t, r]) - v(f, [t_1, r])$$

$$\leq (v_T(f) + \epsilon/2) - \sum_{j=2}^{k} |f(t_j) - f(t_{j-1})|$$

$$= |f(t_1) - f(t)| + \epsilon/2 < \epsilon.$$

Since v_f is an increasing function, this shows that f is right-continuous.   □

Theorem 31.3.4   *(i) The function $v(., \mathbf{R})$ is a norm on $bv_0(\mathbf{R})$; we denote $v(f, \mathbf{R})$ by $\|f\|_{bv}$.*
(ii) If f is an increasing function in $bv_0(\mathbf{R})$, then $\|f\|_{bv} = f(+\infty) = \|f\|_\infty$.
(iii) If $f \in bv_0(\mathbf{R})$ then $\|f\|_{bv} = \left\|v_f^+\right\|_{bv} + \left\|v_f^-\right\|_{bv}$.

Proof   Since $v(f, \mathbf{R}) \geq \|f\|_\infty$, so that $v(f, \mathbf{R}) = 0$ if and only if $f = 0$, this follows immediately from Theorem 31.3.1.   □

Theorem 31.3.5   *The mapping $F : \sigma \to F_\sigma$ is a linear isometry of $(ca_\mathbf{R}(\mathbf{R}, \mathcal{B}), \|.\|_{ca})$ onto $(bv_0(\mathbf{R}), \|.\|_{bv})$, with inverse mapping $\mu : f \to \mu_f$, where $\mu_f = \mu_{v_f^+} - \mu_{v_f^-}$. If $\sigma = \pi - \nu$ is the Jordan decomposition of σ then $F_\pi = v_f^+$ and $F_\nu = v_f^-$.*

Proof   If $\sigma \in ca_\mathbf{R}(\mathbf{R})$, with Jordan decomposition $\sigma = \pi - \nu$, then $F_\sigma = F_\pi - F_\nu$ so that $F_\sigma \in bv_0(\mathbf{R})$, by Proposition 30.5.1. If $F_\sigma = 0$ then $F_\pi = F_\nu$. It therefore follows from Theorem 30.5.2 that $\pi = \nu$, so that F is injective. If $f \in bv_0(\mathbf{R})$, then $f = v_f^+ - v_f^-$. Then $\mu_f = \mu_{v_f^+} - \mu_{v_f^-} \in ca_\mathbf{R}(\mathbf{R})$. If $\sigma \in ca_\mathbf{R}(\mathbf{R})$ then $\sigma = \mu_{F_\sigma}$, so that F is bijective; the mapping $f \to \mu_f$ is the inverse of the mapping F. If μ is a positive measure, then $\|F_\mu\|_{bv} = \mu(\mathbf{R}) = \|\mu\|_{ca}$. Thus if $\sigma \in ca_\mathbf{R}(\mathbf{R})$, with Jordan decomposition $\sigma = \pi - \nu$, then

$$\|F_\sigma\|_{bv} = \|F_\pi - F_\nu\|_{bv} \leq \|F_\pi\|_{bv} + \|F_\nu\|_{bv} = \|\pi\|_{ca} + \|\nu\|_{ca} = \|\sigma\|_{ca},$$

so that F is norm-decreasing. Similarly, if $f \in bv_0(\mathbf{R})$, then

$$\|\mu_f\|_{ca} = \left\|\mu_{v_f^+} - \mu_{v_f^-}\right\|_{ca} \leq \left\|\mu_{v_f^+}\right\|_{ca} + \left\|\mu_{v_f^-}\right\|_{ca}$$

$$= \left\|v_f^+\right\|_{bv} + \left\|v_f^-\right\|_{bv} = \|f\|_{bv},$$

so that F^{-1} is also norm decreasing. Thus F is an isometry. Further,

$$\|F_\sigma\|_{bv} = \|\sigma\|_{ca} = \|\pi\|_{ca} + \|\nu\|_{ca} = \|F_\pi\|_{bv} + \|F_\nu\|_{bv},$$

so that $F_\pi = v_f^+$ and $F_\nu = v_f^-$, by Theorem 31.3.1 (iv). $\qquad \square$

It is a straightforward matter to establish corresponding results for complex Borel measures on $\mathbf{R}$ (Exercise 31.3.2).

Exercises

31.3.1 A *partially ordered vector space* $(E, \leq)$ is a real vector space E together with a partial order $\leq$ on E which satisfies
- if $x \leq y$ then $x + z \leq y + z$, and
- if $x < y$ and $\lambda \geq 0$ then $\lambda x \leq \lambda y$.

Show that $ca(X, \Sigma)$ is a partially ordered vector space when we set $\sigma \leq \tau$ if $\tau - \sigma$ is a positive measure. Show that if $\sigma = \pi - \nu$ is the Jordan decomposition of σ then

$$\pi = \inf\{\mu : \mu \text{ positive}, \mu \geq \sigma.\}.$$

Show that $bv_0(\mathbf{R})$ is a partially ordered vector space when we set $f \leq g$ if $g - f$ is an increasing function. Show that

$$v_f^+ = \inf\{g \in bv_0 : g \text{ increasing}, g \geq f\}.$$

Show that the mapping $\sigma \to F_\sigma$ is an order-preserving mapping of $ca(\mathbf{R}, \mathcal{B})$ onto $bv_0(\mathbf{R})$.

31.3.2 Suppose that $f \in bv_0(\mathbf{R})$ and that $f = g - h$, where g and h are increasing functions in $bv_0(\mathbf{R})$. Show that $g \geq v_f^+$ and $h \geq v_f^-$. Show that equality holds if and only if $\|f\|_{bv} = \|g\|_{bv} + \|h\|_{bv}$.

31.3.3 Define the cumulative distribution function F_σ of a complex Borel measure σ on $\mathbf{R}$, and the total variation $v(f, \mathbf{R})$ of a complex-valued function on $\mathbf{R}$. Define the vector space $bv_0(\mathbf{C})$. If $f \in bv_0(\mathbf{C})$, let $\|f\|_{bv} = v(f, \mathbf{R})$. Show that $\|.\|_{bv}$ is a norm on $bv_0(\mathbf{C})$. Show that the mapping $\sigma \to F_\sigma$ is a linear isometry of $(ca_{\mathbf{C}}(\mathbf{R}, \mathcal{B}), \|.\|_{ca(\mathbf{C})})$ onto $(bv_0(\mathbf{C}), \|.\|_{bv})$.

32

Measures on metric spaces

Lebesgue measure λ was defined on the real line $\mathbf{R}$, and properties of λ are closely connected to the topology of $\mathbf{R}$. In fact, almost all the measures spaces that are met in analysis are defined on a Hausdorff topological space, and, more particularly, on a metric space. In this chapter we consider a metric space (X, d). The *Borel σ-field* $\mathcal{B}$ is the σ-field generated by the open subsets of X (or by the closed subsets of X). We call a measure defined on $\mathcal{B}$ a *Borel measure* on X. Such measures necessarily have good approximation properties.

32.1 Borel measures on metric spaces

Theorem 32.1.1 *A finite Borel measure on a metric space (X, d) is closed-regular: if B is a Borel set then*

$$\mu(B) = \inf\{\mu(O) : O \text{ open, } B \subseteq O\}$$
$$= \sup\{\mu(C) : C \text{ closed, } C \subseteq B\}. \qquad (*)$$

There exist an increasing sequence $(A_n)_{n=1}^\infty$ of closed sets and a decreasing sequence $(U_n)_{n=1}^\infty$ of open sets such that $\mu(A_n) \to \mu(B)$ and $\mu(U_n) \to \mu(B)$ as $n \to \infty$.

Proof The proof uses Dynkin's π-λ theorem in a rather standard way. Let $\mathcal{G}$ be the set of those elements of $\mathcal{B}$ for which $(*)$ holds. We show that the collection $\mathcal{C}$ of closed subsets of X, which is a π-system, is contained in $\mathcal{G}$. We then show that $\mathcal{G}$ is a λ-system; consequently $\mathcal{G} = \mathcal{B}$.

First suppose that C is a closed subset of X. Let

$$C_n = \{x \in X : d(x, C) < 1/n\} = \cup_{c \in C} N_{1/n}(c).$$

Then $(C_n)_{n=1}^{\infty}$ is a decreasing sequence of open sets, whose intersection is C. Thus $\mu(C_n) \to \mu(C)$ as $n \to \infty$, by downwards continuity, so that

$$\mu(C) = \inf\{\mu(O) : O \text{ open}, C \subseteq O\}.$$

Since, trivially, $\mu(C) = \sup\{\mu(A) : A \text{ closed}, A \subseteq C\}$, it follows that $C \in \mathcal{G}$.

Certainly $X \in \mathcal{G}$.

Suppose that $(H_n)_{n=1}^{\infty}$ is an increasing sequence in $\mathcal{G}$, with union H. For each $n \in \mathbf{N}$ there exist a closed set C_n and an open set O_n with $C_n \subseteq H_n \subseteq O_n$, for which

$$\mu(C_n) > \mu(H_n) - 1/2^n \text{ and } \mu(O_n) < \mu(H_n) + 1/2^n.$$

Let $A_n = \cup_{j=1}^{n} C_j$ and $U_n = \cup_{m=n}^{\infty} O_m$. Then $(A_n)_{n=1}^{\infty}$ is an increasing sequence of closed subsets of H, and $(U_n)_{n=1}^{\infty}$ is a decreasing sequence of open sets containing H. Then

$$\lim_{n\to\infty} \mu(A_n) \le \mu(H) = \lim_{n\to\infty} \mu(H_n) \le \lim_{n\to\infty} (\mu(A_n) + 1/2^n) = \lim_{n\to\infty} \mu(A_n),$$

so that $\mu(H) = \lim_{n\to\infty} \mu(A_n)$.

Similarly, since $U_n \setminus H \subseteq \cup_{m=n}^{\infty}(O_m \setminus H_m)$,

$$0 \le \mu(U_n) - \mu(H) \le \mu(U_n \setminus H) \le \sum_{m=n}^{\infty} \mu(O_m \setminus H_m) \le 2/2^n.$$

Thus $\mu(U_n) \to \mu(H)$ as $n \to \infty$. Thus $H \in \mathcal{G}$.

Finally, suppose that $G, H \in \mathcal{G}$ and that $G \subseteq H$. Suppose that $\epsilon > 0$. There exist open sets U and V such that $G \subseteq U$, $H \subseteq V$, $\mu(U) < \mu(G) + \epsilon/2$ and $\mu(V) < \mu(H) + \epsilon/2$. Similarly, there exist closed sets A and B such that $A \subseteq G$, $B \subseteq H$, $\mu(A) > \mu(G) - \epsilon/2$ and $\mu(B) > \mu(H) - \epsilon/2$. Then $V \setminus A$ is open, $H \setminus G \subseteq V \setminus A$ and $\mu(V \setminus A) < \mu(H \setminus G) + \epsilon$. Similarly, $B \setminus U$ is closed, $B \setminus U \subseteq H \setminus G$ and $\mu(B \setminus U) > \mu(H \setminus G) - \epsilon$. Thus $H \setminus G \in \mathcal{G}$, so that $\mathcal{G}$ is a λ-system. $\square$

Corollary 32.1.2 *If σ is a signed Borel measure or complex Borel measure on a metric space (X, d) and $B \in \mathcal{B}$ then there exist an increasing sequence $(A_n)_{n=1}^{\infty}$ of closed subsets of B and a decreasing sequence $(U_n)_{n=1}^{\infty}$ of open sets containing B such that if $(C_n)_{n=1}^{\infty}$ is a sequence in $\mathcal{B}$ with $A_n \subseteq C_n \subseteq B$ then $\sigma(C_n) \to \sigma(B)$ as $n \to \infty$, and if $(D_n)_{n=1}^{\infty}$ is a sequence in $\mathcal{B}$ with $B \subseteq D_n \subseteq U_n$ then $\sigma(D_n) \to \sigma(B)$ as $n \to \infty$.*

Proof There exist an increasing sequence $(A_n)_{n=1}^{\infty}$ of closed subsets of B and a decreasing sequence $(U_n)_{n=1}^{\infty}$ of open sets containing B such that

$|\sigma|(A_n) \to |\sigma|(B)$ and $|\sigma|(U_n) \to |\sigma|(B)$ as $n \to \infty$. If $(C_n)_{n=1}^{\infty}$ is a sequence in Σ with $A_n \subseteq C_n \subseteq B$ then

$$|\sigma(B) - \sigma(C_n)| = |\sigma(B \setminus C_n)| \leq |\sigma|(B \setminus C_n) \leq |\sigma|(B \setminus A_n) \to 0$$

as $n \to \infty$. A similar argument establishes the result for the sequence $(D_n)_{n=1}^{\infty}$. $\qquad\square$

Exercises

32.1.1 Give a proof of Theorem 32.1.1 using the monotone class theorem.

32.1.2 Suppose that μ and ν are finite Borel measures on a metric space (X, d). Show that $\mu = \nu$ if and only if $\mu(U) = \nu(U)$ for each open set U.

32.1.3 Suppose that μ and ν are finite Borel measures on a metric space (X, d). Show that $\mu = \nu$ if and only if $\int_X f \, d\mu = \int_X f \, d\nu$ for each bounded continuous real-valued function f on X.

32.2 Tight measures

In general, compact sets are better behaved than closed sets, and it is important to be able to approximate sets from the inside by compact sets. A finite Borel measure μ on a metric space (X, d) is said to be *tight*, or *regular*, if whenever B is a Borel subset of X then

$$\mu(B) = \inf\{\mu(O) : O \text{ open, } B \subseteq O\}$$
$$= \sup\{\mu(K) : K \text{ compact, } K \subseteq B\}.$$

Proposition 32.2.1 *A finite Borel measure μ on a metric space (X, d) is tight if and only if there exists an increasing sequence $(K_n)_{n=1}^{\infty}$ of compact subsets of X such that $\mu(K_n) \to \mu(X)$ as $n \to \infty$.*

Proof The condition is necessary. Suppose that μ is tight. For each $n \in \mathbf{N}$ there exists a compact subset L_n of X such that $\mu(L_n) > \mu(X) - 1/n$. Let $K_n = \cup_{j=1}^{n} L_j$. Then the sequence $(K_n)_{n=1}^{\infty}$ satisfies the condition.

Conversely, suppose that the condition is satisfied, and that B is a Borel subset of X. Suppose that $\epsilon > 0$. Since μ is closed-regular, there exists a closed set A such that $A \subseteq B$ and $\mu(A) > \mu(B) - \epsilon/2$. There exists $n \in \mathbf{N}$ such that $\mu(K_n) > \mu(X) - \epsilon/2$. Then $A \cap K_n$ is a compact subset of B and

$$\mu(A \cap K_n) = \mu(A) - \mu(A \setminus K_n) \geq \mu(A) - \mu(X \setminus K_n) > \mu(B) - \epsilon. \qquad\square$$

Recall that a topological space is *σ-compact* if it is the union of a sequence of compact subsets.

Corollary 32.2.2 *A finite Borel measure μ on a σ-compact metric space (X, d) is tight.*

Here is a more remarkable result.

Theorem 32.2.3 (Ulam's theorem) *A finite Borel measure μ on a complete separable metric space (X, d) is tight.*

Proof   Let $(x_j)_{j=1}^\infty$ be an enumeration of a countable dense subset of (X, d). For each $n \in \mathbf{N}$ and $k \in \mathbf{N}$, let $A_{n,k} = \cup_{j=1}^k M_{1/n}(x_j)$, (where $M_{1/n}(x_j)$ is the closed $1/n$-neighbourhood of x_j). For fixed $n \in \mathbf{N}$, the sequence $(A_{n,k})_{k=1}^\infty$ is an increasing sequence of closed subsets of X whose union is X, and so $\mu(A_{n,k}) \to \mu(X)$ as $k \to \infty$. Thus there exists k_n such that $\mu(X \setminus A_{n,k_n}) = \mu(X) - \mu(A_{n,k_n}) < 1/2^n$.

Let $K_n = \cap_{m=n}^\infty A_{m,k_m}$. Then K_n is a closed subset of X. It is also totally bounded, since, for each $m \geq n$, $K_n \subseteq A_{m,k_m}$, and is therefore contained in finitely many open balls of radius $2/m$. Since (X, d) is complete, K_n is compact. Further,

$$\mu(X \setminus K_n) = \mu(\cup_{m=n}^\infty (X \setminus A_{m,k_m})) \leq \sum_{m=n}^\infty 2^{-m} = 2^{1-n},$$

and so $(K_n)_{n=1}^\infty$ is an increasing sequence of compact subsets of X for which $\mu(K_n) \to \mu(X)$. The result therefore follows from Proposition 32.2.1.   □

A *Polish space* is a separable topological space (X, τ) for which there is a complete metric on X which defines the topology.

Corollary 32.2.4 *A finite Borel measure μ on a Polish space is tight.*

Proof   For tightness is a topological property.   □

Thus a finite Borel measure on the space I of irrational numbers is tight.

Exercises

32.2.1 Show that a finite Borel measure on a countable metric space (X, d) is tight.

32.2.2 Give an example of a σ-compact metric space (X, d) which is not a Polish space.

32.2.3 Give an example of a Polish space which is not σ-compact.

32.3 Radon measures

We now consider a σ-finite Borel measure μ on a metric space (X, d).

Proposition 32.3.1 *Suppose that μ is a σ-finite Borel measure on a metric space (X, d).*

(i) If $x \in X$ then $\mu(\{x\}) < \infty$.
(ii) There exists an increasing sequence $(C_n)_{n=1}^\infty$ of closed subsets of X of finite measure for which $\mu(X \setminus C) = 0$, where $C = \cup_{n=1}^\infty C_n$.

Proof   There exists an increasing sequence $(A_n)_{n=1}^\infty$ of Borel sets of finite measure for which $\cup_{n=1}^\infty A_n = X$.

(i) $x \in A_n$ for some $n \in \mathbf{N}$, and $\mu(\{x\}) \leq \mu(A_n) < \infty$.
(ii) If B is a Borel subset of X and $n \in \mathbf{N}$, let $\mu_n(B) = \mu(B \cap A_n)$.

Then μ_n is a finite Borel measure on X, and so is closed-regular. Thus there exists a closed set D_n contained in A_n with

$$\mu_n(D_n) > \mu_n(A_n) - 1/2^n = \mu(A_n) - 1/2^n.$$

Let $C_n = \cup_{j=1}^n D_j$, and let $C = \cup_{n=1}^\infty C_n$. Then $(C_n)_{n=1}^\infty$ is an increasing sequence of closed subsets of X of finite measure. Further, if $p > m$ then

$$\mu((X \setminus C) \cap A_m) = \mu(A_m \setminus C) \leq \mu(A_m \setminus D_p) \leq \mu(A_p \setminus D_p) < 1/2^p.$$

Since this holds for all $p > m$, $\mu((X \setminus C) \cap A_m) = 0$, and so

$$\mu(X \setminus C) = \lim_{m \to \infty} \mu((X \setminus C) \cap A_m) = 0. \qquad \square$$

Corollary 32.3.2   *If B is a Borel subset of X then*

$$\mu(B) = \sup\{\mu(D) : D \text{ closed}, D \subseteq B, \mu(D) < \infty\}.$$

Proof

$$\mu(B) = \mu(B \cap C) = \lim_{n \to \infty} \mu(B \cap C_n).$$

Arguing as above, $\mu(B \cap C_n) = \sup\{\mu(D) : D \text{ closed}, D \subseteq B \cap C_n\}$, and so the result follows. $\qquad \square$

Let us consider an example. Let $\overline{\mathbf{R}}$ be the extended real line $\{-\infty\} \cup \mathbf{R} \cup \{\infty\}$ with its usual compact metrizable topology. If B is a Borel subset of $\overline{\mathbf{R}}$, let $\overline{\lambda}(B) = \lambda(B \cap \mathbf{R})$, where λ is Lebesgue measure on $\mathbf{R}$. Then $\overline{\lambda}$ is a σ-finite measure on $\overline{\mathbf{R}}$. $\overline{\lambda}(\{-\infty\}) = 0$, but if N is any open neighbourhood of $-\infty$, then $\overline{\lambda}(N) = \infty$, and the compact set $\overline{\mathbf{R}}$ has infinite measure. This is clearly not very satisfactory. A σ-finite Borel measure on a metric space (X, d) is *locally finite* if each element of X has an open neighbourhood of finite measure.

Proposition 32.3.3   *If μ is a locally finite Borel measure on a metric space (X, d) and K is a compact set of X, then $\mu(K) < \infty$.*

Proof   For each $x \in K$ there exists an open neighbourhood N_x of x with $\mu(N_x) < \infty$. These neighbourhoods cover K, and so there exists a finite subset F of K such that $K \subseteq \cup_{x \in F} N_x$. Thus

$$\mu(K) \leq \sum_{x \in F} \mu(N_x) < \infty.$$

$\square$

A σ-finite measure on a metric space (X, d) is called a *Radon measure* if it is locally finite, and if $\mu(B) = \sup\{\mu(K) : K \text{ compact}, K \subseteq B\}$ for each Borel subset B of X.

Proposition 32.3.4   *A locally finite σ-finite measure μ on a σ-compact metric space (X, d) is a Radon measure.*

Proof   There exists an increasing sequence $(K_n)_{n=1}^{\infty}$ of compact subsets of X whose union is X. If B is a Borel subset of X, then, by Corollary 32.3.2, $\mu(B) = \sup\{\mu(D) : D \text{ closed}, D \subseteq B\}$. But if D is closed, then $\mu(D) = \lim_{n \to \infty} \mu(D \cap K_n)$, so that $\mu(D) = \sup\{\mu(K) : K \text{ compact}, K \subseteq B\}$, and the result follows from this. $\square$

Theorem 32.3.5   *A locally finite σ-finite measure μ on a metric space (X, d) is a Radon measure if and only there exists a σ-compact subset Y of X such that $\mu(X \setminus Y) = 0$.*

Proof   If the condition is satisfied, and B is a Borel subset of X, then $B \cap Y$ is a Borel subset of Y, and

$$\mu(B) = \mu(B \cap Y) = \sup\{\mu(K) : K \text{ compact}, K \subseteq B \cap Y\}$$
$$\leq \sup\{\mu(K) : K \text{ compact}, K \subseteq B\} \leq \mu(B);$$

thus all the terms are equal, and μ is a Radon measure.

Conversely, suppose that μ is a Radon measure. By Proposition 32.3.1. there exists an increasing sequence (C_n) of closed subsets of X of finite measure for which $\mu(X \setminus C) = 0$, where $C = \cup_{n=1}^{\infty} C_n$. Since μ is a Radon measure, for each $n \in N$ there exists a compact subset K_n of C_n for which $\mu(K_n) > \mu(C_n) - 1/2^n$. Let $Y = \cup_{n=1}^{\infty} K_n$. Then Y is σ-compact, and arguing as in Proposition 32.3.1, $\mu(X \setminus Y) = 0$. $\square$

Corollary 32.3.6 *A locally finite σ-finite Borel measure on a Polish space is a Radon measure.*

Proof For each $n \in \mathbf{N}$, the closed set C_n is a Polish subspace of (X, d), and the restriction of μ to the Borel subsets of C_n is a finite measure, which is tight, by Ulam's theorem. There therefore exists a compact subset K_n of C_n with $\mu(K_n) > \mu(C_n) - 1/2^n$. Let $Y = \cup_{n=1}^{\infty} K_n$. Then Y is σ-compact, and once again, $\mu(X \setminus Y) = 0$. $\qquad\square$

33

Differentiation

In this chapter, we compare two measures defined on the same measurable space, and in particular, compare a finite Borel measure on **R** with Lebesgue measure. This involves further properties of integrable functions and of monotonic functions on **R**.

33.1 The Lebesgue decomposition theorem

We consider a σ-finite measure space (X, Σ, μ), and a finite measure ν on Σ. We use the Fréchet–Riesz representation theorem to prove a fundamental theorem of measure theory.

Theorem 33.1.1 (The Lebesgue decomposition theorem) *Suppose that (X, Σ, μ) is a σ-finite measure space, and that ν is a finite measure on Σ. Then there exists a non-negative $f \in L^1(\mu)$ and a set $B \in \Sigma$ with $\mu(B) = 0$ such that $\nu(A) = \int_A f \, d\mu + \nu(A \cap B)$ for each $A \in \Sigma$.*

Two measures μ and ν on the same σ-field Σ are said to be *mutually singular* if there exists $A \in \Sigma$ such that $\mu(A) = 0$ and $\nu(X \setminus A) = 0$. (If so, this is frequently written as $\mu \perp \nu$.) If we define $\nu_B(A) = \nu(A \cap B)$ for $A \in \Sigma$, then ν_B is a measure, and $\nu = f.d\mu + \nu_B$. The measures μ and ν_B are mutually singular.

Proof Let $\pi(A) = \mu(A) + \nu(A)$; π is a σ-finite measure on Σ. Suppose that $g \in L^2_{\mathbf{R}}(\pi)$. Let $L(g) = \int g \, d\nu$. Then, by the Cauchy–Schwarz inequality,

$$|L(g)| \le (\nu(X))^{1/2} \left(\int |g|^2 \, d\nu \right)^{1/2} \le (\nu(X))^{1/2} \|g\|_{L^2_{\mathbf{R}}(\pi)},$$

so that L is a continuous linear functional on $L^2_{\mathbf{R}}(\pi)$. By the Fréchet–Riesz theorem (Volume II, Theorem 14.3.7), there exists an element $h \in L^2_{\mathbf{R}}(\pi)$

such that $L(g) = \langle g, h \rangle$, for each $g \in L^2_{\mathbf{R}}(\pi)$; that is, $\int_X g \, d\nu = \int_X gh \, d\mu + \int_X gh \, d\nu$, so that

$$\int_X g(1 - h) \, d\nu = \int_X gh \, d\mu. \qquad (*)$$

Taking g as an indicator function I_A, we see that

$$\nu(A) = L(I_A) = \int_A h \, d\pi = \int_A h \, d\mu + \int_A h \, d\nu$$

for each $A \in \Sigma$.

Now let $N = (h < 0)$, $G_n = (0 \le h \le 1 - 1/n)$, $G = (0 \le h < 1)$ and $B = (h \ge 1)$. Then

$$\nu(N) = \int_N h \, d\mu + \int_N h \, d\nu \le 0,$$

so that $\nu(N) = 0$. But then $\int_N h \, d\mu = 0$, and so $\mu(N) = 0$. Similarly,

$$\nu(B) = \int_B h \, d\mu + \int_B h \, d\nu \ge \nu(B) + \mu(B),$$

so that $\mu(B) = 0$.

Let $f(x) = h(x)/(1 - h(x))$ for $x \in G$, and let $h(x) = 0$ otherwise. Note that if $x \in G_n$ then $0 \le f(x) \le 1/(1 - h(x)) \le n$. If $A \in \Sigma$, then, using $(*)$,

$$\nu(A \cap G_n) = \int_X \frac{1 - h}{1 - h} I_{A \cap G_n} \, d\nu = \int_X f I_{A \cap G_n} \, d\mu = \int_{A \cap G_n} f \, d\mu.$$

Applying the monotone convergence theorem, we see that $\nu(A \cap G) = \int_{A \cap G} f \, d\mu = \int_A f \, d\mu$. Thus

$$\nu(A) = \nu(A \cap G) + \nu(A \cap B) + \nu(A \cap N) = \int_A f \, d\mu + \nu(A \cap B).$$

Taking $A = X$, we see that $\int_X f \, d\mu < \infty$, so that $f \in L^1(\mu)$. $\qquad \square$

This beautiful proof is due to von Neumann.

Suppose that (X, Σ, μ) is a measure space. Our aim now is to recognize when a complex measure ν on X is of the form $f.d\mu$, where $f \in L^1(X, \Sigma, \mu)$. Suppose that ϕ is a real- or complex-valued function on Σ. We say that ϕ is *absolutely continuous* with respect to μ if whenever $\epsilon > 0$ then there exists $\delta > 0$ such that if $A \in \Sigma$ and $\mu(A) < \delta$ then $|\phi(A)| < \epsilon$; if so, we write $\phi \ll \mu$.

Proposition 33.1.2 *If (X, Σ, μ) is a finite or σ-finite measure space, and ν is a complex measure on Σ, then ν is absolutely continuous with respect to μ if and only if whenever $\mu(A) = 0$ then $\nu(A) = 0$.*

Proof Suppose that ν is absolutely continuous with respect to μ, and that $\mu(A) = 0$. Then $\mu(A) < \delta$ for all $\delta > 0$, so that $|\nu(A)| < \epsilon$ for all $\epsilon > 0$, and so $\nu(A) = 0$. For the converse, suppose first that ν is a finite positive measure and that ν is not absolutely continuous with respect to μ. Then there exists $\epsilon > 0$ such that for each $n \in \mathbf{N}$ there exists $A_n \in \Sigma$ with $\mu(A_n) < 1/2^n$ and $\nu(A_n) \geq \epsilon$. Then $\mu(\limsup_{n\to\infty} A_n) = 0$, by the first Borel–Cantelli lemma, while $\nu(\limsup_{n\to\infty} A_n) \geq \epsilon$. Thus the condition does not hold.

Suppose next that ν is a signed measure, and that $\nu(A) = 0$ whenever $\mu(A) = 0$. Let $X = P \cup N$ be the partition of X in the Jordan decomposition of ν. If $\mu(A) = 0$ then $\mu(A \cap P) = 0$, so that $\nu^+(A) = \nu(A \cap P) = 0$. Thus ν^+ is absolutely continuous with respect to μ; similarly, ν^- is absolutely continuous with respect to μ, and so therefore is ν.

Finally the result follows for complex measures by considering real and imaginary parts. $\qquad\square$

Theorem 33.1.3 (The Radon–Nikodým theorem) *Suppose that (X, Σ, μ) is a σ-finite measure space, and that ν is a finite measure on Σ. Then ν is absolutely continuous with respect to μ if and only if there exists a non-negative $f \in L^1(\mu)$ such that $\nu(A) = \int_A f \, d\mu$ for each $A \in \Sigma$.*

Proof If $f \in L^1(\mu)$, then $f.d\mu$ is absolutely continuous with respect to μ by Proposition 33.1.2. Conversely, suppose that ν is absolutely continuous with respect to μ, and that $\nu = f.d\mu + \nu_B$ is the Lebesgue decomposition of ν. Since $\mu(B) = 0$, $\nu(B) = 0$, and so $\nu_B = 0$. Thus $\nu = f.d\mu$. $\qquad\square$

The Radon–Nikodým theorem clearly extends to signed measures ν, by considering the Jordan decomposition of ν, and to complex measures ν, by considering the real and imaginary parts of ν. The Radon–Nikodým theorem throws light on the relationship between a complex measure σ and the positive measure $|\sigma|$.

Theorem 33.1.4 *Suppose that σ is a complex measure on a measurable space (X, Σ). There exists a complex measurable function ϕ on X, with $|\phi| = 1$, such that $\sigma(A) = \int_A \phi \, d|\sigma|$ for all $A \in \Sigma$.*

In other words, $\sigma = \phi.d|\sigma|$. The function ϕ is the *phase function* of σ.

Proof The complex measure σ is clearly absolutely continuous with respect to $|\sigma|$, and so by the Radon–Nykodým theorem there exists $\phi \in$

$L^1_{\mathbb{C}}(X, \Sigma, |\sigma|)$ such that $\sigma(A) = \int_A \phi \, d|\sigma|$ for all $A \in \Sigma$. We show that $|\phi| = 1$ almost everywhere.

First, let $A_n = (\Re\phi > 1 + 1/n)$. Then

$$|\sigma|(A_n) \geq \Re(\sigma(A_n)) = \int_{A_n} \Re\phi \, d|\sigma| \geq (1 + 1/n)|\sigma|(A_n),$$

so that $|\sigma|(A_n) = 0$. Thus if $A = (\Re\phi > 1)$, then

$$|\sigma|(A) = \lim_{n \to \infty} |\sigma|(A_n) = 0.$$

Next, let $(\theta_n)_{n=1}^{\infty}$ be a dense sequence in $[0, 2\pi)$. Let

$$B_n = (\Re(e^{i\theta_n}\phi) > 1).$$

Then, as above, $|\sigma|(B_n) = 0$. If $B = (|\phi| > 1)$, $B = \cup_{n=1}^{\infty} B_n$, and so $|\sigma|(B) = |\sigma|(\cup_{n=1}^{\infty} B_n) = 0$. Thus $|\phi| \leq 1$ almost everywhere.

Finally, let $C_n = (|\phi| \leq 1 - 1/n)$. Suppose that $D_1, \ldots, D_k$ is a partition of C_n by sets in Σ. Then

$$\sum_{j=1}^{k} |\sigma(D_j)| = \sum_{j=1}^{k} |\int_{D_j} \phi \, d|\sigma|| \leq \sum_{j=1}^{k} \int_{D_j} |\phi| \, d|\sigma| \leq (1 - 1/n)|\sigma|(C_n).$$

Taking the supremum over all partitions, $|\sigma|(C_n) \leq (1 - 1/n)|\sigma|(C_n)$, so that $|\sigma|(C_n) = 0$. Thus $|\sigma|((|\phi| < 1)) = |\sigma|(\cup_{n=1}^{\infty} C_n) = 0$; $|\phi| \geq 1$ almost everywhere. We can change ϕ on a null set so that $|\phi| = 1$.   □

Exercises

33.1.1 Use the fact that if $f \in L^1(X, \Sigma, \mu)$ then $f.d\mu$ is absolutely continuous with respect to μ, and Egorov's theorem, to give another proof of the theorem of dominated convergence.

33.2 Sublinear mappings

We now establish a result which enables us to use approximation arguments to establish results about convergence almost everywhere. First we need some definitions.

Suppose that (X, Σ, μ) is a measure space. A mapping T from a normed space $(E, \|\cdot\|_E)$ into the space $L^0(X, \Sigma, \mu)$ is *subadditive* if $T(f + g) \leq T(f) + T(g)$ for $f, g \in E$, is *positive homogeneous* if $T(\lambda f) = \lambda T(f)$ for $f \in E$ and λ real and positive, and is *sublinear* if it is both subadditive and positive homogeneous. We say that T is of *weak type* (E, q) if there exists $L < \infty$ such that $\mu(|T(f)| > \alpha) \leq L^q \|f\|_E^q / \alpha^q$ for all $f \in E$, $\alpha > 0$. The least constant L for which the inequality holds for all $f \in E$ is called the *weak type* (E, q) *constant*. When $E = L^p(X', \Sigma', \mu')$, we say that T is of *weak type* (p, q).

Weak type is important, when we consider convergence almost everywhere.

Theorem 33.2.1 *Suppose that* $(T_r)_{r \geq 0}$ *is a family of linear mappings from a normed space* E *into* $L^0(X, \Sigma, \mu)$, *and that* M *is a non-negative sublinear mapping of* E *into* $L^0(X, \Sigma, \mu)$, *of weak type* (E, q) *for some* $0 < q < \infty$, *such that*

(i) $|T_r(g)| \leq M(g)$ *for all* $g \in E$, $r \geq 0$, *and*
(ii) *there is a dense subspace* F *of* E *such that* $T_r(f) \to T_0(f)$ *almost everywhere, for* $f \in F$, *as* $r \to 0$.

Then $T_r(g) \to T_0(g)$ *almost everywhere, as* $r \to 0$, *for each* $g \in E$.

Proof We use the first Borel–Cantelli lemma. For each n there exists $f_n \in F$ with $\|g - f_n\|_E \leq 1/2^n$. Let

$$B_n = (M(g - f_n) > 1/n) \cup (T_r(f_n) \nrightarrow T_0(f_n)).$$

Then

$$\mu(B_n) = \mu(M(g - f_n) > 1/n) \leq \frac{Ln^q}{2^{nq}}.$$

Let $B = \limsup(B_n)$. Then $\mu(B) = 0$, by the first Borel–Cantelli lemma.

If $x \notin B$, there exists n_0 such that $x \notin B_n$ for $n \geq n_0$, so that

$$|T_r(g)(x) - T_r(f_n)(x)| \leq M(g - f_n)(x) \leq 1/n, \text{ for } r \geq 0.$$

Thus if $n \geq n_0$, then

$$|T_r(g)(x) - T_0(g)(x)| \leq$$
$$|T_r(g)(x) - T_r(f))(x)| + |T_r(f_n)(x) - T_0(f_n)(x)| + |T_0(f_n)(x) - T_0(g)(x)|$$
$$\leq 2/n + |T_r(f_n)(x) - T_0(f_n)(x)| \leq 3/n$$

for small enough r, and so $T_r(g)(x) \to T_0(x)$ as $r \to 0$. □

We can consider other directed sets than $[0, \infty)$; for example $\mathbf{N}$, or the set

$$\{(h, t) : h \in \mathbf{R}^d, t \geq 0 \text{ and } \|h\| < t\},$$

ordered by $(h, t) \leq (k, s)$ if $N_t(h) \subseteq N_s(k)$.

33.3 The Lebesgue differentiation theorem

We now consider finite Borel measures on $\mathbf{R}^d$ and functions in $L^1(\mathbf{R}^d, \mathcal{L}_d, \lambda_d)$. As usual, let $N_r(x)$ denote the open Euclidean ball $\{y : |y - x| < r\}$ of radius r with centre x. Then $\lambda_d(N_r(x)) = r^d \Omega_d$, where Ω_d is the Lebesgue measure of the unit ball in $\mathbf{R}^d$.

If μ is a finite Borel measure on $\mathbf{R}^d$, we set

$$A_r(\mu)(x) = \frac{\mu(N_r(x))}{\lambda_d(N_r(x))} = \frac{\mu(N_r(x))}{r^d \Omega_d}.$$

Proposition 33.3.1   *The function $A_r(\mu)$ is lower semi-continuous.*

Proof   Suppose that $x \in \mathbf{R}^d$, and that $\epsilon > 0$. Let r_n increase to r as $n \to \infty$. By upwards continuity, there exists $n \in \mathbf{N}$ such that $\mu(N_{r_n}(x)) > \mu(N_r(x)) - \epsilon r^d \Omega_d$. If $\|y - x\| < r - r_n$ then $N_{r_n}(x) \subseteq N_r(y)$, so that $A_r(y) > A_r(x) - \epsilon$.   □

We say that μ has a *spherical derivative* $D\mu(x)$ at x if, given $\epsilon > 0$, there exists $r_0 > 0$ such that if $0 < r < r_0$ and $x \in N_r(y)$ then $|A_r(\mu)(y) - D\mu(x)| < \epsilon$. It is important that in this definition we consider spheres to which x belongs, and not just spheres centred at x.

Similarly, if $f \in L^1(\mathbf{R}^d, \mathcal{L}_d, \lambda_d)$, we set

$$A_r(f)(x) = \frac{\int_{N_r(x)} f \, d\lambda_d}{\lambda_d(N_r(x))} = \frac{1}{r^d \Omega_d} \int_{N_r(x)} f \, d\lambda_d.$$

$A_r(f)(x)$ is the average value of f over the ball $N_r(x)$. Again, we say that f has a *spherical derivative* $Df(x)$ at x if, given $\epsilon > 0$, there exists $r_0 > 0$ such that if $0 < r < r_0$ and $x \in N_r(y)$ then $|A_r(f)(y) - Df(x)| < \epsilon$. Thus the spherical derivative of the function f is the same as the spherical derivative of the measure $f.d\lambda_d$.

First we consider a function f in $L^1(\mathbf{R}^d, \mathcal{L}_d, \lambda_d)$. We set

$$m_u(f)(x) = \sup_{r>0} \left(\sup\{A_r(|f|)(y) : y \in N_r(x)\}\right).$$

Theorem 33.3.2   *The function m_u is a lower semi-continuous sublinear operator of weak type $(1, 1)$.*

Proof   Suppose that $m_u(f)(x) < \infty$. If $\epsilon > 0$, there exist $r > 0$ and $y \in \mathbf{R}^d$ such that $x \in N_r(y)$ and $A_r(|f|)(y) > m_u(f)(x) - \epsilon$. Since $N_r(y)$ is open, there exists $\delta > 0$ such that $N_\delta(x) \subseteq N_r(y)$. If $w \in N_\delta(x)$, then $m_u(f)(w) \geq A_r(|f|)(y) > m_u(f)(x) - \epsilon$. A similar argument applies if $m_u(f) = \infty$. Thus

$m_u(f)$ is lower semi-continuous.

It remains to show that m_u is of weak type $(1, 1)$. The key result is the following covering lemma.

Lemma 33.3.3 (Wiener's lemma) *Suppose that G is a finite set of open balls in $\mathbf{R}^d$. Then there is a finite subcollection F of disjoint balls such that*

$$\sum_{U \in F} \lambda_d(U) = \lambda_d\left(\bigcup_{U \in F} U\right) \geq \frac{1}{3^d}\lambda_d\left(\bigcup_{U \in G} U\right).$$

Proof We use a greedy algorithm. If $U = N_r(x)$ is an open ball, let $U^* = N_{3r}(x)$ be the ball with the same centre as U, but with three times the radius.

Let U_1 be a ball of maximal radius in G. Let U_2 be a ball of maximal radius in G, disjoint from U_1. Continue, choosing U_j of maximal radius, disjoint from $U_1, \ldots, U_{j-1}$, until the process stops, with the choice of U_k. Let $F = \{U_1, \ldots, U_k\}$.

Suppose that $U \in G$. There is a least j such that $U \cap U_j \neq \emptyset$. Then the radius of U is no greater than the radius of U_j (otherwise we would have chosen U to be U_j) and so $U \subseteq U_j^*$. Thus $\bigcup_{U \in G} U \subseteq \bigcup_{U \in F} U^*$ and

$$\lambda_d\left(\bigcup_{U \in G} U\right) \leq \lambda_d\left(\bigcup_{U \in F} U^*\right) \leq \sum_{U \in F} \lambda_d(U^*) = 3^d \sum_{U \in F} \lambda_d(U). \qquad \square$$

Proof of Theorem 33.3.2. Let $f \in L^1(\mathbf{R}^d)$. Let $E_\alpha = (m_u(f) > \alpha)$, for $\alpha > 0$. Let K be a compact subset of E_α. For each $x \in K$, there exist $y_x \in \mathbf{R}^d$ and $r_x > 0$ such that $x \in N_{r_x}(y_x)$ and $A_{r_x}(|f|)(y_x) > \alpha$. It follows from the definition of m_u that $N_{r_x}(y_x) \subseteq E_\alpha$. The sets $N_{r_x}(y_x)$ cover K, and so there is a finite subcover G. By the lemma, there is a subcollection F of disjoint balls such that

$$\sum_{U \in F} \lambda_d(U) \geq \frac{1}{3^d}\lambda_d\left(\bigcup_{U \in G} U\right) \geq \frac{\lambda_d(K)}{3^d}.$$

But if $U \in F$, $\alpha\lambda_d(U) \leq \int_U |f| \, d\lambda_d$, so that since $\bigcup_{U \in F} U \subseteq E_\alpha$,

$$\sum_{U \in F} \lambda_d(U) \leq \frac{1}{\alpha}\sum_{U \in F}\int_U |f| \, d\lambda_d \leq \frac{1}{\alpha}\int_{E_\alpha} |f| \, d\lambda_d.$$

Thus $\lambda_d(K) \le 3^d(\int_{E_\alpha} |f| \, d\lambda_d)/\alpha$, and

$$\lambda_d(E_\alpha) = \sup\{\lambda_d(K) : K \text{ compact}, K \subseteq E_\alpha\} \le \frac{3^d}{\alpha} \int_{E_\alpha} |f| \, d\lambda_d.$$

Thus m_u is sublinear and of weak type $(1,1)$. $\square$

Corollary 33.3.4   *Let* $m(f) = \max(m_u(f), |f|)$. *Then* m *is a sublinear mapping of weak type* $(1,1)$.

Proof   For $(m(f) > \alpha) \subseteq (m_u(f) > \alpha) \cup (|f| > \alpha)$, so that

$$\alpha\lambda_d(m(f) > \alpha) \le (3^d + 1) \int_{\mathbf{R}^d} |f| \, d\lambda_d.$$
$\square$

Theorem 33.3.5 (The Lebesgue differentiation theorem)   *Suppose that* $f \in L^1(\mathbf{R}^d, \mathcal{L}_d, \lambda_d)$. *Then* $f(x)$ *is the spherical derivative of* f *at* x, *for almost every* $x \in \mathbf{R}^d$.

Proof   We use Theorem 33.2.1. For each $r > 0$ and $f \in L^1(\mathbf{R}^d, \mathcal{L}_d, \lambda_d)$, $|A_r(f)| \le m(f)$, so that A_r is a linear mapping of $L^1(\mathbf{R}^d, \mathcal{L}_d, \lambda_d)$ into $L^0(\mathbf{R}^d, \mathcal{L}_d, \lambda_d)$, dominated by m. Let $A_0(f) = f$; then A_0 is also dominated by m. Let F be the linear subspace of $L^1(\mathbf{R}^d, \mathcal{L}_d, \lambda_d)$ consisting of continuous functions of compact support. F is dense in $L^1(\mathbf{R}^d, \mathcal{L}_d, \lambda_d)$, and $A_r(f)(x) \to f(x)$ as $r \to 0$, for each $x \in \mathbf{R}^d$, and so the result follows from Theorem 33.2.1. $\square$

Corollary 33.3.6 (The Lebesgue density theorem)   *If* E *is a measurable subset of* $\mathcal{R}^d$ *then*

$$\frac{1}{r^d \Omega_d} \lambda_d(N_r(x) \cap E) = \frac{\lambda_d(N_r(x) \cap E)}{\lambda_d(N_r(x))} \to 1 \text{ as } r \to 0 \text{ for almost all } x \in E$$

and

$$\frac{1}{r^d \Omega_d} \lambda_d(N_r(x) \cap E) = \frac{\lambda_d(N_r(x) \cap E)}{\lambda_d(N_r(x))} \to 0 \text{ as } r \to 0 \text{ for almost all } x \notin E.$$

Proof   Apply the theorem to the indicator functions $I_{E \cap N_k}(0)$, for $k \in \mathbf{N}$.
$\square$

Next we consider a finite measure μ for which μ and λ_d are mutually singular.

Theorem 33.3.7   *Suppose that* μ *is a finite Borel measure on* $\mathbf{R}^d$ *for which* μ *and* λ_d *are mutually singular. Then* μ *has spherical derivative* 0 *at* λ_d-*almost every point of* $\mathbf{R}^d$.

Proof   There exists a Borel λ_d-null set A for which $\mu(\mathbf{R}^d \setminus A) = 0$. Since μ is tight, there exists an increasing sequence $(K_n)_{n=1}^{\infty}$ of compact subsets of A with $\mu(K_n) > \mu(A) - 1/4^n$ for $n \in \mathbf{N}$. Let U_n be the open set $\mathbf{R}^d \setminus K_n$: then $\mu(U_n) < 1/4^n$.

Let

$$H_n = \{(x,r) : x \in U_n, 0 < r < \min(1/2^n, d(x,K_n)) \text{ and } A_r(\mu)(x) > 1/2^n\}$$

and let $V_n = \cup_{(x,r) \in H_n} N_r(x)$. Suppose that L is a compact subset of V_n. There exists a finite subset G of H_n such that $L \subseteq \cup_{(x,r) \in G} N_r(x)$. By Wiener's lemma, there exists a subset F of G such that the sets $\{N_r(x) : (x,r) \in F\}$ are disjoint and

$$\lambda^d(\cup_{(x,r) \in F} N_r(x)) \geq (1/3^d)\lambda_d(\cup_{(x,r) \in G} N_r(x)).$$

Then

$$\lambda_d(L) \leq \lambda_d(\cup_{(x,r) \in G} N_r(x)) \leq 3^d \lambda_d(\cup_{(x,r) \in F} N_r(x))$$

$$= 3^d \sum_{(x,r) \in F} \lambda_d(N_r(x)) \leq 3^d.2^n \sum_{(x,r) \in F} \mu(N_r(x))$$

$$\leq 3^d.2^n \mu(\cup_{(x,r) \in G} N_r(x)) \leq 3^d.2^n \mu(V_n) \leq 3^d.2^{-n}.$$

Since λ_d is tight, $\lambda_d(V_n) \leq 3^d.2^{-n}$.

Let $B = \limsup_{n \to \infty} V_n$. It follows from the first Borel–Cantelli lemma that $\lambda_d(B) = 0$. Consequently $\lambda_d(A \cup B) = 0$. If $x \notin A \cup B$ then there exists N such that $x \notin V_n$, for $n \geq N$. If $n \geq N$, and $0 < r < \frac{1}{2}\min(1/2^n, d(x,K_n))$ then $A_{2r}(\mu)(x) < 1/2^n$. If $d(x,y) < r$ then $N_r(y) \subseteq N_{2r}(x)$, so that $A_r(\mu)(y) \leq 2^d/2^n$. Thus μ has spherical derivative 0 at x. $\square$

Combining these two theorems, we have the following.

Theorem 33.3.8   *Suppose that μ is a finite Borel measure on $\mathbf{R}^d$. Then μ has a spherical derivative $D\mu$ at λ_d-almost every point of $\mathbf{R}^d$. The function $D\mu$ is λ_d-integrable. Set $\nu(A) = \mu(A) - \int_A D\mu \, d\lambda_d$. Then ν is a Borel measure on $\mathbf{R}^d$, ν and λ_d are mutually singular, and $\mu = D\mu.d\lambda_d + \nu$ is the Lebesgue decomposition of μ.*

Proof   Let $\mu = f.d\lambda_d + \nu$ be the Lebesgue decomposition of μ. Then μ has spherical derivative f λ_d-almost everywhere. $\square$

Exercises

33.3.1 This exercise establishes a version of *Vitali's covering theorem*.

Suppose that V is a bounded open subset of $\mathbf{R}^d$. A *Vitali covering* of V is a set $\mathcal{V}$ of open balls contained in V with the property that if F is a finite subset of $\mathcal{V}$ then

$$V \setminus (\cup_{U \in F} \overline{U}) = \cup \{W \in \mathcal{V} : W \cap (\cup_{U \in F} \overline{U}) = \emptyset\}.$$

Use Wiener's lemma to show that there is a disjoint sequence $(U_n)_{n=1}^{\infty}$ in $\mathcal{V}$ such that $\lambda_d(V \setminus (\cup_{n=1}^{\infty} U_n)) = 0$.

33.3.2 Does the result hold for any open subset of $\mathbf{R}^d$?

33.4 Borel measures on R, II

We now apply the Radon–Nikodým theorem to Borel measures on $\mathbf{R}$. A real- or complex-valued function f on $\mathbf{R}$ is *absolutely continuous* if whenever $\epsilon > 0$ there exists $\delta > 0$ such that if $(I_j)_{j=1}^{k} = ((a_j, b_j))_{j=1}^{k}$ is a sequence of disjoint intervals of total length $\sum_{j=1}^{k} l(I_j) = \sum_{j=1}^{k} (b_j - a_j)$ less than δ then $\sum_{j=1}^{k} |f(b_j) - f(a_j)| < \epsilon$. An absolutely continuous function is clearly uniformly continuous.

Theorem 33.4.1   *A positive, signed or complex Borel measure ν on $\mathbf{R}$ is absolutely continuous with respect to Lebesgue measure λ if and only if its cumulative distribution function F_ν is an absolutely continuous function on $\mathbf{R}$.*

Proof   It is clearly enough to consider the case where ν is a positive measure. Suppose first that ν is absolutely continuous with respect to λ. Given $\epsilon > 0$, there exists $\delta > 0$ such that if $A \in \mathcal{B}$ and $\lambda(A) < \delta$ then $\nu(A) < \epsilon$. If $(I_j)_{j=1}^{k} = ((a_j, b_j))_{j=1}^{k}$ is a sequence of disjoint intervals of total length $\sum_{j=1}^{k} l(I_j) = \sum_{j=1}^{k} (b_j - a_j)$ less than δ then $\lambda(\cup_{j=1}^{k}(a_j, b_j]) < \delta$, so that

$$\nu(\cup_{j=1}^{k}(a_j, b_j]) = \sum_{j=1}^{k} |F_\nu(b_j) - F_\nu(a_j)| < \epsilon,$$

and F_ν is an absolutely continuous function.

Suppose conversely that F_ν is an absolutely continuous function. We use Proposition 33.1.2. Suppose that A is a Borel set for which $\lambda(A) = 0$, and that $\epsilon > 0$. There exists $\delta > 0$ for which the absolute continuity condition is satisfied. There then exists an open set U containing A, with $\lambda(U) < \delta$.

Suppose that $U = \cup_{j=1}^{\infty} I_j = \cup_{j=1}^{\infty}(a_j, b_j)$ is a disjoint union of an infinite sequence of open intervals. Then

$$\nu(A) \le \nu(U) = \lim_{k \to \infty} \nu(\cup_{j=1}^{k}(a_j, b_j))$$

$$\le \lim_{k \to \infty} \nu(\cup_{j=1}^{k}(a_j, b_j]) = \lim_{k \to \infty} \sum_{j=1}^{k} (F_\nu(b_j) - F_\nu(a_j)) \le \epsilon.$$

(The case where U is a finite union is even easier.) Since ϵ is arbitrary, $\nu(A) = 0$, and so ν is absolutely continuous with respect to λ.   □

We now apply the results of the previous section to monotonic real-valued functions on **R**. The results generalize in a straightforward way to functions of bounded variation.

Theorem 33.4.2 *Suppose that F is a bounded increasing function on* **R** *and that $F(t) \to 0$ as $t \to -\infty$. Then F is differentiable almost everywhere. If f is the derivative of F, then f is integrable, and $\int_{(-\infty,t]} f\,d\lambda \le F(t)$ for almost all $t \in$ **R**. Equality holds for all $t \in$ **R** if and only if $\int_{\mathbf{R}} f\,d\lambda = \lim_{t \to +\infty} F(t)$, and if and only if F is an absolutely continuous function on* **R**.

Proof   Since F is monotonic, it is continuous except on a countable set J, and the discontinuities are all jump discontinuities. Let $G(t) = F(t+)$, for $t \in$ **R**. Then G is right-continuous, and is equal to F, except on a subset of J; G is continuous except on J, and the discontinuities are all jump discontinuities. G is therefore the cumulative distribution function of a finite Borel measure μ. The measure μ has a spherical derivative $D\mu$ except on a null-set N, which clearly includes J. Thus if $x \notin N$ then

$$\lim_{h,k \searrow 0} \frac{F(x+h) - F(x-k)}{h+k} = \lim_{h,k \searrow 0} \frac{\mu((x-k, x+h])}{h+k} = D\mu(x).$$

Since f is continuous at x,

$$\frac{F(x+h) - F(x-k)}{h+k} \to \frac{F(x+h) - F(x)}{h} \text{ as } k \searrow 0,$$

$$\text{and } \frac{F(x+h) - F(x-k)}{h+k} \to \frac{F(x) - F(x-k)}{k} \text{ as } h \searrow 0.$$

Thus F is differentiable at x, with derivative $f = D\mu$.

By Theorem 33.3.8, f is integrable, and $\mu = f.d\lambda + \nu$, where ν and λ are mutually singular. If $t \notin J$, then

$$F(t) = \mu((-\infty, t]) = \int_{(-\infty, t]} f \, d\lambda + \nu((-\infty, t]) \geq \int_{(-\infty, t]} f \, d\lambda.$$

Equality holds for all t if and only if $\nu = 0$. This happens if and only if F is absolutely continuous, and if and only if

$$\mu(\mathbf{R}) = \lim_{t \to +\infty} F(t) = \int_{\mathbf{R}} f \, d\lambda.$$

$\square$

Finally, let us consider the structure of a finite Borel measure μ on $\mathbf{R}$. By the Lebesgue decomposition theorem, $\mu = f.d\lambda + \nu$, where $f \in L^1(\mathbf{R}, \mathcal{B}, \lambda)$, and ν and λ are mutually singular. The cumulative distribution function $\int_{(-\infty, t]} f \, d\lambda$ of $f.d\lambda$ is absolutely continuous, so that F_ν has the same set J of discontinuities as F_μ.

If $x \in J$, let $j(x)$ be the size of the jump at x. If A is a Borel set, let $\alpha(A) = \sum\{j(x) : x \in A \cap J\}$. Then α is an *atomic* Borel measure: $\alpha(\{x\}) = j(x) > 0$ if $x \in J$, and $\alpha(\mathbf{R} \setminus J) = 0$. Further, α and λ are mutually singular.

Now let $\pi = \nu - \alpha$. Then π is a finite measure, π and λ are mutually singular, as are π and α. The cumulative distribution function F_π has no jumps, and is therefore a continuous function. Since π and λ are mutually singular, F_π is differentiable almost everywhere, and its derivative is 0 almost everywhere. A Borel measure on $\mathbf{R}$ such as π, which has a continuous cumulative distribution function, but for which π and λ are mutually singular, is called a *continuous singular* measure.

Summing up, if μ is a finite Borel measure on $\mathbf{R}$, then μ can be written as the sum of an absolutely continuous measure $f.d\lambda$, an atomic measure α, and a continuous singular measure π. It is easy to see that this decomposition is uniquely determined.

34

Applications

In this chapter, we give examples to show how the theory of measure that we have developed is used.

34.1 Bernstein polynomials

We now use Chebyshev's inequality to show that the polynomial functions are dense in $C[0,1]$. Suppose that f is a continuous real- or complex-valued function on $[0,1]$. The nth Bernstein polynomial $B_n(f)$ is defined as

$$B_n(f)(t) = \sum_{j=0}^{n} f(\frac{j}{n}) \binom{n}{j} t^j (1-t)^{n-j}.$$

Note that $B_n(f)$ is a polynomial of degree n, that $B_n(f)(0) = f(0)$ and that $B_n(f)(1) = f(1)$.

Theorem 34.1.1 *If f is a continuous real- or complex-valued function on $[0,1]$ then $B_n(f)$ converges uniformly to f as $n \to \infty$.*

Proof The proof is usually given in the language of probability theory, but we shall give a purely analytic account. We consider the unit cube $J_n = [0,1]^n$ with Lebesgue measure λ_n. Suppose that $t \in [0,1]$. For $1 \le j \le n$ let

$$C_j = \{x = (x_1, \ldots, x_n) \in J_n : 0 \le x_j \le t\},$$

and let c_j be the indicator function of C_j. Then

$$\int_{J_n} c_j \, d\lambda_n = \int_{J_n} c_j^2 \, d\lambda_n = t \text{ for } 1 \le j \le n,$$

$$\text{and } \int_{J_n} c_i c_j \, d\lambda_n = t^2 \text{ for } 1 \le i < j \le n.$$

Let $a_n = (c_1 + \cdots + c_n)/n$ be the average of $c_1, \ldots, c_n$. a_n takes values $0, 1/n, 2/n, \ldots, 1$, and $a_n(x) = j/n$ if and only if $x \in C_i$ for exactly j values of i. Thus

$$\lambda_n(a_n = \frac{j}{n}) = \binom{n}{j} t^j (1 - t)^{n-j}.$$

Further,

$$\int_{J_n} a_n \, d\lambda_n = \frac{1}{n} \sum_{j=1}^{n} \left(\int_{J_n} c_j \, d\lambda_n \right) = t, \text{ and}$$

$$\int_{J_n} a_n^2 \, d\lambda_n = \frac{1}{n^2} \sum_{j=1}^{n} \left(\int_{J_n} c_j^2 \, d\lambda_n \right) + \frac{2}{n^2} \sum_{1 \le i < j \le n} \left(\int_{J_n} c_i c_j \, d\lambda_n \right)$$

$$= \frac{t}{n} + \frac{n-1}{n} t^2,$$

so that

$$\sigma^2(a_n) = \int_{J_n} a_n^2 d\lambda_n - \left(\int_{J_n} a_n \, d\lambda_n \right)^2 = \frac{t(1-t)}{n} \le \frac{1}{4n}.$$

We now consider $f \circ a_n$. Suppose that $\epsilon > 0$. Since f is uniformly continuous on $[0, 1]$, there exists $\delta > 0$ such that $|f(s) - f(t)| < \epsilon/2$ for $|s - t| < \delta$. Let $L = (|a_n - t| > \delta)$ and let $S = (|a_n - t| \le \delta)$. By Chebyshev's inequality (Proposition 29.6.8),

$$\lambda_n(L) \le \frac{\sigma^2(a_n)}{\delta^2} \le \frac{1}{4n\delta^2}.$$

Now

$$\int_{J_n} f \circ a_n \, d\lambda_n = \sum_{j=0}^{n} f(\frac{j}{n}) \binom{n}{j} t^j (1 - t)^{n-j} = B_n(f)(t),$$

so that if $0 \le t \le 1$ then

$$|f(t) - B_n(f)(t)| = | \int_{J_n} (f(t) - f(a_n(s))) \, d\lambda_n(s)|$$

$$\le \int_{J_n} |f(t) - f(a_n(s))| \, d\lambda_n(s)$$

$$= \int_{S} |f(t) - f(a_n(s))| \, d\lambda_n(s) + \int_{L} |f(t) - f(a_n(s))| \, d\lambda_n(s)$$

$$\le \frac{\epsilon}{2} + \frac{2 \|f\|_\infty}{4n\delta^2},$$

since $|f(t) - f \circ a_n| < \epsilon/2$ on S and $|f(t) - f \circ a_n| \leq 2\|f\|_\infty$ on T. Thus $|f(t) - B_n(f)(t)| < \epsilon$ if $n > \|f\|_\infty / \epsilon \delta^2$, and so $B_n(f)$ converges uniformly to f as $n \to \infty$. $\qquad\square$

Corollary 34.1.2 *Suppose that μ and ν are finite, or signed, or complex Borel measures on $[0,1]$ for which*

$$\int_{[0,1]} t^n \, d\mu(t) = \int_{[0,1]} t^n \, d\nu(t) \text{ for } n \in \mathbf{Z}^+.$$

Then $\mu = \nu$.

Proof For if $f \in C[0,1]$ then

$$\int_{[0,1]} f \, d\mu = \lim_{n\to\infty} \int_{[0,1]} B_n(f) \, d\mu = \lim_{n\to\infty} \int_{[0,1]} B_n(f) \, d\nu = \int_{[0,1]} f \, d\nu,$$

and so the result follows from Exercise 32.1.3. $\qquad\square$

Exercises

There are many ways of showing that continuous functions on $[0,1]$ can be approximated uniformly by polynomials. The following exercises provide another proof.

34.1.1 Define a sequence of polynomials $(p_n)_{n=0}^\infty$ by setting $p_0 = 0$ and

$$p_{n+1}(t) = p_n(t) + \tfrac{1}{2}(t^2 - (p_n(t))^2) \text{ for } n \in \mathbf{N}.$$

Show that if $-1 \leq t \leq 1$ then $0 \leq p_n(t) \leq p_{n+1}(t) \leq |t|$.
34.1.2 Use Dini's theorem to show that $p_n(t) \to |t|$ uniformly on $[-1,1]$.
34.1.3 Show that if g is a piecewise linear function on $[0,1]$ then there exists $x_1 \ldots, x_k \in [0,1]$ and constants $c_0, \ldots, c_k$ such that

$$g(x) = c_0 + \sum_{j=1}^k c_j |x - x_j|,$$

and deduce that g can be approximated by polynomials uniformly on $[0,1]$.
34.1.4 Show that a continuous function on $[0,1]$ can be approximated uniformly by polynomials.

34.2 The dual space of $L^p_{\mathbf{C}}(X, \Sigma, \mu)$, for $1 \leq p < \infty$

We now use the Radon–Nikodým theorem to determine the dual space of $L^p_{\mathbf{C}}(X, \Sigma, \mu)$, where μ is a finite or σ-finite measure, and $1 \leq p < \infty$. Recall that if $(E, \|.\|_E)$ is a normed space then the dual space E' is the space of continuous linear functionals on E. The quantity $\|\phi\|' = \sup\{|\phi(x)| : \|x\|_E \leq 1\}$ is then a complete norm on E'. Recall also that if $1 < p < \infty$ then $p' = p/(p-1)$ is the conjugate index of p; we also set $1' = \infty$.

Theorem 34.2.1 *Suppose that (X, Σ, μ) is a finite or σ-finite measure space and that $1 \leq p < \infty$. If $g \in L^{p'}_{\mathbf{C}}(X, \Sigma, \mu)$ and $f \in L^p_{\mathbf{C}}(X, \Sigma, \mu)$, let $\phi_g(f) = \int_X fg \, d\mu$. Then the mapping $\phi : g \to \phi_g$ is a linear isometry of $L^{p'}_{\mathbf{C}}(X, \Sigma, \mu)$ onto $(L^p_{\mathbf{C}}(X, \Sigma, \mu)', \|.\|')$.*

A corresponding result holds in the real case, and the proof is essentially the same.

Proof   Theorem 29.6.6 shows that $fg \in L^1_{\mathbf{C}}(X, \Sigma, \mu)$, so that ϕ_g is defined, and that ϕ is a linear isometry of $L^{p'}_{\mathbf{C}}(X, \Sigma, \mu)$ into $(L^p_{\mathbf{C}}(X, \Sigma, \mu)', \|.\|')$. We must show that ϕ is surjective.

First we consider the case where μ is a finite measure. Suppose that $\psi \in (L^p_{\mathbf{C}}(X, \Sigma, \mu)'$. If $A \in \Sigma$, let $\nu_\psi(A) = \psi(I_A)$. Then

$$|\nu_\psi(A)| \leq \|\psi\|' \cdot \|I_A\|_p = \|\psi\|' \mu(A)^{1/p}.$$

We show that ν_ψ is a signed measure on Σ. Suppose that $(A_n)_{n=1}^\infty$ is a sequence of disjoint elements of Σ, with union A. If $n \in \mathbf{N}$ then

$$|\nu_\psi(A) - \sum_{j=1}^n \nu_\psi(A_j)| = |\nu_\psi(\cup_{m=n+1}^\infty A_m)| \leq \|\psi\|' \cdot \mu(\cup_{m=n+1}^\infty A_m)^{1/p}.$$

Since $\mu(\cup_{m=n+1}^\infty A_m)^{1/p} \to 0$ as $n \to \infty$, $\nu_\psi(A) = \sum_{j=1}^\infty \nu_\psi(A_j)$, and so ν_ψ is a complex measure. If $\mu(A) = 0$ and $B \in \Sigma$ is a subset of A, then $\nu_\psi(B) = 0$, and so $|\nu_\psi|(A) = 0$. Thus $|\nu_\psi|$ is absolutely continuous with respect to μ, by Proposition 33.1.2. By the Radon–Nikodým theorem, there exists $g \in L^1_{\mathbf{C}}(X, \Sigma, \mu)$ such that $\nu_\psi = g.d\mu$. Thus $\psi(I_A) = \int_A g \, d\mu$ for $A \in \Sigma$, and so $\psi(f) = \int_X fg \, d\mu$ when f is a simple function.

Next we show that $g \in L^{p'}(X, \Sigma, \mu)$. First we consider the case $p = 1$. Let $B = (\Re(g) > \|\psi\|')$. If $\mu(B) > 0$ then $\Re(\psi(I_B)) > \|\psi\|' \cdot \|I_B\|_1$, giving a contradiction. Thus $\mu(B) = 0$. In the same way, if $\theta \in (0, 2\pi]$ then $\mu(\Re(e^{i\theta}g) > \|\psi\|') = 0$. Considering a dense sequence $(\theta_n)_{n=1}^\infty$ in $(0, 2\pi]$, it follows that $g \in L^\infty_{\mathbf{C}}(X, \Sigma, \mu)$, and $\|g\|_\infty \leq \|\psi\|'$.

Next, suppose that $1 < p < \infty$. For $n \in \mathbf{N}$, let $G_n = (|g| \le n)$, let $g_n = gI_{G_n}$ and let $f_n = \overline{\text{sgn } g_n}|g_n|^{p'-1}$. Then

$$\int_X |f_n|^p \, d\mu = \int_X |g_n|^{p'} \, d\mu, \text{ so that } \|f_n\|_p = \|g_n\|_{p'}^{p'/p},$$

and

$$|\psi(f_n)| = |\int_X f_n g \, d\mu| = |\int_X f_n g_n \, d\mu| = \int_X |g_n|^{p'} \, d\mu,$$

so that

$$\|g_n\|_{p'}^{p'} \le \|\psi\|' \cdot \|f_n\|_p = \|\psi\|' \cdot \|g_n\|_{p'}^{p'/p}.$$

Thus $\|g_n\|_{p'} \le \|\psi\|'$. It then follows from the monotone convergence theorem that $\int_X |g|^{p'} \, d\mu \le (\|\psi\|')^{p'}$, so that $g \in L_{\mathbf{C}}^{p'}(X, \Sigma, \mu)$ and $\|g\|_{p'} \le \|\psi\|'$. If $f \in L_{\mathbf{C}}^p(X, \Sigma, \mu)$, it follows, by approximating f by simple functions, that $\psi(f) = \int_X fg \, d\mu$.

If μ is σ-finite, there exists an increasing sequence $(C_n)_{n=1}^\infty$ of sets in Σ of finite measure, with $\cup_{n=1}^\infty C_n = X$. The result follows easily by considering the restriction of ψ to the spaces $L_{\mathbf{C}}^p(C_n, \Sigma, \mu)$, and letting n tend to infinity. $\qquad\square$

A similar result does not hold for $L_{\mathbf{C}}^\infty(X, \Sigma, \mu)$. In general, the mapping $g \to \phi_g$ is a linear isometry of $L_{\mathbf{C}}^1(X, \Sigma, \mu)$ onto a proper subspace of the dual of $L^\infty(X, \Sigma, \mu)$.

Exercises

34.2.1 Suppose that (X, Σ, μ) is a finite measure space and that $1 \le p \le 2$. Use the fact that $L^2(X, \Sigma, \mu) \subseteq L^p(X, \Sigma, \mu)$ and that the inclusion is continuous, to show that any continuous linear functional on $L^p(X, \Sigma, \mu)$ can be represented by an element of $L^{p'}(X, \Sigma, \mu)$, without using the Radon–Nikodým theorem.

34.3 Convolution

We have seen in Theorem 31.2.1 that if (X, Σ) is a measurable space, then $(ca_{\mathbf{C}}(X, \Sigma), \|.\|_{ca})$ is a complex Banach space. We now consider the case where $(X, \Sigma) = (\mathbf{T}, \mathcal{B})$. We write $(\mathcal{M}(\mathbf{T}), \|.\|)$ for $(ca_{\mathbf{C}}(\mathbf{T}, \mathcal{B}), \|.\|_{ca(\mathbf{C})})$, and $L^p(\mathbf{T})$ for $L_{\mathbf{C}}^p(\mathbf{T}, \mathcal{B}, m)$ (where m is Haar measure), for $1 \le p \le \infty$. First, we show that we can define an associative multiplication $\star$ on $\mathcal{M}(\mathbf{T})$ which makes it into a Banach algebra: that is to say, $\mathcal{M}(\mathbf{T})$ is an algebra, and $\|\mu \star \nu\| \le \|\mu\|.\|\nu\|$, for $\mu, \nu \in \mathcal{M}(\mathbf{T})$. The essential fact that we use is that

$\mathbf{T}$ is a compact topological group: the mappings $\psi : (e^{i\theta}, e^{i\phi}) \to e^{i(\theta+\phi)}$ from $\mathbf{T} \times \mathbf{T}$ to $\mathbf{T}$ and $j : e^{i\phi} \to e^{-i\phi}$ from $\mathbf{T}$ to itself are continuous. In fact, similar results hold for any locally compact group, and in particular for the additive group of Euclidean space (see Exercise 34.3.3).

If $\theta \in (-\pi, \pi]$ and A is a Borel set in $\mathbf{T}$, let $T_\theta(A) = e^{-i\theta} A$, and if $\mu \in \mathcal{M}(\mathbf{T})$, let $T_\theta(\mu)$ be defined by setting $T_\theta(\mu)(A) = \mu(T_\theta(A))$. Then T_θ is a norm-preserving linear isomorphism of $\mathcal{M}(\mathbf{T})$ onto itself.

Suppose that μ and ν are complex Borel measures on $\mathbf{T}$. Then the product measure $\mu \otimes \nu$ is a Borel meaure on $\mathbf{T} \times \mathbf{T}$. We define the *convolution product* $\mu \star \nu$ to be the push forward measure $\psi_*(\mu \otimes \nu)$:

$$(\mu \star \nu)(A) = (\mu \otimes \nu)(\psi^{-1}(A)) = (\mu \otimes \nu)(\{(e^{i\theta}, e^{i\phi}) : e^{i(\theta+\phi)} \in A\}).$$

Using the definition of the product measure, it follows that

$$(\mu \star \nu)(A) = \int_{\mathbf{T}} \nu(T_\theta(A)) \, d\mu(\theta) = \int_{\mathbf{T}} \mu(T_\phi(A)) \, d\nu(\phi),$$

and that

$$\int_{\mathbf{T}} f \, d(\mu \star \nu) = \int_{\mathbf{T}} \left(\int_{\mathbf{T}} f(e^{i(\theta+\phi)}) \, d\mu(\theta) \right) d\nu(\phi)$$

$$= \int_{\mathbf{T}} \left(\int_{\mathbf{T}} f(e^{i(\theta+\phi)}) \, d\nu(\phi) \right) d\mu(\theta).$$

Proposition 34.3.1 *Suppose that $\mu, \nu, \pi \in \mathcal{M}(\mathbf{T})$, and that $\alpha, \beta \in \mathbf{C}$.*

(i) $\mu \star \nu = \nu \star \mu$.
(ii) $(\mu \star \nu) \star \pi = \mu \star (\nu \star \pi)$.
(iii) $(\alpha\mu + \beta\nu) \star \pi = \alpha(\mu \star \pi) + \beta(\nu \star \pi)$.
(iv) $\|\mu \star \nu\| \leq \|\mu\| \cdot \|\nu\|$, *with equality if both are positive measures.*

Proof (i)–(iii) follow from the definitions. If $B_1, \ldots, B_k$ are disjoint Borel sets in $\mathbf{T} \times \mathbf{T}$, then

$$\sum_{j=1}^{k} |(\mu \otimes \nu)(B_j)| \leq \sum_{j=1}^{k} (|\mu| \otimes |\nu|)(B_j) \leq (|\mu| \otimes |\nu|)(\mathbf{T} \times \mathbf{T}) = \|\mu\| \cdot \|\nu\|.$$

Hence, if $A_1, \ldots, A_k$ are disjoint Borel sets in $\mathbf{T}$ then $\sum_{j=1}^{k} |(\mu \star \nu)(A_j)| \leq \|\mu\| \cdot \|\nu\|$, and so $\|\mu \star \nu\| \leq \|\mu\| \cdot \|\nu\|$. If μ and ν are positive measures, then the inequality becomes equality. $\square$

Thus $(\mathcal{M}(\mathbf{T}), \|.\|)$ is indeed a Banach algebra, the *measure algebra* of $\mathbf{T}$.

Example 34.3.2 Let δ_θ be the atomic measure which gives mass 1 to $\{e^{i\theta}\}$. Then $\delta_\theta \star \mu = T_\theta(\mu)$.

In particular, $\delta_0 \star \mu = \mu$; δ_0 is the multiplicative identity of the algebra.

Example 34.3.3 Let $m = \lambda/2\pi$ be Haar measure on $\mathbf{T}$. Thus m is an invariant probability measure on $\mathbf{T}$. Then $m \ast \mu = \mu(\mathbf{T}).m$.

For if A is a Borel set in $\mathbf{T}$ and $\theta \in (-\pi, \pi]$ then $m(T_\theta(A)) = m(A)$, so that

$$m \star \mu = \int_{\mathbf{T}} m(T_\theta(A)) \, d\mu(\theta) = \mu(\mathbf{T})m(A).$$

Thus $\operatorname{span} m$ is an ideal in $\mathcal{M}(\mathbf{T})$, and the mapping $\mu \to m \star \mu$ is a norm-decreasing projection of $\mathcal{M}(\mathbf{T})$ onto $\operatorname{span} m$.

The measure algebra $(\mathcal{M}(\mathbf{T}), \|.\|)$ is very large and complicated, and its properties are still not well understood. It does however provide a good framework for considering the convolution of functions.

We have seen that the space $L^1(\mathbf{T})$ can be identified with the closed subspace of $(\mathcal{M}(\mathbf{T}), \|.\|)$ consisting of measure which are absolutely continuous with respect to m. We can say more.

Theorem 34.3.4 *If $f \in L^1(\mathbf{T})$ and $\mu \in \mathcal{M}(\mathbf{T})$ then $f.dm \star \mu \in L^1(\mathbf{T})$.*

Proof We use the Radon–Nikodým theorem. If $m(A) = 0$ then $m(T_\theta(A)) = 0$, so that

$$(f.dm \star \mu)(A) = \int_{\mathbf{T}} \left(\int_{T_\theta(A)} f \, dm \right) d\mu(\theta) = 0.$$

Consequently $f.dm \star \mu$ is absolutely continuous with respect to m, and so belongs to $L^1(\mathbf{T})$. $\qquad\square$

We write $f \star \mu$ for the measure $f.dm \star \mu$.

Thus $L^1(\mathbf{T})$ is a closed ideal in the measure algebra $(\mathcal{M}(\mathbf{T}), \|.\|)$, and is therefore a Banach algebra (without identity element).

Let us consider the convolution product of two absolutely continuous measures. Suppose that $f, g \in L^1(\mathbf{T})$. If $A \in \mathcal{B}$, then, using Fubini's theorem,

$$(f.dm \star g.dm)(A) = \int_{\mathbf{T}} \left(\int_{T_\theta(A)} f \, dm \right) g(e^{i\theta}) \, dm(\theta)$$

$$= \int_{\mathbf{T}} \left(\int_A f(e^{i(\phi-\theta)}) \, dm(\phi) \right) g(e^{i\theta}) \, dm(\theta)$$

$$= \int_A \left(\int_{\mathbf{T}} f(e^{i(\phi-\theta)}) g(e^{i\theta}) \, dm(\theta) \right) dm(\phi).$$

Thus $f.dm \star g.dm = h.dm$, where

$$h(e^{i\phi}) = \int_{\mathbf{T}} f(e^{i(\phi-\theta)}) g(e^{i\theta}) \, dm(\theta).$$

We therefore define the convolution product of two elements f and g of $L^1(\mathbf{T})$ by setting

$$(f \star g)(e^{i\phi}) = \int_{\mathbf{T}} f(e^{i(\phi-\theta)}) g(e^{i\theta}) \, dm(\theta) = \frac{1}{2\pi} \int_0^{2\pi} f(e^{i(\phi-\theta)}) g(e^{i\theta}) \, d\theta.$$

By Fubini's theorem, the integral exists for almost all ϕ, and, as we have seen, the product is in $L^1(\mathbf{T})$; convolution is a bilinear operator on $L^1(\mathbf{T})$. Since $m(\mathbf{T}) = 1$, it follows that $L^q(\mathbf{T}) \subseteq L^p(\mathbf{T})$ for $1 \le p < q \le \infty$, and that the inclusion mapping is norm-decreasing. How does this relate to convolution?

Theorem 34.3.5 *Suppose that $f \in L^1(\mathbf{T})$ and $g \in L^p(\mathbf{T})$, where $1 < p < \infty$. Then $f \star g \in L^p(\mathbf{T})$, and $\|f \star g\|_p \le \|f\|_1 \cdot \|g\|_p$.*

Proof Suppose that $h \in L^{p'}(\mathbf{T})$, where $p' = p/(p-1)$ is the conjugate index. Then

$$\int_{\mathbf{T}} |(f \star g)h| \, dm \le \int_{\mathbf{T}} (|f| \star |g|)|h| \, dm$$

$$= \frac{1}{2\pi} \int_0^{2\pi} \left(\frac{1}{2\pi} \int_0^{2\pi} |g(e^{i(\phi-\theta)})| . |f(e^{i\theta})| \, d\theta \right) |h(e^{i\phi})| \, d\phi$$

$$= \frac{1}{2\pi} \int_0^{2\pi} \left(\frac{1}{2\pi} \int_0^{2\pi} |g(e^{i(\phi-\theta)})| . |h(e^{i\phi})| \, d\phi \right) |f(e^{i\theta})| \, d\theta$$

$$\leq \frac{1}{2\pi} \int_0^{2\pi} \|g\|_p \cdot \|h\|_{p'} |f(e^{i\theta})| \, d\theta$$

$$= \|f\|_1 \cdot \|g\|_p \cdot \|h\|_{p'} < \infty.$$

Thus $(f \star g)h \in L^1(\mathbf{T})$, and $\|(f \star g)h\|_1 \leq \|f\|_1 \cdot \|g\|_p \cdot \|h\|_{p'}$. Hence the mapping $h \to \int_{\mathbf{T}} |(f \star g)h| \, dm$ is a continuous linear functional on $L^{p'}(\mathbf{T})$, with norm at most $\|f\|_1 \cdot \|g\|_p$. It now follows from Theorem 34.2.1 that $f \star g \in L^p(\mathbf{T})$, and $\|f \star g\|_p \leq \|f\|_1 \cdot \|g\|_p$. □

We can say more.

Theorem 34.3.6 *Suppose that $f \in L^p(\mathbf{T})$ and $g \in L^{p'}(\mathbf{T})$, where $1 \leq p < \infty$, and $p' = p/(p-1)$ is the conjugate index. Then $f \star g \in C(\mathbf{T})$, and $\|f \star g\|_\infty \leq \|f\|_p \cdot \|g\|_{p'}$.*

Proof Suppose first that f is continuous and that $g \in L^1(\mathbf{T})$. Suppose that $\epsilon > 0$. Since f is uniformly continuous, there exists $\delta > 0$ such that if $|e^{i\phi} - e^{i\phi'}| < \delta$ then $|f(e^{i\phi}) - f(e^{i\phi'})| < \epsilon/(\|g\|_1 + 1)$. It then follows that if $|e^{i\phi} - e^{i\phi'}| < \delta$ then

$$|(f \star g)(e^{i\phi}) - (f \star g)(e^{i\phi'})| \leq \frac{1}{2\pi} \int_0^{2\pi} |f(e^{i(\phi-\theta)}) - f(e^{i(\phi'-\theta)})| \cdot |g(e^{i\theta})| \, d\theta < \epsilon,$$

so that $f \star g$ is continuous.

Now consider the general case. It follows from Hölder's inequality that

$$|(f \star g)(e^{i\phi})| \leq \frac{1}{2\pi} \int_0^{2\pi} |f(e^{i(\phi-\theta)})| \cdot |g(e^{i\theta})| \, d\theta \leq \|f\|_p \cdot \|g\|_{p'},$$

so that $f \star g \in L^\infty(\mathbf{T})$, and $\|f \star g\|_\infty \leq \|f\|_p \cdot \|g\|_{p'}$. If $\epsilon > 0$, there exists $f' \in C(\mathbf{T})$ for which $\|f - f'\|_p < \epsilon/(\|g\|_{p'} + 1)$. Then

$$|(f \star g)(e^{i\phi}) - (f' \star g)(e^{i\phi})| \leq \epsilon \text{ for all } e^{i\phi} \in \mathbf{T},$$

so that $\|f \star g - f' \star g\| < \epsilon$. Thus the function $f \star g$ can be approximated uniformly by continuous functions, and so it is continuous. □

Exercises

34.3.1 Suppose that $1 < p < \infty$ and that p' is the conjugate index. Let $f(e^{i\theta}) = |\cot \theta|^{1/p}$ and let $g(e^{i\theta}) = |\cot \theta|^{1/p'}$, for $e^{i\theta} \in \mathbf{T}$. Show that $f \in L^q(\mathbf{T})$ for $1 \leq q < p$ and that $g \in L^q(\mathbf{T})$ for $1 \leq q < p'$. Show that $f \star g$ is unbounded.

34.3.2 Construct a non-negative element f of $L^1(\mathbf{T})$, for which $f \star f$ is unbounded.

34.3.3 Define the convolution product of two complex Borel measures on $\mathbf{R}^d$. Use this to define the convolution of a function in $L^1(\mathbf{R}^d) = L^1(\mathbf{R}^d, \mathcal{B}, \lambda_d)$ with a measure. Define the convolution of two functions in $L^1(\mathbf{R}^d)$, and show that it is a bounded continuous function. Problems occur when we consider functions in $L^p(\mathbf{R}^d)$, for $p > 1$, since $L^p(\mathbf{R}^d)$ is not contained in $L^1(\mathbf{R}^d)$. Extend other definitions and results of this section by considering approximations $f I_R$, where I_R is the indicator function of $\{x : \|x\| \le R\}$, and letting $R \to \infty$.

34.4 Fourier series revisited

In Volume I, we established some fundamental properties of Fourier series, using the Riemann integral. We now have the Lebesgue integral available, and can take the theory further. We shall only prove a few results from an enormous subject; these are intended as an introduction, and also as an illustration of how results from measure theory are used in practice. In particular, we restrict attention to Fourier series, and do not consider the Fourier transform on Euclidean space, or Fourier analysis on more general groups.

We continue with the notation of the previous section. If μ is a complex Borel measure on $\mathbf{T}$, we define its Fourier coefficients, by setting

$$\hat{\mu}_n = \int_{\mathbf{T}} e^{-in\theta} \, d\mu(\theta) = (\gamma_n \star \mu)(0),$$

where $n \in \mathbf{Z}$ and $\gamma_n(e^{i\theta}) = e^{in\theta}$. Since $\|\gamma_n\|_\infty = 1$, $|\hat{\mu}_n| \le \|\mu\|$, and so $(\hat{\mu}_n)_{n=-\infty}^\infty$ is a bounded sequence.

Example 34.4.1

(i) $\hat{\mu}_0 = \mu(\mathbf{T})$.

(ii) $(\hat{\delta}_\theta)_n = e^{-in\theta}$ for $n \in \mathbf{Z}$. In particular, $(\hat{\delta}_0)_n = 1$ for all n.

(iii) $(\hat{m})_0 = 1$ and $(\hat{m})_k = 0$ if $k \ne 0$. (Recall that m is Haar measure on $\mathbf{T}$.)

These results all follow immediately from the definition.

Proposition 34.4.2 *If μ and ν are complex Borel measures on $\mathbf{T}$, then* $(\widehat{\mu \star \nu})_n = \hat{\mu}_n \hat{\nu}_n$.

Proof   For

$$(\widehat{\mu \star \nu})_n = \int_{\mathbf{T} \times \mathbf{T}} e^{-in(\theta + \phi)} \, d(\mu \otimes \nu)(\theta, \phi)$$

$$= \left(\int_{\mathbf{T}} e^{-in\theta} \, d\mu(\theta) \right) \cdot \left(\int_{\mathbf{T}} e^{-in\phi} \, d\nu(\phi) \right) = \hat{\mu}_n \hat{\nu}_n.$$

$\square$

Theorem 34.4.3   *If μ is a complex Borel measure on $\mathbf{T}$ for which $\hat{\mu}_n = 0$ for all $n \in \mathbf{Z}$, then $\mu = 0$.*

Proof   If p is a trigonometric polynomial, then $\int_{\mathbf{T}} p \, d\mu = 0$. Since the trigonometric polynomials are dense in $C(\mathbf{T})$, it follows that if $f \in C(\mathbf{T})$, then $\int_{\mathbf{T}} f \, d\mu = 0$. If U is an open subset of $\mathbf{T}$, there exists an increasing sequence $(f_n)_{n=1}^\infty$ of non-negative functions in $C(\mathbf{T})$ which converges point-wise to the indicator function of U. It follows from the theorem of bounded convergence that $\mu(U) = 0$. If $\mu = \mu^+ - \mu^-$ is the Jordan decomposition of μ, then $\mu^+(U) = \mu^-(U)$. Since Borel measures on $\mathbf{T}$ are regular, it follows that $\mu^+ = \mu^-$, and so $\mu = 0$. $\square$

We now consider the Fourier coefficients of integrable functions. If $f \in L^1(\mathbf{T})$, we set

$$\hat{f}_n = (\widehat{f.dm})_n = \frac{1}{2\pi} \int_{-\pi}^{\pi} f(e^{i\theta}) e^{-in\theta} \, d\theta,$$

so that $\hat{f}_n = (f \star \gamma_n)(0)$.

In fact, we begin by considering functions in $L^2(\mathbf{T})$. The proofs of Bessel's inequality (Volume I, Theorem 9.3.1), and of Parseval's equation (Volume I, Corollary 9.4.7), given in Volume I, can be applied to functions in $L^2(\mathbf{T})$. Parseval's equation has the following important consequence.

Theorem 34.4.4   *The mapping $\mathcal{F} : f \rightarrow (\hat{f}_n)_{n=-\infty}^\infty$ is an isometric linear isomorphism of $L^2(\mathbf{T})$ onto $l_2(\mathbf{Z})$.*

Proof   Parseval's equation implies that $\mathcal{F}$ is an isometric linear isomorphism of $L^2(\mathbf{T})$ into $l_2(\mathbf{Z})$. On the other hand, $(\gamma_n)_{n=-\infty}^\infty$ is an orthonormal sequence in $L^2(\mathbf{T})$. Thus if $a = (a_n)_{n=-\infty}^\infty \in l_2(\mathbf{Z})$ then $\left(\sum_{j=-n}^n a_j \gamma_j \right)_{n=1}^\infty$ is a Cauchy sequence in $L^2(\mathbf{T})$, which, since $L^2(\mathbf{T})$ is complete, converges to an element $f \in L^2(\mathbf{T})$. Further, $\hat{f}_n = \langle f, \gamma_n \rangle = a_n$, so that $\mathcal{F}(f) = a$: $\mathcal{F}$ is surjective. $\square$

This result illustrates the importance of measure theory in constructing complete normed spaces.

Theorem 34.4.5 (The Riemann–Lebesgue theorem) *If $f \in L^1(\mathbf{T})$, then $\hat{f}_n \to 0$ as $|n| \to \infty$.*

Proof If $k \in \mathbf{N}$, let $f^{(k)} = f.I_{(|f| \le k)}$. Suppose that $\epsilon > 0$. Since $\left\| f - f^{(k)} \right\|_1 \to 0$, by the theorem of dominated convergence, there exists k such that $\left\| f - f^{(k)} \right\|_1 < \epsilon/2$. Since $f^{(k)}$ is bounded, it is in $L^2(\mathbf{T})$, so that $\mathcal{F}(f^{(k)}) \in l_2(\mathbf{Z})$. Hence there exists n_0 such that $|(\widehat{f^{(k)}})_n| < \epsilon/2$ for $|n| \ge n_0$. Thus if $|n| \ge n_0$ then

$$|\hat{f}_n| \le |(\widehat{f - f^{(k)}})_n| + |(\widehat{f^{(k)}})_n| < \epsilon/2 + \epsilon/2 = \epsilon.$$

$\square$

If $f \in L^1(\mathbf{T})$, we set $s_n(f) = \sum_{j=-n}^n \hat{f}_j \gamma_j$. Since an element of $L^1(\mathbf{T})$ is an equivalence class of functions, it is appropriate to express Dini's test in the following terms.

Theorem 34.4.6 (Dini's test) *Suppose that $f \in L^1(\mathbf{T})$, that $\alpha \in \mathbf{C}$ and that $e^{it} \in \mathbf{T}$. Let*

$$\phi_t(f)(e^{is}) = \tfrac{1}{2}(f(e^{i(t+s)}) + f(e^{i(t-s)})) - \alpha,$$

and let $\theta_t(f)(e^{is}) = \phi_t(f)(e^{is}) \cot(s/2)$. If $\theta_t(f) \in L^1(\mathbf{T})$ then $s_n(f)(t) \to \alpha$ as $n \to \infty$.

Proof Note that $\phi_t(f)$ is an even function and $\theta_t(f)$ is an odd function. Recall that it follows from the form of the Dirichlet kernel that

$$s_n(f)(e^{it}) - \alpha = \frac{1}{2\pi} \int_{-\pi}^{\pi} \theta_t(f)(s) \sin ns\, ds + \frac{1}{2\pi} \int_{-\pi}^{\pi} \phi_t(f)(s) \cos ns\, ds$$

$$= -i(\widehat{\theta_t(f)})_n + (\widehat{\phi_t(f)})_n.$$

The conditions ensure that $\phi_t(f)$ and $\theta_t(f)$ are in $L^1(\mathbf{T})$, and so the result follows from the Riemann–Lebesgue theorem. $\square$

The proof of Riemann's localization theorem that was given in Theorem 9.6.3 of Volume I does not extend to unbounded functions in $L^1(\mathbf{T})$; but we can now give an easier proof.

Theorem 34.4.7 (Riemann's localization theorem) *Suppose that $f \in L^1(\mathbf{T})$ and that $f(e^{it}) = 0$ for $a \le t \le b$. If $\delta < (b-a)/2$, then $s_n(f) \to 0$ uniformly on $I_\delta = \{e^{it} : a + \delta \le t \le b - \delta\}$.*

Proof If $h \in L^1(\mathbf{T})$, the mapping $t \to T_t(h)$ from $[-\pi, \pi]$ to $L^1(\mathbf{T})$ is continuous, and so $\{T_t(h) : t \in [-\pi, \pi]\}$ is a compact subset of $L^1(\mathbf{T})$. It follows from this that $\{\phi_t(f) : t \in [a + \delta, b - \delta]\}$ is a compact subset of $L^1(\mathbf{T})$. Let

$$g(e^{is}) = \begin{cases} \cot(s/2) & \text{if } \delta/2 \leq |s| \leq \pi \\ 0 & \text{otherwise.} \end{cases}$$

If $a + \delta \leq t \leq b - \delta$, then $\theta_t(f) = \phi_t(f).g$, so that

$$K = \{\theta_t(f) : a + \delta \leq t \leq b - \delta\}$$

is a compact subset of $L^1(\mathbf{T})$.

Suppose now that $\epsilon > 0$. There exists a finite subset F in $[a + \delta, b - \delta]$ such that $\{\theta_u(f) : u \in F\}$ is an $\epsilon/3$-net in K and $\{\phi_u(f) : u \in F\}$ is an $\epsilon/3$-net in $\{\phi_t(f) : t \in [a + \delta, b - \delta]\}$. By Dini's test, there exists n_0 such that $|s_n(f)(e^{iu})| < \epsilon/3$ for $n \geq n_0$ and $u \in F$. If $t \in [a + \delta, b - \delta]$ there exists $u \in F$ such that $\|\theta_t(f) - \theta_u(f)\|_1 < \epsilon/3$ and $\|\phi_t(f) - \phi_u(f)\|_1 < \epsilon/3$. Then $|s_n(f)(e^{it}) - s_n(f)(e^{iu})| \leq 2\epsilon/3$ for $n \geq n_0$, and so $|s_n(f)(e^{it})| \leq \epsilon$ for $n \geq n_0$. $\square$

Exercises

34.4.1 Let $(K_n)_{n=1}^{\infty}$ be the sequence of Fejér kernels. Show that if $f \in L^p(\mathbf{T})$, where $1 \leq p < \infty$, then $K_n \star f \to f$ in L^p-norm, as $n \to \infty$.

34.4.2 Suppose that f is an absolutely continuous function on $\mathbf{T}$. Show that $\hat{f}_n = o(1/n)$ as $|n| \to \infty$.

34.5 The Poisson kernel

The Poisson kernel in d-dimensional Euclidean space was defined in Volume II, Section 19.8, and was used to solve the Dirichlet problem for the unit sphere. Here we restrict attention to the two-dimensional case. In this case, ideas and results are more transparent, since the Poisson kernel is the real part of a holomorphic function. We give a fairly self-contained account.

Let $m(z) = (1+z)/(1-z)$. m is a Möbius transformation which maps the unit disc $\mathbf{D}$ onto the right-hand half-plane $H_r = \{z : \Re(z) > 0\}$, with $m(-1) = 0$, $m(i) = i$ and $m(-i) = -i$. Writing $z = x + iy = re^{i\theta}$, we have

$$m(z) = \frac{(1+z)(1-\bar{z})}{1-z)(1-\bar{z})} = \frac{1 - z\bar{z}}{1 - (z + \bar{z}) + z\bar{z}} + \frac{z - \bar{z}}{1 - (z + \bar{z}) + z\bar{z}}$$

$$= \frac{1 - r^2}{1 - 2x + r^2} + i\frac{2y}{1 - 2x + r^2}$$

$$= \frac{1 - r^2}{1 - 2r \cos\theta + r^2} + i\frac{2r \sin\theta}{1 - 2r \cos\theta + r^2}$$

$$= P(re^{i\theta}) + iQ(re^{i\theta}).$$

$P(re^{i\theta}) = P_r(e^{i\theta})$ is the two-dimensional *Poisson kernel* and $Q(re^{i\theta}) = Q_r(e^{i\theta})$ is the *conjugate Poisson kernel*.

Proposition 34.5.1   *The Poisson kernel has the following properties.*

(i) $P_r(e^{i\theta}) = P_r(e^{-i\theta}) > 0$ *for* $0 \le r < 1$;
(ii) $P_r(e^{i\theta}) \le P_r(e^{i\delta})$ *for* $0 < \delta \le |\theta| \le \pi$;
(iii) $P_r(e^{i\theta}) \to 0$ *uniformly for* $0 < \delta \le |\theta| \le \pi$, *as* $r \nearrow 1$;
(iv) $\frac{1}{2\pi}\int_{-\pi}^{\pi} P_r(e^{i\theta})\, d\theta = 1$.

Proof   (i), (ii) and (iii) follow from the formula for P. By Cauchy's integral formula,

$$1 = m(0) = \frac{1}{2\pi i}\int_{T_r(0)} \frac{m(z)}{z}\, dz = \frac{1}{2\pi}\int_{-\pi}^{\pi} (P_r(e^{i\theta}) + iQ_r(e^{i\theta}))\, d\theta,$$

so that (iv) follows by taking the real part.   □

We can also consider the Taylor series expansion of m:

$$\frac{1+z}{1-z} = (1+z)(1 + z + z^2 + \cdots)$$

$$= 1 + 2z + 2z^2 + \cdots.$$

Thus

$$\Re\left(\frac{1+z}{1-z}\right) = 1 + (z + \bar{z}) + (z^2 + \bar{z}^2) + \cdots;$$

hence

$$P_r(e^{i\theta}) = 1 + r(e^{i\theta} + e^{-i\theta}) + r^2(e^{2i\theta} + e^{-2i\theta}) + \cdots$$

$$= \sum_{-\infty}^{\infty} r^{|n|}e^{in\theta},$$

and the convergence is absolute and uniform in $|z| \le r < 1$, for $0 \le r < 1$.

Suppose that μ is a complex Borel measure on $\mathbf{T}$. Let

$$P(\mu)(re^{it}) = \mu_r(e^{it}) = (P_r * \mu)(e^{it}) = \int_{\mathbf{T}} P_r(e^{i(t-s)})\, d\mu(s).$$

Theorem 34.5.2 *Suppose that μ is a positive Borel measure on $\mathbf{T}$. Then $P(\mu)$ is a non-negative harmonic function on $\mathbf{D}$ and*

$$\frac{1}{2\pi} \int_{-\pi}^{\pi} \mu_r(e^{it}) \, dt = \mu(\mathbf{T}).$$

Proof Let $z = re^{it}$ and let

$$P^{(N)}(re^{it}) = \sum_{-N}^{N} r^{|n|} e^{int} = \sum_{0}^{N} z^n + \sum_{1}^{N} \bar{z}^n.$$

Then $P^{(N)}(re^{it}) \to P(re^{it})$ as $N \to \infty$, and $|P^{(N)}(re^{it})| \le (1+r)/(1-r)$. Thus by dominated convergence,

$$(P^{(N)} * \mu)(re^{it}) = \int_{-\pi}^{\pi} P^{(N)}(re^{i(t-s)}) \, d\mu(s)$$

$$\to \int_{-\pi}^{\pi} P(re^{i(t-s)}) d\mu = P(\mu)(re^{it}).$$

But $(P^{(N)} * \mu)(z) = \sum_{0}^{N} \hat{\mu}_n z^n + \sum_{1}^{N} \hat{\mu}_{-n} \bar{z}^n$, and so

$$P(\mu)(z) = \sum_{0}^{\infty} \hat{\mu}_n z^n + \sum_{1}^{\infty} \hat{\mu}_{-n} \bar{z}^n.$$

Since $\sup_n |\hat{\mu}_n| \le \|\mu\|_1$, the two power series have radii of convergence greater than or equal to 1. Thus $P(\mu)$ is harmonic on $\mathbf{D}$. Finally

$$\frac{1}{2\pi} \int_{-\pi}^{\pi} \mu_r(t) \, dt = \frac{1}{2\pi} \int_{-\pi}^{\pi} \left(\int_{\mathbf{T}} P_r(e^{i(t-s)}) \, d\mu(s) \right) dt$$

$$= \int_{\mathbf{T}} \left(\frac{1}{2\pi} \int_{-\pi}^{\pi} P_r(e^{i(t-s)}) \, dt \right) d\mu(s) = \mu(\mathbf{T}).$$

$\square$

We can extend this result to signed measures and complex measures.

Theorem 34.5.3 *Suppose that μ is a signed or complex Borel measure on $\mathbf{T}$. Then $P(\mu)$ is a harmonic function on $\mathbf{D}$, $\frac{1}{2\pi} \int_{-\pi}^{\pi} |\mu_r(t)| \, dt \le |\mu|(\mathbf{T})$ and $\frac{1}{2\pi} \int_{-\pi}^{\pi} |\mu_r(t)| \, dt \to |\mu|(\mathbf{T}) = \|\mu\|_{ca}$ as $r \nearrow 1$.*

Proof We prove this in the case where μ is a signed measure: the complex case is similar, but messier.

First, $P_r(\mu) = P_r(\mu^+) - P_r(\mu^-)$ is harmonic, and $|P_r(\mu)| \le P_r(|\mu|)$, so that $\|P_r(\mu)\|_1 \le \|P_r(|\mu|)\|_1 = \|\mu\|_{ca}$.

Suppose that $\epsilon > 0$. There exist disjoint Borel sets P and N with $\mathbf{T} = P \cup N$ such that μ is positive on P and negative on N. There exist compact sets C and D such that $C \subseteq P$, $D \subseteq N$ and $\mu(C) > \mu(P) - \epsilon$ and $\mu(D) < \mu(N) + \epsilon$. Let $d = d(C, D)$. Let $\delta = d/3$, so that the δ-neighbourhoods C_δ and D_δ are disjoint.

As usual, $\mu^+(A) = \mu(A \cap P)$ and $\mu^-(A) = -\mu(A \cap N)$. Let $\mu_C(A) = \mu(A \cap C)$, and let $\mu_D(A) = -\mu(A \cap D)$. Then $\|\mu^+ - \mu_C\|_{ca} < \epsilon$, so that

$$\int_{C_\delta} P_r(\mu_C) dm - \int_{C_\delta} P_r(\mu^+) dm \le \left\| P_r(\mu^+ - \mu_C) \right\|_1 < \epsilon.$$

Thus

$$\int_{C_\delta} P_r(\mu) \, dm = \int_{C_\delta} P_r(\mu^+) \, dm - \int_{C_\delta} P_r(\mu^-) \, dm$$

$$\ge \int_{C_\delta} P_r(\mu^+) \, dm \ge \int_{C_\delta} P_r(\mu_C) \, dm - \epsilon$$

$$= \int_{\mathbf{T}} P_r(\mu_C) \, dm - \int_{\mathbf{T} \setminus C_\delta} P_r(\mu_C) \, dm - \epsilon.$$

Now $P_r(\mu_C)(e^{it}) \to 0$ uniformly on $\mathbf{T} \setminus C_\delta$, and so there exists $0 \le r_C < 1$ such that $P_r(\mu_C)(e^{it}) \le \epsilon$ for $t \in \mathbf{T} \setminus C_\delta$ and $r_C \le r < 1$. Thus $\int_{\mathbf{T} \setminus C_\delta} P_r(\mu_C) dm < \epsilon$, and so

$$\int_{C_\delta} P_r(\mu) dm \ge \int_{\mathbf{T}} P_r(\mu_C) dm - 2\epsilon$$

$$= \|\mu_C\|_{ca} - 2\epsilon \ge \left\|\mu^+\right\|_{ca} - 3\epsilon$$

for $r_C \le r < 1$. Similarly, there exists $0 \le r_D < 1$ such that $\int_{D_\delta} P_r(\mu) \, dm \ge -\|\mu^-\|_{ca} + 3\epsilon$ for $r_D \le r < 1$. Consequently,

$$\|P_r(\mu)\|_1 \ge \int_{C_\delta} P_r(\mu) \, dm - \int_{D_\delta} P_r(\mu) \, dm - \int_{\mathbf{T} \setminus (C_\delta \cup D_\delta)} P_r(\mu) \, dm$$

$$\ge \|\mu\|_{ca} - 7\epsilon,$$

for $\max(r_C, r_D) \le r < 1$. Thus $\|P_r(\mu)\|_1 \to \|\mu\|_{ca}$ as $r \nearrow 1$. $\qquad\square$

We can also consider functions in $L^1(\mathbf{T})$. If $f \in L^1(\mathbf{T})$, we set $P(f) = P(f.dm)$, so that

$$P(f)(re^{i\theta}) = P_r(f)(e^{i\theta}) = \frac{1}{2\pi} \int_{-\pi}^{\pi} \left(\sum_{-\infty}^{\infty} r^{|n|} e^{in(\theta-t)} \right) f(e^{it}) \, dt$$

$$= \sum_{-\infty}^{\infty} \hat{f}_n r^{|n|} e^{in\theta}.$$

Again, $P(f)$ is a harmonic function on $\mathbf{D}$. Let us first consider the case where f is a continuous function.

Theorem 34.5.4 (Solution of the Dirichlet problem)   *Suppose that $f \in C(\mathbf{T})$, where $\mathbf{T} = \{z \in \mathbf{C} : |z| = 1\}$. Let*

$$P(f)(re^{i\theta}) = f_r(e^{i\theta}) = (P_r * f)(e^{i\theta}) = \frac{1}{2\pi} \int_{-\pi}^{\pi} P_r(e^{i(\theta-t)}) f(e^{it}) \, dt.$$

Then $P(f)$ is a harmonic function on D, and $f_r \to f$ uniformly on $\mathbf{T}$.

Further, $P(f)$ is unique: if g is a continuous function on $\bar{D}$ which is harmonic on D and equal to f on $\mathbf{T}$ then $g = P(f)$ on D.

Proof   We have just seen that $P(f)$ is harmonic. Suppose that $\epsilon > 0$. Since f is uniformly continuous, there exists $\delta > 0$ such that $|f(e^{i\theta}) - f(e^{i\phi})| < \epsilon/2$ if $|\theta - \phi| < \delta$. By Theorem 34.5.1(iii), there exists $0 < r_0 < 1$ such that $2 \|f\|_\infty |P_r(e^{i\phi})| < \epsilon/2$ for $r_0 \le r < 1$ and $\delta \le |\phi| \le \pi$. Suppose that $e^{i\theta} \in \mathbf{T}$. Then if $r_0 \le r < 1$,

$$|P_r(f)(e^{i\theta}) - f(e^{i\theta})| = \left| \frac{1}{2\pi} \int_{-\pi}^{\pi} P_r(e^{i(\theta-t)})(f(e^{it}) - f(e^{i\theta})) \, dt \right|$$

$$\le \frac{1}{2\pi} \int_{-\pi}^{\pi} P_r(e^{i(\theta-t)}) |f(e^{it}) - f(e^{i\theta})| \, dt$$

$$= \frac{1}{2\pi} \int_{|\theta-t| \ge \delta} P_r(e^{i(\theta-t)}) |f(e^{it}) - f(e^{i\theta})| \, dt$$

$$+ \frac{1}{2\pi} \int_{|\theta-t| < \delta} P_r(e^{i(\theta-t)}) |f(e^{it}) - f(e^{i\theta})| \, dt$$

$$\le \epsilon/2 + \epsilon/2 = \epsilon.$$

This holds for all $e^{i\theta} \in \mathbf{T}$, and so $\|P_r(f) - f\|_\infty \le \epsilon$ for $r_0 \le r < 1$.

Finally we show that $P(f)$ is unique. The function equal to $P(f) - g$ on $\mathbf{D}$ and zero on $\mathbf{T}$ is continuous on $\bar{\mathbf{D}}$ and harmonic on $\mathbf{D}$, and must therefore be zero (see Exercise 22.6.7). $\qquad \square$

Corollary 34.5.5 *The trigonometric polynomials are dense in $C(\mathbf{T})$.*

Proof Given $f \in C(\mathbf{T})$ and $\epsilon > 0$, there exists $0 < r < 1$ such that $\|f_r - f\|_\infty < \epsilon/2$, and there exists N such that

$$\left\| f_r - \sum_{n=-N}^{N} \hat{f}_n r^{|n|} e^{in\theta} \right\|_\infty < \epsilon/2. \qquad \square$$

We can use this to give another proof that the polynomials are dense in $C[-1,1]$. Suppose that $g \in C[-1,1]$. Let $f(e^{i\theta}) = g(\cos\theta)$. Then f is an even function in $C(\mathbf{T})$, so that

$$\hat{f}_n = \hat{f}_{-n} = \frac{1}{\pi} \int_0^\pi f(e^{it}) \cos nt \, dt, \text{ and } f_r(e^{i\theta}) = \hat{f}_0 + 2 \sum_{n=1}^{\infty} \hat{f}_n r^n \cos n\theta.$$

Now $\cos n\theta = T_n(\cos\theta)$, where T_n is a polynomial of degree n, the n-th *Chebyshev polynomial*. Thus

$$\sum_{n=-N}^{N} \hat{f}_n r^{|n|} e^{in\theta} = \hat{f}_0 + 2 \sum_{n=1}^{N} \hat{f}_n r^n \cos n\theta$$

$$= \hat{f}_0 + 2 \sum_{n=1}^{N} \hat{f}_n r^n T_n(\cos\theta) = p_{r,N}(\cos\theta),$$

where $p_{r,N}$ is a polynomial of degree at most N. Then, arguing as above, $\|g - p_{r,N}\|_\infty < \epsilon$ for suitable r and N.

We now consider the spaces $L^p(\mathbf{T}) = L^p(\mathbf{T}, \mathcal{B}, m)$, for $1 \le p < \infty$. We define

$$h_p(\mathbf{D}) = \{f \text{ harmonic on } \mathbf{D} : \|f\|_{h_p} = \sup_{0<r<1} \|f_r\|_p < \infty\},$$

for $1 \le p < \infty$.

Theorem 34.5.6 *Suppose that $1 \le p < \infty$. If $f \in L^p(\mathbf{T})$ then $P(f) \in h_p(\mathbf{D})$, and $\|P(f)\|_{h_p} = \|f\|_p$. Further, $P_r(f) \to f$ in L^p-norm as $r \nearrow 1$.*

Proof Suppose that p' is the conjugate index. If $f \in L^p$, $g \in L^{p'}$ and $\|g\|_{p'} \leq 1$,

$$
\left| \frac{1}{2\pi} \int_{-\pi}^{\pi} P_r(f)(e^{it}) g(e^{it}) \, dt \right| = \left| \frac{1}{2\pi} \int_{-\pi}^{\pi} \left(\frac{1}{2\pi} \int_{-\pi}^{\pi} f(e^{i(t-s)}) P_r(e^{is}) \, ds \right) g(e^{it}) \, dt \right|
$$

$$
= \left| \frac{1}{2\pi} \int_{-\pi}^{\pi} \left(\frac{1}{2\pi} \int_{-\pi}^{\pi} f(e^{i(t-s)}) g(e^{it}) \, dt \right) P_r(e^{is}) \, ds \right|
$$

$$
\leq \frac{1}{2\pi} \int_{-\pi}^{\pi} \left(\frac{1}{2\pi} \int_{-\pi}^{\pi} |f(e^{i(t-s)}) g(e^{it})| \, dt \right) P_r(e^{is}) \, ds
$$

$$
\leq \frac{1}{2\pi} \int_{-\pi}^{\pi} \|f\|_p \cdot \|g\|_{p'} \, P_r(e^{is}) \, ds \leq \|f\|_p,
$$

so that $P(f) \in h_p(\mathbf{D})$, and $\|P(f)\|_{h_p} \leq \|f\|_p$. Given $\epsilon > 0$ there exists $g \in C(\mathbf{T})$ with $\|f - g\|_p < \epsilon/3$, and there exists $0 < r_0 < 1$ such that $\|P_r(g) - g\|_\infty < \epsilon/3$ for $r_0 < r < 1$. Thus $\|P_r(g) - g\|_p < \epsilon/3$ for $r_0 < r < 1$. If $r_0 < r < 1$ then

$$
\|f - P_r(f)\|_p \leq \|f - g\|_p + \|g - P_r(g)\|_p + \|P_r(g - f)\|_p < \epsilon.
$$

Thus $P_r(f) \to f$ in L^p-norm as $r \nearrow 1$. Consequently, $\|f\|_p \leq \|P(f)\|_{h_p}$. $\square$

When $1 < p < \infty$, we can say more.

Theorem 34.5.7 *Suppose that $1 < p < \infty$. The mapping $f \to P(f)$ is a linear isometry of $L^p(\mathbf{T})$ onto $h_p(\mathbf{D})$.*

Proof Theorem 34.5.6 shows that the mapping is a linear isometry of $L^p(\mathbf{T})$ into $h_p(\mathbf{D})$. We must show that it is surjective. Suppose that $f \in h_p(\mathbf{D})$. Suppose that $0 < r < s < 1$. $f_s \in C(\mathbf{T})$, and so $f_s \sim \sum_{n=-\infty}^{\infty} c_n \gamma_n$, where $c_n = (\hat{f}_s)_n$. Let $a_n = c_n s^{-|n|}$. It then follows that

$$
f_r = P_{r/s}(f_s) = \sum_{n=-\infty}^{\infty} a_n r^{|n|} \gamma_n.
$$

This holds for all $0 < r < s < 1$, so that a_n does not depend on s, and

$$
f_r = \sum_{n=-\infty}^{\infty} a_n r^{|n|} \gamma_n \text{ for all } 0 < r < 1.
$$

If $g \in L^{p'}$, let $\phi_r(g) = \int_{\mathbf{T}} f_r g \, dm$. Then ϕ_r is a continuous linear functional on $L^{p'}(\mathbf{T})$, and $\|\phi_r\|' = \|f_r\|_p \leq \|f\|_{h_p}$. Let T be the vector space of trigonometric polynomials. If $g = \sum_{-N}^{N} g_n \gamma_n \in T$, let $\phi(g) = \sum_{n=-N}^{N} a_n g_n$. Then

$\phi_r(g) = \sum_{n=-N}^{N} r^{|n|} a_n g_n \to \phi(g)$, as $r \nearrow 1$, and $|\phi(g)| \leq \|f\|_{h_p} \cdot \|g\|_{p'}$. Note that $\phi(\gamma_n) = a_n$. Thus ϕ is a continuous linear functional on the dense linear subspace T of $L^{p'}(\mathbf{T})$, and so it extends to a continuous linear functional, which we again denote by ϕ, on $L^{p'}(\mathbf{T})$. By Theorem 34.5.6, there exists $h \in L^p$ such that $\phi(g) = \int_{\mathbf{T}} hg\, dm$. Since $\phi(\gamma_n) = \hat{h}_n$, it follows that $a_n = \hat{h}_n$, and so $f = P(h)$. $\qquad\qquad\qquad\qquad\qquad\qquad\qquad\qquad\qquad\square$

This result does not extend to $L^1(\mathbf{T})$. Indeed, the following theorem holds.

Theorem 34.5.8   *The mapping $\mu \to P(\mu)$ is a linear isometry of $ca_C(\mathbf{T})$ onto $h_1(\mathbf{D})$.*

The proof of this theorem is beyond the scope of this book[1].

Exercises

34.5.1 How would you prove Theorem 34.5.3 for complex measures?

34.5.2 Show that the Chebyshev polynomials satisfy the recurrence relation $T_{n+1}(x) = 2xT_n(x) - T_{n-1}(x)$ for $n \in \mathbf{N}$, and deduce that if $|x| < 1$ and $|t| < 1$ then

$$\sum_{n=0}^{\infty} T_n(x)t^n = \frac{1 - tx}{1 - 2tx + t^2}.$$

34.6 Boundary behaviour of harmonic functions

What can we say about the behaviour of the values of an element $f(re^{it})$ of $h_1(\mathbf{D})$ as $r \nearrow 1$?

Theorem 34.6.1   *Suppose that μ is a complex Borel measure on $\mathbf{T}$ with Lebesgue decomposition $\mu = f.dm + \nu$. Then $P(\mu)(re^{it}) \to f(e^{it})$ for almost all t as $r \nearrow 1$.*

Proof   By considering real and imaginary parts, and positive and negative parts, we can suppose that μ is a positive measure. If I is an open interval in $\mathbf{T}$, let $A_I(\mu) = \mu(I)/m(I)$, let

$$m_u(\mu)(e^{it}) = \sup\{A_I(\mu) : I \text{ an open interval}, e^{it} \in I\},$$

and let

$$m_\delta(\mu)(e^{it}) = \sup\{A_I(\mu) : I \text{ an open interval}, l(I) \leq 2\delta, e^{it} \in I\}.$$

---

[1] See P.L. Duren, *Theory of H^p Spaces*, Dover, 2000.

Then, as in Theorem 33.3.2, m_u is a an operator of weak type $(\mathcal{M}(\mathbf{T}), 1)$, and so therefore is m_δ.

Let $s_r(t) = P_r(e^{it})$. Then s_r is a continuous even function on $[-\pi, \pi]$ which is strictly decreasing on $[0, \pi]$. Let $\gamma_r(u) = m(P_r > u)$. Then $\gamma_r(u) = 1$ for $0 \le u \le s_r(\pi)$, $\gamma_r(u)/2$ is the function inverse to s_r for $s_r(\pi) \le u \le s_r(0)$, and $\gamma_r(u) = 0$ for $u > s_r(0)$.

Suppose that $e^{it} \in \mathbf{T}$, that $0 < \delta \le \pi$ and that $0 < r < 1$. Let $J = (t - \delta, t + \delta)$. Then

$$\int_J P_r(e^{i(t-u)}) \, d\mu(u) = \int_J \left(\int_0^{s_r(t-u)} dv \right) d\mu(u)$$

$$= \int_0^{s_r(0)} \mu(J \cap (s_r(t-u) > v)) \, dv$$

$$\le m_\delta(\mu)(e^{it}) \int_0^{s_r(0)} m(J \cap (|t - u| < \gamma_r(v)/2)) \, dv$$

$$\le m_\delta(\mu)(e^{it}) \int_0^{s_r(0)} 2m(P_r > u) \, du = 2m_\delta(\mu)(e^{it}).$$

Suppose first that $f \in L^1(\mu)$. Then, taking $\delta = \pi$,

$$\frac{1}{2\pi} \int_{-\pi}^{\pi} P_r(f)(e^{i(t-s)}) \, ds \le 3m_u(f)(e^{it}).$$

Since the continuous functions are dense in $L^1(\mathbf{T})$, the result therefore follows from Theorem 33.2.1.

Next suppose that ν and m are mutually singular. By Theorem 33.3.7, ν has spherical derivative 0 almost everywhere. Suppose that ν has spherical derivative 0 at e^{it}. Suppose that $\epsilon > 0$. There exists $0 < \delta < \pi$ such that $m_\delta(\nu) < \epsilon$, and there exists r_0 such that $P_r(e^{i\delta}) < \epsilon$ for $r_0 < r < 1$. If $r_0 < r < 1$ then

$$P_r(\nu)(e^{it}) = \int_{(|t-s|<\delta)} P(re^{i(t-s)}) \, d\nu(s) + \int_{(|t-s|\ge\delta)} P(re^{i(t-s)}) \, d\nu(s)$$

$$\le 3\epsilon + \epsilon \|\nu\|,$$

which establishes the result. □

Index

Contents

Volume I

Printed in the United States
by Baker & Taylor Publisher Services